T0189757

Intelligent Systems Reference Library

Volume 119

Series editors

Janusz Kacprzyk, Polish Academy of Sciences, Warsaw, Poland
e-mail: kacprzyk@ibspan.waw.pl

Lakhmi C. Jain, University of Canberra, Canberra, Australia;
Bournemouth University, UK;
KES International, UK
e-mails: jainlc2002@yahoo.co.uk; jainlakhmi@gmail.com
URL: http://www.kesinternational.org/organisation.php

About this Series

The aim of this series is to publish a Reference Library, including novel advances and developments in all aspects of Intelligent Systems in an easily accessible and well structured form. The series includes reference works, handbooks, compendia, textbooks, well-structured monographs, dictionaries, and encyclopedias. It contains well integrated knowledge and current information in the field of Intelligent Systems. The series covers the theory, applications, and design methods of Intelligent Systems. Virtually all disciplines such as engineering, computer science, avionics, business, e-commerce, environment, healthcare, physics and life science are included.

More information about this series at http://www.springer.com/series/8578

Anthony Lewis Brooks · Sheryl Brahnam
Bill Kapralos · Lakhmi C. Jain
Editors

Recent Advances in Technologies for Inclusive Well-Being

From Worn to Off-body Sensing, Virtual Worlds, and Games for Serious Applications

 Springer

Editors
Anthony Lewis Brooks
Aalborg University Esbjerg
Esbjerg
Denmark

Sheryl Brahnam
Computer Information Systems
Missouri State University
Springfield, MO
USA

Bill Kapralos
Faculty of Business and Information
  Technology
University of Ontario Institute
  of Technology
Oshawa, ON
Canada

Lakhmi C. Jain
University of Canberra
Canberra
Australia

and

Bournemouth University
Poole
UK

ISSN 1868-4394   ISSN 1868-4408   (electronic)
Intelligent Systems Reference Library
ISBN 978-3-319-84263-9   ISBN 978-3-319-49879-9   (eBook)
DOI 10.1007/978-3-319-49879-9

This Springer imprint is published by Springer Nature
The registered company is Springer International Publishing AG
The registered company address is: Gewerbestrasse 11, 6330 Cham, Switzerland

Foreword

We are being bombarded with messages of doom about our inability to deliver the required level of healthcare. In the western world we are currently experiencing an increase in life expectancy but not necessarily of years of active life. Thanks to advances in medicine and public health, babies can survive being born at an early stage of gestation and those with a condition who would not have survived long past birth are living into middle age and beyond. The consequent challenge to our health care systems is not restricted to the western world. Less developed countries have been the first to suffer from the effects of climate change and virulent pandemics and frequently live long distances from health services.

Luckily, innovators have been pointing out that the solution is not just "more of the same" but is to be found in radically different ways of solving the problem. The astounding rate of change caused by more recent technological developments has enabled us to do more things faster but has also had more profound changes on old models of delivery. After a period of being astounded by the rate of development of digital technology, we are now witnessing its transition into a qualitatively different era. Whereas many of the original developments were very much fixed in physical space, tethered to a desktop computer, miniaturisation and mobile technology has moved on from the need to be located in one space and ownership is passing from the hands of experts. This in turn has led to the binary between experts and the rest of us becoming fuzzy. Taken further, this could result in a redistribution of power and devolvement of knowledge.

This collection of chapters from a wide range of disciplines, illustrates the march of digital technology out of the laboratory and out of the hands of the experts to be located in the end users' own space. Head mounted VR displays mean that they can be accessed where and when the user wishes and, in the case of a therapeutic application, the user can take back control of their own rehabilitation.

And it's not just this boundary that is being broken down. It is also possible that in addition to challenging the location of expertise, new technology will challenge the binary conception of health versus illness and disability versus lack of disability. The concept of assistive technology started off being seen as something created for people with a disability by others such that it enabled them to achieve a function previously denied to them. Wheelchairs could be seen to act as exemplars but however much we tinkered with their design and appearance, they marked the owner out as different, disabled as against able bodied and could contribute to their stigmatisation.

The phenomenon that the editors refer to as "an ecosystem that encompasses such tools as microcontrollers, sensors for just about everything, and personal 3D printing" stands to challenge this stigmatisation. Using a smartphone to remind you in which order to carry out self care procedures no longer marks you out as different if everyone around you is also holding a smartphone. Hearing aids do not look so unusual when every other person you pass in the street is wearing high end headphones for their smart device.

Mention of hearing aids reminds me that almost two decades ago, there was speculation (Cromby and Standen 1999) about the parallels between using assistive devices such as wheelchairs and being a cyborg. Cyborgs were originally the stuff of science fiction, Robocop type characters whose physical abilities were extended beyond normal human limitations by mechanical elements built into their bodies. However, in a 2014 article on the BBC Future website http://www.bbc.com/future/story/20140924-the-greatest-myths-about-cyborgs.

Frank Swain used this concept to describe his use of a hearing aid: "So I am now a cyborg out of necessity, not choice. Being part machine is my resting state".

This use of the term illustrates how technology could soon end the binary distinction between being disabled and being described as able bodied. Kevin Warwick, Professor of Cybernetics at Reading University Warwick installed a microchip in his arm that allowed him to operate doors, lights, heaters and other computers remotely as he moved from room to room. He wasn't seeking to supplement any identified lack of ability but I bet he was able to move around the environment much more easily when carrying a cup of coffee. Think what an advantage such a microchip could give someone who uses a wheelchair or other mobility device that required the use of their hands. In chapter 10 participants wore a vibrotactile vest that transmitted the heart beats and respiration rates of the vest wearers to a public display (the "wall") and any tactile stimuli applied to the wall were transmitted back to the vest wearer whenever the wall is swiped or knocked upon, these stimuli are vibrotactically transmitted to the vest wearers. Having achieved this, it is only a small step to developing smart wearables that look no different from normal clothing. Use of them would not mark you out as different and their employment could support the wearer in a whole range of activities from just making our day to day activities easier to providing prompts to those whose cognition has deteriorated.

At first sight, this innovation seems to be revolutionary, new and exciting but, as illustrated by our concept of cyborgs, the merging of human and technology has been going on for centuries. Transhumanism is a cultural and intellectual movement that believes we can, and should, improve the human condition through the use of advanced technologies. In the forward to their book *The Transhumanist Reader* (2013) More and Vita-More see transhumanism as promoting an interdisciplinary approach to understanding and evaluating the opportunities for enhancing the human condition and the human organism opened up by the advancement of technology.

After reading the contributions to this volume, some of you will be inspired to take some of them forward. However, it is equally important for us to read about these developments and reflect on their implications in terms of the human condition. At least two of the contributors raise potential concerns about these innovations. In the words of More and Vita-More "The same powerful technologies that can transform human nature for the better can also be used in ways that, intentionally or unintentionally, cause direct damage or more subtly undermine our lives". It was ever thus: no advances have come without the potential for harm. That has never stopped us from moving forward.

Penny Standen is Professor in Health Psychology and Learning Disabilities at Nottingham University. Her main area of research is developing and evaluating new technologies, robotics, multimedia and gaming for people with intellectual disabilities to help the acquisition of independence skills and improve cognition and quality of life. This work has resulted from collaboration with computer scientists and the involvement of users at all stages of the research process.

Prof. Penny Standen
Nottingham University, UK

References

1. Cromby J, Standen PJ (1999) Cyborgs and stigma: technology, disability, subjectivity. In: Gordo-López AJ, Parker I (eds) Cyberpsychology. Macmillan, pp 95–112
2. More M, Vita-More N (2013) The Transhumanist Reader. Wiley, Blackwell

Preface

This preface briefly introduces Recent Advances in Technologies of Inclusive Well-Being: Wearables, Virtual Interactive Spaces (VIS)/Virtual Reality, Authoring tools, and Games (Serious/Gamification). The book, which builds upon an initial Springer volume focused on technologies of inclusive well-being, came about through the amassed positive response attributed to the earlier publication edited by the first three co-editors herein. Inclusive well-being would seem a hot topic and as associated technologies continue to advance alongside adoptions and applied practices, it was clear the demand to expand to include topics as per title. Virtual worlds and games for serious applications is a growing field and this is a reason behind the book, the editors' motivation, the subject matter chosen, and its scope and aims to impact the field and associated. The earlier book's metrics, at time of writing, is near ten thousand downloads and progressively increasing citations. Readers of the first volume will notice an additional co-editor to the team in the form of Associate Professor Bill Kapralos from the University of Ontario Institute of Technology, Canada: It is a pleasure to have Bill onboard and his chapters are an important contribution.

The publication covers wide ground as introduced in the first chapter. Authors covering a gamut of disciplines come together under the inclusive well-being theme and it is anticipated that there is something for everyone, be they academic, student, or otherwise interested party. The main aim of the book is to disseminate this growing field through a combined effort that predicts to inform, educate, evoke—or even provoke, at least in thought—responses and discussions. While not the sole purpose, the editors along with the authors believe it important to bring such work presented out from behind the walls of establishments into the public specter so as to impact from a societal level.

The challenge of bringing together a collection of seminal work relating to technology subject of encroachment is just that—things move fast. We have been aware of this challenge and need to publish a contemporary volume within a schedule considering the prerequisite for up-to-date(ness) of presented research. The initial timeline had to be extended due to counterbalancing to the editors' different time zones, work and family commitments, and busy lives and distractions of the real world—for the delay we apologize to authors. However, in stating this, we believe that the extension has resulted in an even stronger contribution, realized in a form to credit all involved.

Acknowledgements are given to all authors for their submitted works and patience and understanding in the editorial team's challenges to realize what is anticipated as an impactful volume. We thank Springer's publishing team for their input to realize the volume. The editors thank their own families whose tolerance in supporting us in tackling such endeavors to publish is often tested; we are indebted for their support. The last acknowledgement is given to you, the reader, who we thank for coming onboard from your specific individual perspective; in thanking you for the interest in the work, we anticipate your curiosity being stimulated by individual texts so as to read not only chapters labeled in line with your position but also to stray and explore chapters not aligned to your discipline. In line with this latter statement, we offer no suggestions about how to read the book; there is no special structure, however, contents are split to theme in order to assist your reading adventure.

In closing we, the editors, extend our warmest regards encouraging you to explore the texts herein toward whetting your appetite to then dive further into the body of work and possible being stimulated to even visit the earlier volume—enjoy!

Best wishes from the team
Sheryl, Lakhmi, Bill and Anthony (aka Tony)

Contents

About the Editors

Dr. Anthony Brooks is Associate Professor at Aalborg University, Denmark where he is director/founder of the "SensoramaLab". He was a founding team-member of the Medialogy education wherein he is section leader, lecturer, coordinator, supervisor, and study board member. Originating from Wales, he created a bespoke volumetric invisible sensing system for unencumbered gesture-control of media to stimulate interactions and thus, human performance potentials. Initially titled Handi-MIDI, in the mid 1990's he renamed the maturing body of work as SoundScapes. Virtual Worlds for creative expression, Games for Serious Applications, and Virtual Interactive Space (VIS) are core aspects of his human-centered research targeting alternative channeling of biofeedback to realize human afferent-efferent neural feedback loop closure as a supplement for traditional rehabilitation, healthcare, and Quality of Life interventions. Making music, digital painting and robotic light control via available function/residual movements are selectable. He originated the ZOOM model (Zone of Optimized Motivation) for in-action intervention and an on-action analysis and assessment model under a custom-made synthesized methodology of Action Research integrated with Hermeneutics—resulting the Recursive Reflection model.

He has approximately 200 publications. His research is responsible for—sizable externally funded national and international (European) projects; –a games industry start-up, and patented commercial product (Patent US6893407 "Communication Method and

Apparatus"). He is acknowledged as a third culture thinker and "world pioneer in digital media and its use with the disabled community". He is an active keynote speaker at international events and has presented SoundScapes globally (including The Olympics and Paralympics 1996/2000 and numerous others). His work targets societal impact and benefit in respect of future demographics and service industries through applied ICT and optimized motivation of system use through inclusive intervention strategies: He is an ambassador for accessibility. He is Danish representative for UNESCO's International Federation for Information Processing (IFIP) Technical committee (WG14) on "Entertainment Computing"– specifically under work groups WG14.7 "Art and Entertainment"; WG14.8 "Serious Games", and WG 14.9 "Game Accessibility". He is appointed by the European Commission as EU expert examiner, rapporteur, (Serious Games/Gamification, Human-Computer Confluence, Presence and Future Emerging Technologies) and panel reviewer of funded projects (e.g. Presenccia – an Integrated Project funded under the European Sixth Framework Program, Future and Emerging Technologies (FET), Contract Number 27731). As vice-chair of the European Alliance for Innovation's Wellbeing (SIB-WellB) Market Trends and Society segment, he is also steering person for the international conference "ArtsIT, Interactivity and Game Creation". He achieved his PhD under the University of Sunderland, Great Britain.

Dr. Sheryl Brahnam is the Director/Founder of Missouri State University's infant COPE (Classification Of Pain Expressions) project. Her interests focus on face recognition, face synthesis, medical decision support systems, embodied conversational agents, computer abuse, and artificial intelligence. Dr. Brahnam has published articles related to medicine and culture in such journals as *Artificial Intelligence in Medicine, Expert Systems with Applications, Journal of Theoretical Biology, Amino Acids, AI & Society, Bioinformatics, Pattern Recognition, Human Computer Interaction, Neural Computing and Applications, and Decision Support Systems.*

Dr. Bill Kapralos is an Associate Professor in (and Program Director of) the Game Development and Entrepreneurship Program at the University of Ontario Institute of Technology. His current research interests include: serious games, multi-modal virtual environments/reality, the perception of auditory events, and 3D (spatial) sound generation for interactive virtual environments and serious games. He has led several large interdisciplinary and international serious gaming research projects that have included experts from medicine/surgery, and medical education with funding from a variety of government and industry sources. He is currently leading the serious gaming theme within the *Social Sciences and Humanities Research Council of Canada* (SSHRC) Interactive and *Multi-Modal Experience Research Syndicate* (IMMERSe) initiative. Bill chaired the 2014 *IEEE Games, Entertainment, and Media* (GEM) conference, and the *ACM FuturePlay International Conference on the Future of Game Design and Technology* from 2007-2010. He co-chaired the 2015 IEEE GEM conference, and the *ACM Virtual Reality Software and Technology Conference* in 2012. He is a past recipient of an IBM Centers for Advanced Studies Faculty Award, a past co-recipient of a Google Faculty Award, and a past recipient of a *Natural Sciences and Engineering Research Council of Canada* (NSERC) and *Japan Society for the Promotion of Science* (JSPS) Fellowship to conduct research in Japan. He recently completed a two-month stay at Shizuoka University in Hamamatsu, Japan, as a Visiting Research Fellow and Guest Professor as part of this Fellowship.

Lakhmi C. Jain BE (Hons), ME, PhD Fellow (Engineers Australia) is with the Faculty of Education, Science, Technology & Mathematics at the University of Canberra, Australia and Bournemouth University, UK.

Professor Jain founded the KES International for providing a professional community the opportunities for publications, knowledge exchange, cooperation and teaming. Involving around 10,000 researchers drawn from universities and companies world-wide, KES facilitates international cooperation and generate synergy in teaching and research. KES regularly provides networking opportunities for professional community through one of the largest conferences of its kind in the area of KES www.kesinternational.org

His interests focus on the artificial intelligence paradigms and their applications in complex systems, security, e-education, e-healthcare, unmanned air vehicles and intelligent agents.

Chapter 1
An Overview of Recent Advances
in Technologies of Inclusive Well-Being

**Anthony Lewis Brooks, Sheryl Brahnam, Bill Kapralos
and Lakhmi C. Jain**

Abstract This chapter reflects on luminary work on the use of digital media for the
therapeutic benefit and well-being of a wide range of people. It also discusses the
importance of multidisciplinarity, end user participation in design, and the future of
new technologies of inclusive well-being. A summary of the chapters included in
this book is also presented.

Keywords Alternative realities · Serious games · Rehabilitation · Therapy ·
Virtual environments · CVAA · Multidisciplinarity · Ethics · Multimodal
sensory environments

1.1 Introduction

This book is the second volume on the "Recent Advances in Technologies of
Inclusive Well-Being" and is published under the Intelligent Systems Reference
Library. The first volume was published in a series under the umbrella of Springer's

A.L. Brooks (✉)
Aalborg University, Esbjerg, Denmark
e-mail: tb@create.aau.dk

S. Brahnam
Department of Computer Science, Missouri State University, Springfield, MO, USA
e-mail: sbrahnam@missouristate.edu

B. Kapralos
Faculty of Business and Information Technology, University of Ontario Institute
of Technology, Oshawa, ON, Canada
e-mail: bill.kapralos@uoit.ca

L.C. Jain
Bournemouth University, Poole, UK
e-mail: Jainlc2002@yahoo.co.uk; jainlakhmi@gmail.com

L.C. Jain
University of Canberra, Canberra, Australia

© Springer International Publishing AG 2017

A.L. Brooks et al. (eds.), *Recent Advances in Technologies
for Inclusive Well-Being*, Intelligent Systems Reference Library 119,
DOI 10.1007/978-3-319-49879-9_1

"Studies in Computational Intelligence" (SCI) portfolio that brings together the topics of serious games, alternative realities, and play therapy.

The focus on this specific sequence of volumes is to share luminary work on the use of digital media for the therapeutic benefit and well-being of a wide range of people—spanning those with special needs to the elderly to entire urban neighborhoods. This book builds upon the first volume to further bring together these topics to demonstrate the increasing trans/inter/multi-disciplinary initiatives apparent today in science, medicine, and academic research—interdisciplinary initiatives that are already profoundly impacting society. In line with this, the editors believe it important to promote the transcending of what is researched, developed and realized behind the walls of academia into applied societal bearing for the benefit of individuals and communities. This second volume thus shares such work reflecting contemporary impactful designs, learning/practices, and innovations in physical therapies and multimodal experiences.

As was pointed out in the foreword in the previous volume, the last decade has presented a dramatic increase in the global adoption of innovative digital technologies. This can be seen in the rapid acceptance and growing demand for high-speed network access, mobile devices/wearable displays, smart televisions, social media, hyper realistic computer games, novel interactions, and numerous behavioral sensing devices. There are new ways to sense the world around us with alternative reality (virtual reality, augmented reality, mixed reality, and so on—) devices and content becoming available at affordable prices for all with related software tools available for creating one's own content without the need for academic background or complex laboratory facilities. Thus, consumer-driven and consumer-created interactive technologies have truly arrived and evolved beyond a simple maker-movement to now realize an ecosystem that encompasses such tools as microcontrollers, sensors for just about everything, and personal 3D printing, and so on—. These advances in technology enable a driving of innovation in manufacturing, engineering, industrial design, hardware technology and education for benefitting society. This volume thus presents work by ingenious 'makers/creators' utilizing innovative technologies toward investigating well-being and quality of life across differences of ability, age, and situation.

It is clear from these texts that what is being witnessed in the fields targeted is that healthcare researchers, providers, and related professionals, have adopted customizable (i.e., made for task) and adapted (i.e., not made for task) systems, apparatus, and methods. Advent of new game control interfaces, commercially available since the first decade of the twenty-first century, has probably had greatest impact in such adoption. Such systems have contextual added value in that whilst being known to many end users through home use, data can be generated of the human input that informs treatment programmes and other interventions. Software that enables generic use (vis., specific to game platforms) is commonly available so that the interface devices can capture human input and map to desired content that can be tailored for each individual, task, and situation. Analysis tools are also used to support progressions in design and re-design iterations and thus optimally support progressions. However, such work is ongoing with numerous challenges innate to the genre. Such challenges include how to optimize accessibility and inclusion, and this is partly addressed via development of (and upon) the United States Twenty-First Century Communications and Video Accessibility Act (CVAA), as pointed out in this volume by the lead editor's contribution. Thus, the emergence and growing adoption of clinical applications that both leverage off-the-shelf technology and push the boundaries of new technologic development should eventually benefit from enforcement of the act such that home-training is increasingly accessible for 'all'. As professor Skip Rizzo stated in his foreword to the first volume—"electric typewriters gave way to word processors and handwritten letters to email, we are now witnesses to a wave of technological innovation that is driving how healthcare is accessed and delivered, and how clinical users engage with it". With increased technology 'buy-in' by the general public such that most have a powerful personal computer that can link to a home-based entertainment system, home-based activities using such technological apparatus and methods as presented herein suggest optimization toward supporting self-responsibility for healthcare, well-being and quality of life aligned with what this volume presents.

The context and content of the series in the form of the two volumes published to date offers much to digest for a variety of readers from scientists, academics, and students, to healthcare professionals, providers, and end-users. In this context end-users are important due to the availability, affordability and usability of the current spate of systems and adaptations on the market. Whilst the CVAA is currently targeting the United States, it is envisioned by the editors that it is a first step in making visible issues on accessibility and inclusion that should be addressed globally for the benefit of the majority. We, the editors, envision continued implementations, developments and improvements of such acts that will support the accessibility and inclusion argument that certain industries have avoided addressing focusing instead solely on the economic strategies that target the maximum financial gain versus maximum human benefit.

We strongly support and wish to reiterate in this Introduction what Rizzo stated in his foreword to the first volume:

While 20 years ago the title, "Technologies of Well-Being: Serious Games, Alternative Realities, and Play Therapy", would raise the specter of Star Trek, Lawnmower Man, and Super Mario Brothers, in the current context it instead evokes a sense that new possibilities are within our reach as we harness technology to create user experiences that promote human well-being. The use of games and simulation technology and play for engaging users with health care has passed through its initial phase of scientific doubt and quizzical skepticism. These concepts are now seen as vibrant options that bring together both art and science for a useful human purpose. No longer seen as harebrained schemes, we see respected scientific journals like Nature, American Psychologist, and JAMA publishing research that probes these concepts. Papers in this area are routinely presented at long-established scientific venues in addition to the more specialized homegrown conferences that our community has now evolved. Major funding agencies are now earnestly supporting R&D in these areas. And, when you describe what you do for a living to your neighbor, they get it right away and seem genuinely impressed! In essence, the science and technology has caught up with the vision in clear and demonstrative ways.

This second volume is entitled "Recent Advances in Technologies of Inclusive Well-Being: Wearables; Virtual Interactive Spaces (VIS)/Virtual Reality; Emotional Robots; Authoring tools; and Games (Serious/Gamification)." It evolves the first volume in part through the title's expansion of topics to reflect current trends and activities.

The first volume in this series had sections on (1) *Technologies for Rehabilitation* —including titles such as "Design Issues for Vision-Based Motor-Rehabilitation Serious Games"; "Development of a Memory Training Game"; "Assessing Virtual Reality Environments as Cognitive Stimulation Method for Patients with MCI"; "Adaptive Cognitive Rehabilitation"; "A Body of Evidence: Avatars and the Generative Nature of Bodily Perception"; "Virtual Teacher and Classroom for Assessment of Neurodevelopmental Disorders"; "Engaging Children in Play Therapy: The Coupling of Virtual Reality Games with Social Robotics"; (2) *Technologies for Music Therapy and Expression*—including titles "Instruments for Everyone: Designing New Means of Musical Expression for Disabled Creators"; "Designing for Musical Play"; (3) *Technologies for Well-Being* —including titles "Serious Games as Positive Technologies for Individual and Group Flourishing"; "Spontaneous Interventions for Health: How Digital Games May Supplement Urban Design Projects"; "Using Virtual Environments to Test the Effects of Lifelike Architecture on People"; (4) *Technologies for Education and Education for Rehabilitative Technologies*—including titles "An Overview of Virtual Simulation and Serious Gaming for Surgical Education and Training"; "The Ongoing Development of a Multimedia Educational Gaming Module"; and the final section, (5) *Disruptive Innovation*—with a single title of "Disruptive Innovation in Healthcare and Rehabilitation". This second volume furthers through contributions relating to Wearables; Virtual Interactive Spaces (VIS)/Virtual Reality; Emotional Robots; Authoring tools; and once again Games (Serious/Gamification). It is clear from the texts in both volumes to date that the field has many perspectives and can be considered as multidisciplinary, interdisciplinary, and transdisciplinary. Whilst these terms are increasingly used in the literature, they are ambiguously defined and interchangeably used.

Choi and Pak [1] state that "Multidisciplinarity draws on knowledge from different disciplines but stays within their boundaries. Interdisciplinarity analyzes, synthesizes and harmonizes links between disciplines into a coordinated and coherent whole. Transdisciplinarity integrates the natural, social and health sciences in a humanities context, and transcends their traditional boundaries. The objectives of multiple disciplinary approaches are to resolve real world or complex problems, to provide different perspectives on problems, to create comprehensive research questions, to develop consensus clinical definitions and guidelines, and to provide comprehensive health services." He states further that "The three terms refer to the involvement of multiple disciplines to varying degrees on the same continuum. The common words for multidisciplinary, interdisciplinary and transdisciplinary are additive, interactive, and holistic, respectively."

1.2 Contributions in This Book

The chapters in this book are divided into five parts that reflect major reviews and themes in recent advances in technologies of inclusive well-being:

Part 1: Literature Reviews and Taxonomies
Part 2: Physical Therapy
Part 3: Touch and Wearables
Part 4: Special Needs
Part 5: Ethics and Accessibility

Below we provide a synopsis of each of the chapters in this book.

Part 1: Literature Reviews and Taxonomies

In Chap. 2, entitled "An Overview of Serious Game Engines and Frameworks," Brent Cowan and Bill Kapralos examined the tools (game engines and frameworks) that are commonly used to develop serious games. They present the results of two surveys that were conducted to determine the most widely used tools for serious game development and summarize their features. The motivation for this review is to provide those seeking tools to develop serious games with the appropriate knowledge and insight regarding the tools that are currently available to them so that they can make an informed decision. The chapter ends with a brief discussion regarding a framework that Cowan and Kapralos are currently developing specifically for the development of serious games that aims to strike a balance between ease of use and functionality, while providing the user with the necessary options and tools to ideally develop effective serious games.

In Chap. 3, "A Review of and Taxonomy for Computer Supported Neuro-Motor Rehabilitation Systems," Lucas Stephenson and Anthony Whitehead, as the title of the chapter suggest, provide a review of the literature of computer supported neuro-motor rehabilitation. Unlike other forms of physical therapy rehabilitation that focus on rebuilding muscle strength and flexibility, neuro-motor rehabilitation rebuilds and strengthens neural pathways to affected motor systems. As nuero-motor rehabilitation requires intense repetition of therapeutic exercises that lead to cognitive fatigue, research into computer supported systems and serious games has grown considerably the last couple of decades. The focus of the chapter, however, is primarily on providing a taxonomy for organizing the literature, a taxonomy that is carefully developed and validated by cognitively inspecting new input and output devices and verifying via a thought experiment that these I/O mechanisms could be represented in the taxonomy.

Part 2: *Physical Therapy*

In Chap. 4, "A Customizable Virtual Reality Framework for the Rehabilitation of Cognitive Functions," Gianluca Paravati, Valeria Maria Spataro, Fabrizio Lamberti, Andrea Sanna, and Claudio Giovanni Demartini present a virtual reality rehabilitation system, which they call VR2, for cognitive therapy along with a framework that was developed in collaboration with neuropsychology experts for adapting training scenarios, the main novelty offered by VR2 being a focus on system adaptation to the abilities of individuals or groups. Accordingly clinicians are provided with a set of tools for determining the number of stimuli present in rehabilitative scenarios and for testing and fine tuning parameters that personalize training sessions for individuals. Care is taken as well in the system to eliminate the effect of learning by generating randomly distributed stimuli and characteristics. This chapter not only describes the VR2 framework but it also presents a set of preliminary usability tests on a group of participants with cerebral lesions who varied widely in age and chronicity of the disorder. The training scenarios (named *library* and *supermarket*) focused on selective attention, i.e., the ability to focus on a stimulus or some relevant information in an environment containing competing stimuli and information. Patient acceptance of the system and its ability to produce cognitive improvement were examined, and results demonstrate VR2's usability and potential effectiveness. Included in this chapter is a short review of some state-of-the-art VR systems for rehabilitation involving motor and cognitive activities.

In Chap. 5, entitled "Technology for standing up and balance training in rehabilitation of people with neuromuscular disorders," Imre Cikajlo, Andrej Olenšek, Matjaž Zadravec, and Zlatko Matjaèiæ present a VR support balance training system that provides real-time balance feedback to patients undergoing dynamic balance training. Currently, a number of devices are available for dynamically maintaining balance (DMB); these devices are platforms that hold a patient upright while engaged in limited movement in all directions of the transversal plane. Newer systems are also being develop that use VR displays to provide visual correlates to the platform's movements. The authors of this chapter explore VR training with a DMB that has been modified with several additions: a haptic floor that translates in all directions of the horizontal plane and postural response assessment and progression from sit-to-stand with controlled active support. The results of several pilot studies using this system with both healthy and impaired participants are reported. These studies demonstrate not only the feasibility of using their system but also superior clinical results when compared to standard treatments. Some potential benefits of their VR balance training system that the authors point out include enabling the patient to train at home and a single physical therapists to supervise therapeutic procedures without an assistant and using only the DMB device. In addition, their system would allow the therapist to update difficulty levels and strategies that enhance adaptability to a broad range of environments.

In Chap. 6, entitled "Exergaming for Shoulder-Based Exercise and Rehabilitation," Alvaro Uribe-Quevedo and Bill Kapralos present two exergames that were developed specifically for shoulder-based exercise and rehabilitation. The first exergame focuses on lateral and medial rotations, which are important movements for strength and functionality. The goal of the game is to paddle from an origin position in a lake to the finish line. Correct paddling results in swift navigation that allows collecting power bars to restore energy depletion caused by movement. The second exergame, Rapid Recovery makes use of the Spincore Helium 6 baton, a unique stand-alone fitness and rehabilitation device. The player's kayak paddling motions of the Helium 6 baton are captured at interactive rates (captured with the Microsoft Kinect), and mapped to movements in the game world. Rapid Recovery supports multiple players, allowing the opportunity for players to race each other. In the process of describing the two exergames, Uribe-Quevedo and Kapralos provide a detailed description of the upper limb (including the shoulder), to characterize ranges of motion, the body parts involved, and the exercises suitable to be implemented as exergames. This characterization provides information that serves as an initial step in determining the hardware and software development requirements. They also outline their experience in dynamic design, development, and preliminary user testing.

In Chap. 7, entitled "Development of an Occupational Health Care Exergaming Prototype Suite," Alvaro Uribe-Quevedo, Sergio Valdivia, Eliana Prada, Mauricio Navia, Camilo Rincon, Estefania Ramos, Saskia Ortiz, and Byron Perez leverage the engaging and motivational aspects of exergames to promote preventive exercise routines in the workplace. With the widespread use of computers and a variety of mobile devices in the workplace, computer-related health problems arising from prolonged and incorrect use of a computer (and/or mobile device) are increasing at a rapid pace. Preventive exercises can help reduce the risk of developing computer-related health problems and many employers encourage employees to perform various preventive exercises typically outlined in printed material and/or online. To overcome some of the issues associated with such an approach and most importantly, the lack of motivation amongst employees to start and maintain such exercise routines, the authors of this chapter outline an exergaming prototype suite comprised of four exergames developed specifically to address common work-related preventive exercises associated with the lower-limb, upper-limb, hands, and eyes. The exergames are designed as casual games intended to be played in short bursts, thereby allowing them to be incorporated into an employee's daily work routine.

In Chap. 8, "Game-Based Stroke Rehabilitation" by Mehran Kamkarhaghighi, Pejman Mirza-Babaei, and Khalil El-Khatib discuss and analyze advancements made in the areas of human computer interaction and games specifically for stroke-based rehabilitation. The chapter begins with an overview of stroke and its associated consequences and a discussion regarding the problems and drawbacks inherent in traditional stroke rehabilitation approaches. This is followed by an overview regarding the benefits of game-based stroke rehabilitation by linking the efforts and advancements made of the fields of human-computer interaction and games research to medical research. More specifically, a thorough discussion regarding increased patient motivation and engagement and lower access barriers to rehabilitation is provided. The chapter then proceeds to outline the requirements for

efficient game-based stroke rehabilitation which include the need to provide rehabilitation exercises for different parts of the body, monitor patient progress and
provide feedback, provide diversity and flexibility to accommodate the wide range
of stroke-based effects, and the need to allow for patient autonomy and the ability
for patients to connect with other patients as part of their rehabilitation. A thorough
literature review regarding game-based stroke rehabilitation is provided. The review
provides historical context describing the most current efforts in the context of prior
advancements. A summary of various games developed for stroke rehabilitation
along with their target body part(s) and input devices employed, is provided in
tabular form.

Part 3: Touch and Wearables
In Chap. 9, entitled "Multi-sensory environmental stimulation for users with multiple disabilities," Cristina Manresa-Yee, Ann Morrison, Joan Jordi Muntaner, and
Maria Francesca Roig-Maimó present their multi-sensory environment which they
call SINASense that combines a number of technologies, such as computer vision
and hand recognition and vibratory and other tactile stimuli, with the aim of
inducing the execution of particular body movements in people with multiple
disabilities, including cognitive disabilities. After describing their system, the
authors report a two-stage preliminary evaluation on users with cerebral palsy.
Included in this chapter is a short review of the literature on interactive systems
used in multi-sensory environments, defined as spaces that stimulate a person's
senses in a way that does not require much high level cognitive processing, and
vibrotactile interfaces for people with disabilities, esp. vision and cognitive
disabilities.

In Chap. 10, entitled "Interactive Furniture: Bi-directional Inter-action with a
Vibrotactile Wearable Vest in an Urban Space," Ann Morrison, Jack Leegaard,
Cristina Manresa-Yee, Walther Jensen, and Hendrik Knoche investigate the experience of wearing a vibrotactile vest that interacts with a vibroacourtic wall, called
The Humming Wall, located in a public urban park setting. The heart beats and
breath rates of the vest wearers are publically and vibroacoustically displayed on
The Humming Wall via an android API; whenever the wall is swiped or knocked
upon, these stimuli are vibrotactically transmitted to the vest wearers. The authors
situate their vest and huggable wall within both the huggable toy industry and the
literature on health care wearables. One goal of the project is to reduce stress.
Participants were asked questions about their experiences and a large amount of
data was collected and analyzed regarding the interactions with the wall.

In Chap. 11, "The Acceptance, Challenges, and Future Applications of Wear-
able Technology and Virtual Reality to Support People with Autism Spectrum
Disorders", authors Nigel Newbutt, Connie Sung, Hung Jen Kuo, and Michael.
J Leahy present a review regarding the application of virtual reality technologies
(VRTs), within a research context, to support those with Autism spectrum disorder.
The chapter begins with a thorough review of the literature with an emphasis on the
potentially useful application of VRTs to train and support those with an ASD with
the development social skills, job skills, independent living skills (i.e., life skills).

The authors then outline the unique characteristics of VR that make it an effective intervention for those with autism. More specifically, the interactive nature of VR leads to an increase in user motivation, VR provided a safe and controlled environment allowing for repeated practice without real-world consequences when a mistake is made, VR provides the ability for immediate feedback and to track the users progress over time, a VR intervention can be customized to meet the unique needs of each user and this includes the manipulation of the presented stimuli and any distractors, and finally, a VR intervention allows for hierarchical learning where a scenario can differ slightly each time a user encounters it, forcing the user to think about different ways to solve the same problem. The paper then shifts focus to the use of head-mounted display- (HMD-) based VR for autism support given the gaining popularity of HMDs along with the potential benefits HMDs offer. Through a case study that involved the use of HMD-based VR amongst 29 participants with autism, the authors set out to examine the acceptance of wearing a HMD and more specifically, whether there are any adverse effects from the use of the HMD, and to gauge the 3D immersive experience and thus provide insight into whether HMD-based VR could be a potential intervention for the ASD population. The results of the study show promise regarding the use of HMD-based VR to support those with autism spectrum disorder, the study also revealed a number of potential concerns warranting further study.

Part 4: *Special Needs*
In Chap. 12, "Nursing Home Residents versus Researcher: Establishing Their Needs while Finding Your Way," Jon Ram Bruun-Pedersen describes problems encountered when designing exercise technology that uses virtual reality for nursing home residents. The author spent several years investigating, interviewing, and following the daily activities of residents in a local nursing home. The first part of this paper motivates the need for such technology. The author takes a detailed look at the elderly's reluctance to engage in physical exercise and offers a solution design that motivates this target group to engage in physical activities. Two studies are presented that examine the usefulness of the design solution that was derived from the author's careful investigations of the residents' needs. The second part of this chapter provides some insightful comments concerning the difficulties researchers encounter when doing research specific to this population.

Part 5: Ethics and Accessibility
In Chap. 13, "DigitalEthics: 'Ethical' considerations of post-research ICT impact", Anthony Brooks introduces a case study from the turn of the century conducted at the Centre for Advanced Visualization and Interactivity (CAVI) Aarhus University, Denmark. The subjects were four non-verbally capable children averaging five years of age diagnosed as suffering PMLD (profound and multiple learning disabilities). The children, attended by their personal professional care-providers, experienced unencumbered gesture control via the author created bespoke infrared sensor system (enabling movement of residual function in free space as controlling interface). CAVI's separate studio facilities were involved, i.e., the active 3D stereoscopic Virtual Reality Panorama studio, the Holobench studio, and the

Experimentarium. Multimodal stimuli in the studios were mapped from the author's sensor system enabling control by residual gesture within volumetric sensor spaces, which were scaled and otherwise tailored and fine-tuned according to differences of each child's ability. A specific focus presents how one of the children was deeply positively affected through interacting using his residual head motion to control a 'super-toy' (a VR model of a Star Wars fighter) in the Panorama studio and how this may have contributed to a post-research condition and outcome. A posited conclusion (from this speculated position) is how additional consideration for post-research impact in such instances of sensitivity need to be discussed and where necessary actioned such that where possible, funding and budgets address continued access aligned with training initiatives for facilitators to be empowered to optimize experiences and treatment.

In Chap. 14, "Accessibility: Definition, Labeling, and CVAA impact", Brooks, also the author of the previous chapter, elaborates on an act passed in USA regarding communication accessibility. The act updates a previous version to include encompassing game platforms and software. The contribution highlights the adoption of games with serious application in healthcare and beyond and points out some potential challenges and impact the act may have to this growing segment of support tool arsenal seen within various therapies. These are linked to the defined access challenges already in place for blind, deaf and cognitively impaired, and direct sources of information are introduced and shared in order to clarify issues involved. Such issues will align in the enforcing of the act, which many outside of the accessibility networks are not aware of. The chapter acts as an information messenger due to the dense data that has been sourced directly from selected main players such as AbleGamers Charity, PEGI (Pan European Game Information) age ratings and content labeling system), and other Special Interest Groups (SIGs). The contribution foresees challenges in enforcement of the CVAA aligned with the challenges of developers and industry to comply with requirements as expected and defined within game genres. A goal of the chapter is to promote discussions and awareness on accessibility issues and activities to realize increased inclusion that relates to technologies for inclusive wellbeing. An extended appendixes section adjoins this contribution by sharing the full CVAA and the full associated waiver granted to the games software industry, the latter presenting arguments associated to the innate challenges of compliance. The appendixes inclusion is considered an optimal means to inform the reader in printed form in one place the aligned materials.

1.3 Conclusion

These two volumes to date represent the referenced continuum starting with content sections as shared above, continued and advanced through the contributions in this volume where chapters are covering a wider gambit of topics to inform in and across fields relating to technologies of inclusive well-being.

However, as was asked in the first volume, what might a reader consider as they study this book (and the series)? Again, we believe Rizzo was eloquent in his preface to the first volume and insightful in his vision:

> We sometimes observe that great insights into the present turn up in the words of those who lived and died in a not-too-distant past. Such insights, derived from a relatively recent yet sufficiently disjointed past, can deliver a vision of the future that illuminates our present in surprising ways. This can be nicely illustrated in the words of the French author, poet, and aviator, Antoine de Saint-Exupery (1900–1944) with his comment, "The machine does not isolate man from the great problems of nature but plunges him more deeply into them." While interpretations may vary, in one sentence from a writer who lived exclusively in the first half of the twentieth century, I see the exquisite juxtaposition of apprehension and engagement that always looms in our pursuit of technology solutions that address the problems of nature. This is not a bad thing. With whatever technology tools we have available, we plunge deeper into the nature of problems, and hopefully come close to where the solutions lie. I see this book in much the same fashion; a fascinating collection of visionary works by a diverse collection of scientists and practitioners who implicitly acknowledge the same struggle. The many ideas presented in these pages for using digital technology to help change the course of challenged lives in ways unthinkable in just the last century is bold and provocative. And to do this requires a team of scientists, artists, programmers, clinicians, users, among others, who are willing to plunge deeply into the struggle, rather than to use technology to become isolated from the reality of the challenges that we aim to address. The authors in this book have successfully done this and these writings will play a significant part in further illuminating the bright future in this area.

In closing this opening statement, the editors reflect on this growing field and the advancements and impact of increased digital media adoption across associated practices. Today's pioneers and ambassadors, who are pushing the boundaries and limits previously conceived in this wide-ranging field, realize it is not solely about technology, importantly, it is increasingly about the thinking behind the technic that targets human performance and the conviction to put words into practice in such a way to impact society and benefit others. Enjoy the read, enjoy the journey!

Reference

1. Choi BC, Pak AW (2006) Multidisciplinarity, interdisciplinarity and transdisciplinarity in health researchservices, education and policy: 1. Definitions, objectives, and evidence of effectiveness. Clin Invest Med 29(6):351–364

Part I
Literature Reviews and Taxonomies

Chapter 2
An Overview of Serious Game Engines and Frameworks

Brent Cowan and Bill Kapralos

Abstract Despite the growing popularity and widespread use of serious games, the development of effective serious games is difficult, requiring an appropriate balance between game design and instructional design. Although there are fundamental differences between games developed purely for entertainment compared to those developed for "serious" purposes, there are currently no standard development tools specifically intended for serious game design and development available that encourage developers to follow a set of best practices. Rather, developers of serious games often rely on existing game engines and frameworks that are specific to entertainment-based game development. Given the availability of a large number of game engines and frameworks, deciding on which one to use to develop a serious game may be difficult, yet the choice of engine or framework can play a significant role in the development process. In this paper we present the results of a literature review that examined the frameworks and game engines that are used to develop serious games. We provide a list of the most commonly used frameworks and game engines and summarize their features. Knowledge of the frameworks and game engines that are most popular and details regarding why they are popular may prove to be useful to serious games developers seeking such tools. The chapter ends with a brief discussion regarding a framework that is currently being developed specifically for the development of serious games. Through consultation with the potential users of the framework (serious games developers), the framework aims to strike a balance between ease of use and functionality, while providing the user with the necessary options and tools to ideally develop effective serious games.

Keywords Serious gaming · Virtual simulation · Game engine · Game development · Framework · Review

B. Cowan
Faculty of Science, University of Ontario Institute of Technology, Oshawa, ON, Canada

B. Kapralos (✉)
Faculty of Business and Information Technology, University of Ontario Institute of Technology, Oshawa, ON, Canada
e-mail: bill.kapralos@uoit.ca

© Springer International Publishing AG 2017
A.L. Brooks et al. (eds.), *Recent Advances in Technologies for Inclusive Well-Being*, Intelligent Systems Reference Library 119, DOI 10.1007/978-3-319-49879-9_2

2.1 Introduction

The idea of using games for purposes other than entertainment was first formulated in a book titled 'Serious Games' by Clark C. Abt in 1975. Abt introduces the concept of serious games and defines them by stating: "We are concerned with serious games in the sense that these games have an explicit and carefully thought-out educational purpose and are not intended to be played primarily for amusement" [1]. The examples discussed by Abt were limited to table-top ("board") games as video games were still in their infancy in 1975. In 2002, motivated by Clark Abt's book 'Serious Games', David Rejeski from the Woodrow Wilson Center for Scholars added the term serious games to a report Ben Sawyer prepared titled "Improving Public Policy through Game-based Learning and Simulation" [47]. The expression "serious game" may be seen as a contradiction, or a tautology [7]. More specifically, if games are fun, how can they also be serious? It could even be argued that games have an evolutionary background as instruments for survival training [7]. Although no particularly clear definition of the term is currently available, "serious games" usually refer to games that are used for training, advertising, simulation, or education, and are designed to run on personal computers or video game consoles [55]. Serious games have also been referred to as "games that do not have entertainment, enjoyment, or fun as their primary purpose" [34]. A serious game can more formally be defined as an interactive computer application, with or without a significant hardware component that (i) has a challenging goal, (ii) is fun to play and/or engaging, (iii) incorporates some concept of scoring, and (iv) imparts to the user a skill, knowledge, or attitude that can be applied to the real world [5]. The terms "serious game" and "educational game" are often used interchangeably. However, the primary purpose of a serious game is not necessarily educational. For example, the America's Army series of games serve as a recruitment tool designed to persuade young people to consider a career in the U. S. military [59]. Educational games are typically viewed as a subset of serious games, and are mainly developed for use within kindergarten to grade twelve (K-12) education [42]. Education generally refers to the acquisition of knowledge, while training refers to the acquisition of both skills and knowledge. Educational games generally focus on the acquisition of knowledge while using entertainment as a motivator.

Serious games "leverage the power of computer games to captivate and engage players for a specific purpose such as to develop new knowledge or skills" [13] and greater engagement within an educational setting has been linked to higher academic achievement [50]. In addition to promoting learning via interaction, there are various other benefits to serious games. More specifically, they allow users to experience situations that are difficult (even impossible) to achieve in reality due to a number of factors including cost, time, safety, and ethical concerns [54]. Serious games may also support the development of various skills including analytical and spatial, strategic, recollection, and psychomotor skills, as well as visual selective attention [36]. Further benefits of serious games may include improved self-monitoring, problem recognition and solving, improved short- and long-term memory, and increased social skills [36]. Serious games are being applied within a wide range of applications including, medical/surgical skills development [3], military strategy training [43], and interpersonal skills development [8]. Given the ubiquity of video game use across a large demographic (i.e., males, females, youth, and the elderly), and their ability to engage and motivate learners, the popularity of serious games has seen a recent surge across many areas of education and training.

Despite the growing popularity of serious games and the benefits they afford, the development of effective serious games is a difficult and time consuming process, requiring an appropriate balance between game design and instructional design. It has been suggested that a lack of proper instructional design will lead to ineffective serious games or "instructional games" [4, 26]. Further complicating matters, there are currently no standard development tools available that emphasize, and encourage developers of serious games to follow best practices. Many serious game developers are using tools that were developed for the creation of commercial entertainment games instead of tools designed specifically for education and training. Serious games are often developed by developers with limited knowledge regarding a game's educational content and possess limited, if any knowledge of instructional design.

In order to understand how serious games are currently being developed, we begin by examining the development tools used to create them. Here, we present the results of two literature reviews that were conducted across three databases. The first survey compiled a list of development tools that were frequently mentioned in academic writing. A search was performed for each framework using three separate search terms. By measuring the number of search results for each framework in relation to each search term, we determined which development tool or framework is the "most discussed academically". In addition to measuring the notoriety of each framework/engine, separating the results by search term allows us to examine the context in which the framework/engine was frequently discussed. The second survey was conducted to determine which tools were utilized most often in the creation of a serious game. By knowing which tools are popular among the developers of serious games, we are able to investigate the features that are most important to developers. The toolset utilized to create a game may also provide insight regarding the size of the team, their skill level, and the project's budget.

2.1.1 Paper Organization

The remainder of this paper is organized as follows. Section 2.2 provides background information, and more specifically, an overview of existing frameworks and game engines. The results of a literature review that examined the tools (game engines and frameworks), used to develop serious games are also presented. This review led to a compiled list of the frameworks and game engines that are the most popular among serious games developers. A discussion of the survey results and concluding remarks are provided in Sect. 2.3. This includes a summary of the most popular game engines and frameworks along with common features that are important to developers of serious games in addition to a description of our ongoing work that is seeing the development of a serious game framework for medical education. This framework is intended to bridge the gap between the specific needs of serious game developers and the development tools they currently have available to them.

2.2   Game Engines and Frameworks

With respect to entertainment-based video games, early video games such as Pong were designed to run on hardware that was not well suited for video game development. With little in the way of processing power or memory to work with, games were typically written completely from scratch in assembly language (a low-level language). The close link between the game "code" and the hardware prevented the code from being re-used [6]. As hardware capabilities improved, higher level languages such as C/C++ and Java gradually replaced assembly language for game programming, leading to greater code re-use. Over time, many game companies accumulated a library of well-tested reusable code. To further reduce production time, simplified Application Programming Interfaces (APIs) and external tools such as level editors were developed. id software began licensing their Quake engine to other companies as an additional source of revenue in 1996 [31]. Other companies such as Epic Games soon began licensing their game engines (Unreal) too. There are now hundreds of commercial game engines competing for licensees. Modern game engines provide advanced rendering technologies, simple tools for content creation, allow game developers to reuse code, and decrease development time and costs [31]. Sherrod [49] defines a game engine as "a framework comprised of a collection of different tools, utilities, and interfaces that hide the low-level details of the various tasks that make up a video game". The terms "game engine" and "framework" are often used interchangeably. For the purpose of this paper, the term game engine refers to the functionality and features that become part of the completed game. A framework includes a game engine in addition to tools and interfaces that simplify the process of game development (this is depicted graphically in Fig. 2.1).

Listed below are the main features provided by most modern game engines. The code responsible for providing this functionality becomes a part of the finished game.

- **Scripting**: Simple code that can be written to control game objects and events.
- **Rendering**: The speed and accuracy by which a three-dimensional (3D) scene is generated, as well as the visual effects provided.
- **Animation**: The movement and deformation of objects, such as characters.

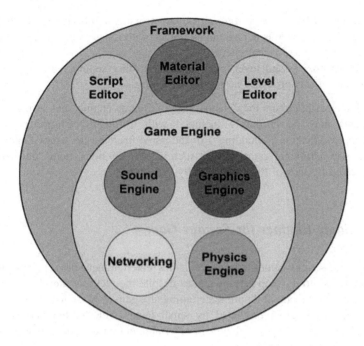

Fig. 2.1 The relationship between game framework, game engine, and their constituent components

- **Artificial Intelligence**: Steering behaviors, such as pursuing, dodging, and fleeing are combined with path finding.
- **Physics**: Objects respond accurately to applied forces or pressures (e.g., when colliding).
- **Audio**: Spatial rendering of audio allows sounds to have a location within the environment. Digital Signal Processing (DSP) is used to add variation as well as environmental cues such as reverberation.
- **Networking**: Allows players to interact with other players within the game by sharing data through a network.

Game creation frameworks generally provide a Graphical User Interface (GUI) which often ties together several editors. Listed below are some of the editing tools commonly included with game creation frameworks:

- **Level Editor**: Also known as a "world editor", this tool aids in the creation of two-dimensional (2D) or three-dimensional (3D) environments (game levels, maps, etc.).
- **Script Editor**: Scripts can be attached to objects selected in the level editor to customize their behavior.

- **Material Editor**: Shader (programs that provide a procedure for rendering surface details), code is edited and combined with images to form the surface of objects or to create visual effects.
- **Sound Editor**: Sound level, attenuation and other settings can be combined with filter effects provided by the sound engine.

Level and script editors may also be part of a game engine, and, at times, these components may be purposely included with a finished (shipped) game to encourage modification ("modding") of the game itself by the users. Game modding refers to additional game content such as new missions, items, and environments created by fans of the game, not by the developer.

2.2.1 Game Engines for Serious Games

Developers of serious games often use existing (entertainment-based) game engines and frameworks to develop serious games despite the fact that they have different needs than developers of "pure entertainment" games [41]. More specifically, serious games are often produced by small teams with limited budgets when compared to companies specializing in entertainment-based games whose budget often times is millions of dollars.

Although the development of games for entertainment often requires considerable planning, there are fewer design restrictions when compared to the development of serious games. Entertainment games must be fun and engaging; the game's content may be designed around the chosen gameplay mechanic and the designers are free to modify the design throughout the development process. Developers of serious games often begin with content that exists outside of the game world and, therefore, this content cannot be changed [4, 26]. For example, a serious game for surgical skills training must strictly adhere to the process and the equipment used in the surgical procedure in order for the skills acquired in the game to translate into real world application. Being games, serious games should also be fun and engaging, though this may not necessarily be their primary purpose.

A serious game development team may not include team members skilled in every area of game development. For example, an engine capable of rendering very large, highly detailed environments might not appeal to a team that lacks the art and modeling skills necessary to construct such an environment. In addition, some frameworks and engines are specialized for building a specific type of game, such as a first person shooter (FPS). Selecting the right engine is dependent on the type of game being developed and the skill set of the development team.

2.2.1.1 Engine Survey 1: Engines/Frameworks Most Discussed in Academic Literature

Although having knowledge of the frameworks and game engines that are popular amongst the developers of serious games will allow for the features most important to developers to be easily examined, very little prior work has considered this. As a result, two literature surveys to determine the frameworks and engines most commonly used by serious games developers (i.e., the most popular frameworks and engines) were recently conducted. The first survey (Survey 1) involved searching the following three data bases for game engines and frameworks that are most popular among serious games development: (i) Google Scholar, (ii) the ACM Digital Library, and (iii) the IEEE Xplore Digital Library. The Google Scholar search was conducted over a period of three days (August 2–5, 2015), while the ACM and IEEE Digital Library searches were conducted over a period of four days (August 3–7, 2015). The search terms "serious game", "educational game", and "simulator" were used to reveal more than 200 academic publications related to serious games and game development. Each of these publications were then scanned for the names of game engines without regard for the context in which the engine was discussed. By counting the number of publications that mention a framework by name, each framework was given a score signifying its notoriety among academic publications. The top 20 most mentioned game engines and frameworks were deemed worthy of further investigation (this initial investigation was conducted to limit any potential personal bias).

For each of the top 20 game engines, Google Scholar, the IEEE Xplore Digital Library, and the ACM Digital Library were used to survey the engine's popularity based on the number of search results returned. In each case, the engine name was combined with the search terms described above ("educational game", "serious game", or "simulator"). Many game engines, including Unity, Unreal, Torque, Half-Life, and Flash, have names that are common words in the English language and whose meaning may not necessarily have any relation to game engines or frameworks. For example, the term "half-life" is a commonly used in the physics domain to describe the decay rate of radioactive elements, but has no gaming reference. "Half-life" could also refer to a game by the same name, and not the game engine itself. To separate the search results specific to game engines and frameworks from those that did not relate to game development, the first 30 documents for each search term were manually reviewed. The approximate percentage of search results relating to the game engine was determined by dividing the number of documents found that referred to the engine by the total number of documents reviewed. The results were then normalized to ensure that each of the three search terms were equally weighted. For example, the search term "simulator" returned many more search results than the term "serious game".

The results of Survey 1 are summarized in Fig. 2.2 where the *x-axis* is a list of the top ten most discussed (academically) engines and frameworks, and the *y-axis* represents the normalized occurrence (the number of search results) for each of the three search terms ("educational game", "serious game", and "simulator").

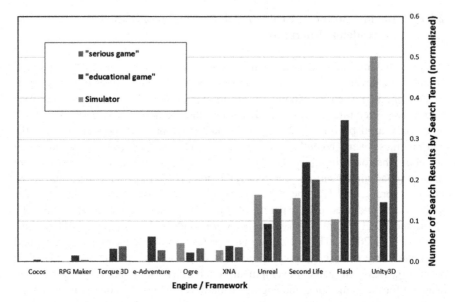

Fig. 2.2 Game engine/framework versus normalized occurrence (number of search results) by search term

The results of this survey provide an indication of the game engines/frameworks that are mentioned most frequently in academic writing. The results do not take into account whether or not the authors view the framework/engine favorably, and do not directly relate to the number of games made with each engine. For example, many of the publications that mention Second Life do not use it as a framework for their own content creation. Instead, Second Life is frequently discussed in relation to the psychological or social impact of virtual worlds. Adobe Flash is often discussed in papers relating to educational games. Unity and Unreal are frameworks capable of rendering highly realistic 3D graphics, so it is not surprising that they were frequently mentioned in papers relating to simulators.

2.2.1.2 Survey 2: Engines/Frameworks Most Utilized by Game Developers

After the completion of Survey 1, Survey 2 was conducted (August 7–11, 2015) to determine the engines or frameworks that serious game developers are using to create their serious games. This survey consisted of conducting a search within Google Scholar, IEEE Xplore Digital Library, and ACM Digital Library, using the search term "serious game". Five hundred peer reviewed papers were downloaded from the three databases. These papers were manually reviewed to determine which

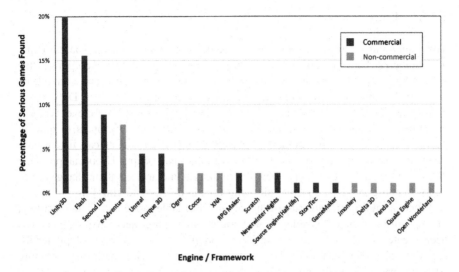

Fig. 2.3 Game engine/framework versus the percentage of serious games that were created with it

game engine the authors used to create a serious game. The majority of the papers that were downloaded (approximately 82 %) were disqualified since there was no mention regarding the framework/engine used, described more than one serious game, or described a serious game that was developed from "scratch" without the use of a specific framework/engine. Papers were also removed from the list if they featured a game that had been discussed in a previous paper, or if the first author matched an author of a previous paper. Only 90 out of 500 papers satisfied the above requirements and were used to determine the most popular engines utilized by serious game developers. The results of Survey 2 are summarized in Fig. 2.3 where, for each framework/engine, the percentage of games developed with it is provided. The frameworks/engines shown in green are freely available, while the frameworks/engines shown in blue are commercial products (although many of them are free for non-commercial use).

2.2.2 Notable Frameworks and Engines

A detailed description of the top ten most utilized frameworks/engines (as revealed by Survey 2) is provided below. The descriptions are not intended to be a review or evaluation the frameworks/engines, but rather an objective look at the main features that each framework/engine has to offer.

2.2.2.1 Unity

The Unity game creation framework was developed by Denmark-based Unity Technologies. Unity combines a powerful rendering engine that incorporates the nVidia PhysX physics engine [15]. Many successful commercial games have been developed using Unity, including various AAA titles such as Call of Duty Strike Team, and Rain. A "What You See Is What You Get" (WYSIWYG) level editor is combined with an intuitive terrain sculpting tool allowing developers to quickly generate realistic 3D environments. Objects in the environment can be visually selected, and changes can be made to their variables and scripts. Users can press the "Play" button at any time to preview the running game and test their changes [23]. Unity is non-game-type specific, allowing for 2D or 3D rendering [58].

Efficient multiplatform publishing at an affordable price has made Unity a favorite for independent developers and non-commercial projects. After a game has been created once in Unity, it can be exported to a wide variety of platforms, including all major game consoles and many mobile device platforms including Android and iOS, with only minor changes (such as reducing the file size or complexity of 3D assets) [58]. Figure 2.4 illustrates a sample screenshot taken from "Sort 'Em Up", a serious game that was commissioned by the Regional Municipality of Durham (Ontario, Canada) to allow residents to test their knowledge about recycling in a first-person shooter manner.

Access to the Unity engine source code is not provided with any of the standard licenses. All game logic is written in one of three scripting languages: JavaScript, C#, or Boo [23]. Unity simplifies programming by providing built-in visual effects such as particles, water, and post-processing. Artificial intelligence routines such as

Fig. 2.4 Sort 'Em Up, a serious game created by Squabble Studios to teach children about recycling [53]

path finding and steering behaviors are also provided. These routines allow amateur programmers to easily create believable non-player characters (NPCs) [58]. A single-player game can be converted into a multiplayer experience with minimal changes to the existing code. Unity also allows for easy integration with social media sites such as Facebook. Massively multiplayer games can be created by way of third-party solutions that must be purchased at additional cost through the Unity Asset Store [58]. Unity's asset store also contains many sound effects, images, 3D models, scripts, and complete games that can be used as a "starting point". However, the majority of the content found in the asset store is not free of charge. Community members are encouraged to upload assets to the store and set their own prices. Members receive 70 % of the sale price for each item sold through the store (Unity Technologies receives a 30 % commission on all sales).

Support is primarily provided by community members through online forums. A Premium Support package may be purchased granting the user direct access to the support team at Unity Technologies. Support packages cost between $500 USD and $1,000 USD a month, and there are limits to the number of times support can be requested and the number of users within a development team who may request support [58]. Unity does not charge on a per-title basis, and there are no annual maintenance fees. A free, Personal Edition of Unity is available that has a limited subset of available features. The Personal Edition of Unity can be used to create commercial projects royalty free provided that the licensee does not receive revenue or funding exceeding $100,000 USD in the previous year [58]. Projects (games) developed with the free version can be upgraded once a Unity Pro license has been purchased. A Unity Pro license costs $1,500 USD and permits publication of content created using Unity on desktop computers and the web (browser-based content). An additional license must be purchased at a cost of $1,500 USD for each mobile device (iOS, Android, BlackBerry) one wishes to publish content for [58]. Licenses may be provided at a reduced cost for developers of serious games or non-commercial projects. The licensing fee for serious game developers is negotiated on a case-by-case basis by the Unity sales team. Prices for console licenses are not posted and must be negotiated. Licensing the Unity source code is also possible.

2.2.2.2 Adobe Flash

Adobe Flash (formerly Macromedia Flash) began as a multimedia platform allowing 2D vector animations to be played in a web browser or as standalone applications. The free Flash player is included pre-installed on most computers and many mobile devices. It is estimated that the Flash player is currently installed on more than 1.3 billion computers and more than 500 million portable devices worldwide [2]. The Flash Builder framework combines a simple scripting language (Action Script) with a graphical editor used to position and animate objects. With Flash Builder, developers are able to easily release games on a variety of platforms including mobile devices. There are currently more than 50,000 mobile games and applications developed using Flash [2]. Some of the more notable Flash games include Machinarium, a version of Angry Birds, and FarmVille. At its peak, FarmVille alone attracted more than 80 million active monthly players [9]. Flash now supports High Definition (HD) video playback, full screen viewing, and 3D GPU accelerated rendering [2].

Support is primarily provided by other community members through the online forums. Expert support packages are available for purchase from Adobe. There are also several books specific to Flash development currently available including "Flash Game Development by Example [19]. Flash builder is available for Windows and Mac OS. The content created with Flash Builder can be viewed on Windows, Mac OS, Linux, iOS, and Android. The Flash Builder software (standard edition) sells for $249 USD. Games created with Flash Builder can be sold without incurring additional license fees [2].

Plug-ins such as Flash are no longer needed to create rich interactive content online. The HyperText Markup Language (HTML) is the standard language used to create web pages. Until recently, the HTML standard did not support real-time interaction. Interactive elements and other forms of multimedia were added to web pages using web browser plug-ins such as Flash. The new HTML5 standard now includes video and audio playback, vector animation, and support for WebGL applications which allow for hardware-accelerated 3D graphics [56]. However, browser-based game development frameworks such as Flash may persist based on the strength of their development tools and a large user base.

2.2.2.3 Second Life

Second Life was launched in 2003 by Linden Labs, located in San Francisco, CA [33]. Second Life is not a game, and not a game engine. Instead, it is a massively multiplayer (supports a large number of players simultaneously) 3D virtual world; a social space where people can meet, chat, play, explore, and also construct their own virtual spaces within the world. Each Second Life account comes with an avatar that the player can customize to look any way they want. Players interact with each other and the world through their avatars. The content for this virtual world is created by the people who inhabit it. Massively Multiplayer Online games (MMOs) such as World of Warcraft (WOW) or EVE Online, provide a themed environment created by the game's developers. The virtual world within an MMO can be viewed as a stage where the game's narrative plays out. Second Life does not provide any theme or narrative. Instead, users are encouraged to create their own themed environments and items, including games [29]. The environments they create may be static or interactive. Users may use their creations to tell a story, or simply provide a space to inhabit and explore. Second Life has managed to harness the creativity of its user base in order to provide an ever expanding world that is as diverse as those who author it.

In Second Life, simple tools are provided for content creation. Original objects can be made by combining many simple shapes such as cubes, spheres, cones, etc.

[29]. Second Life also allows users to import models created with professional 3D modeling software such as Maya or 3D Studio Max. Objects in Second Life can be made interactive with the addition of scripts written in Linden Script (LSL), Second Life's proprietary scripting language. Scripted objects can move, react to being touched, and even respond to chat messages [21].

Linden Labs grants the users with intellectual property rights for the objects and environments they create, implying that users can buy and sell virtual items such as land, buildings, vehicles, fashion accessories, and even in-game pets [21]. Second Life employs its own currency called Linden dollars that can be bought or sold in exchange for real dollars. The rate of exchange fluctuates as it is based on what users are willing to pay. For example, between February 2008 and February 2011, the exchange rate ranged between 250 and 270 Linden dollars for one USD [32].

A standard membership is free of charge. A premium membership costs $9.95 USD per month. Premium members receive 1,200 Linden dollars (L$) a month [48], have the right to purchase land at auctions, and can have a second avatar on the same account [46]. A full region of land measures 65,536 m^2 and costs 295 L$ per month in addition to the purchase price. Land may also be purchased from Linden Labs or other users in smaller sizes. The price of virtual items in Second Life varies and is driven by the market [24].

It has been many years since Second Life was launched, and it currently maintains a large user base with new content added daily. As of June 2011, there were approximately 800,000 logins per month [32], and the average number of users online at any given time was estimated to be around 50,000 [24]. Many colleges and universities around the world have a presence in Second Life, often enabling students to explore a virtual version of the campus. Simulation Linked Object Oriented Dynamic Learning Environment (SLOODLE) is a free and open source project which integrates Second Life with Moodle, a learning management system. With SLOODLE, students can register for classes, attend lectures, and write quizzes within a virtual environment [51]. Real-world tourist attractions such as museums and historical sites are replicated in Second Life to educate and encourage tourism. Government agencies such as the Center for Disease Control (CDC) and the National Aeronautics and Space Administration (NASA) also have a presence in Second Life, with the goal of educating the public, and encouraging an interest in the sciences among students (see Fig. 2.5 for an example).

With Second Life, there are technical limitations due to the underlying architecture which at times could hinder its use for educational purposes. For example, the inclusion of user-created content forces the Second Life server to send every user's avatar and objects to every other user present in the same area. If too many users and unique objects are grouped together, it will lead to an increase in both computational requirements and network traffic (bandwidth) [60]. In addition to reducing immersion, lag caused by excessive network traffic reduces the accuracy of movement within the game and can lead to motion sickness [20]. Therefore, Second Life is not well suited to serious games that require accurate movement, pointing, or timing. Furthermore, the Second Life server is provided by and maintained exclusively by Linden Labs. Institutions may set up an "island"

Fig. 2.5 Large rockets on display at Spaceport Alpha in Second Life [52]

(a privately owned area of land within Second Life where the owner has control over the content and access), on the Linden server. In January 2007, Linden Labs released source code of the Second Life client software (Second Life Viewer). The Open Simulator community was able to create their own server software capable of connecting to Second Life clients [40]. Institutions can now host their own virtual worlds similar to those found on Second Life but with fewer restrictions. Second life restricts the number of objects and the size of scripts depending on the size of the land owned [30].

2.2.2.4   eAdventure

eAdventure was developed as a research project at Complutense University of Madrid (Spain), one of the oldest universities in the world [12]. eAdventure is the only framework in our top-ten survey results that was designed specifically for educational use. The framework has been designed to allow educators to author their own interactive content. eAdventure games can be integrated with learning management systems such as Moodle and Sakai [17].

Unlike most of the frameworks revealed by our literature survey, eAdventure is limited to building one specific type of game, 2D point-and-click adventures. Most step-by step procedures can be easily adapted to a point-and-click style interface.

By limiting development to one type of interaction, the framework is far less complicated than a general purpose game development tool. Games can be developed using the eAdventure language or created with a graphic-based editor which does not require a technical background or any programming skill [57]. eAdventure games can be deployed as a standalone application or as an applet playable in a web browser [17]. eAdventure is freely available and open source. You may use it, modify it, redistribute it, or integrate it into your project even if your project has commercial applications [17]. Documentation and tutorials are provided on the official site. Support is provided by the developers as well as community members by way of an online forum.

2.2.2.5 Unreal

The Unreal Engine was created by Epic Games, Inc., currently based in North Carolina, USA. Unreal Engine 1 was released in 1998 and since then, Unreal has been used to create more than 50 top-selling games, including Gears of War, Borderlands, Mass Effect, and BioShock [18]. In addition to being one of the most widely used commercial game engines, the Unreal engine has also been used to develop many well-known serious games, including the US Department of Defense's America's Army [59] (Fig. 2.6), and NASA's Moonbase Alpha [38]. Unreal Engine 4 features a 64-bit color HDR (High Dynamic Range) rendering pipeline with a wide range of post-processing effects, such as motion blur, depth of field, bloom, and screen space ambient occlusion. Unreal is well known for having some of the most realistic lighting effects, including real-time global illumination and surface interreflections [18].

Fig. 2.6 America's Army serious game [59]

Epic games now provides full access to the Unreal engine's C++ source code. However, most games are developed completely using UnrealScript, a proprietary object oriented scripting language that is similar in syntax to Java. Unreal also provides a visual scripting tool called Kismet that allows events and actions to be connected together by way of a "drag and drop" style interface. Kismet enables level designers to create simple behaviors and gameplay without writing any code. The Unreal Editor (UnrealEd) is a WYSIWYG editing tool. Designers can make changes to the environment and then preview the level directly within the editor. UnrealEd is more than just a level editor; it is a collection of visual tools for sculpting terrain, mesh painting, sound editing, amongst other operations [18]. One common complaint about the Unreal engine is that it is geared toward FPS style games. However, the Unreal engine can be used (and has been used) to develop many different types of games from virtually every genre even though game development with the Unreal engine becomes more difficult the farther you stray from the engine's FPS roots [61].

Support is provided by the large community of users consisting of both professional and amateur developers who provide help to fellow members in the community forums. Licensed developers receive a greater level of support provided by Epic [41]. The Unreal Development Kit (UDK) is free for non-commercial and educational use. For commercial projects, there is a 5 % royalty charged for revenue in excess of $3,000 USD per product per quarter [18].

2.2.2.6 Torque

The technology behind Torque Game Engine was originally developed by Dynamix for the Tribes series of games. Torque is currently maintained by GarageGames and has been re-engineered as a general purpose game-development framework. The Torque game engine was written in the C++ programming language and is now open source. The TorqueScript scripting language is similar syntactically to C++. Most of the game engine's functionality is available through TorqueScript, which makes it possible to build an entire game without the need to recompile the engine. The World Editor is a collection of graphical editors which allow artists to sculpt terrain, lay roads and rivers, place objects, edit materials, and create particle effects without writing any code [22].

Torque 3D supports Windows and all major web browsers. Unofficial support for other platforms may be provided by the community. GarageGames also offers a 2D version of the framework (Torque 2D) which is dedicated to 2D game development and supports most mobile devices. Both Torque 3D and Torque 2D are free under the MIT open source software license. Support is generally provided by the online community. Expert on-site support provided by GarageGames starts at $5,000 USD [22].

2.2.2.7 Ogre

According to our survey, Ogre was the 7th most utilized engine despite the fact that Ogre is a graphics engine only. In other words, if you want to add features such as audio, physics, and artificial intelligence to your game, you must either program such features yourself or integrate third party libraries into the project. Example programs combining Ogre with various physics and audio libraries are freely available thanks to the active community [39]. Ogre is freely available and open source under the MIT license. Supported platforms currently include Windows, Linux, Mac OSX, Android, iOS, and Windows Phone. Support is provided by the community via online forums [39].

2.2.2.8 Cocos

Cocos is a very popular freely available open source game development framework that supports development on all major mobile and desktop platforms. Cocos games can be deployed as a browser-based or stand-alone application [25]. Cocos includes visual editors for building the user interface, animating characters, and placing objects in the scene. There are several versions of Cocos being maintained by the community, some of which are focused on developing games for a specific platform. The most popular is Cocos2D-x which is written in C++ and supports the Javascript and Lua scripting languages. There are versions that support Python, C#, and Objective-C (iOS development) [10]. Cocos3D extends the 2D version by adding full support for 3D rendering based on OpenGL [11].

2.2.2.9 XNA

XNA Game Studio is game engine developed by Microsoft with the goal of making it easier to develop games for Windows, the Xbox, and Windows Phone [35]. It is free to develop XNA games for Windows, Xbox 360 game development requires a small yearly subscription fee. XNA was released in 2006 and quickly attracted a large community of independent game developers because of its simplicity and low cost [28]. Both 3D and 2D games can be created with XNA. XNA games written using the C# programming language. Although there are no official visual editing tools included with XNA, third party editors for UI design and level editing are available. Despite its popularity, Microsoft decided to retire XNA in 2014 [44]. The open source community has picked up where Microsoft left off with the creation of MonoGame. MonoGame is an open source implementation of XNA that has been extended to support additional platforms such as Linux, iOS, Android, and MacOS [37].

2.2.2.10   RPG Maker

RPG Maker was created by Enterbrain located in Tokyo Japan, and as the name suggest, this framework was designed specifically for the development of Role Playing Games (RPGs) only. More specifically, RPG Maker can only be used to create tile based games with separate combat screens viewed from the side (non-tiled). The developer is provided with limited control over the interface or game play, and must therefore focus on crafting a story told through NPC dialogs and quests instead. However, RPG Maker's singular focus makes it one of the simplest frameworks to use. No knowledge of programming is required. The visual editor assists the user in developing animated characters, and the WYSIWYG map editor allows trees, mountains and other objects to be added simply by selecting the object and clicking on one of the grid squares [45].

2.3   Discussion and Conclusions

Two literature surveys were conducted using Google Scholar, ACM Digital Library, and the IEEE Xplore Digital Library. The first survey (Survey 1) compared the number of papers mentioning the name of each engine in relation to search terms such as "serious game," "educational games," and "simulator." The frameworks and engines that scored well in this survey are those that are often discussed in academic publications. It could be argued that some of the frameworks listed here are not game engines or game development frameworks at all. Adobe Flash for example, is a framework for developing web-based applications in general, and is commonly used to make online advertisements. Second Life is a massively multiplayer virtual world, and it is not technically a game or a game engine. However, the content creation tools available to Second Life users, allows them to create their own games and interactive content. The serious games created with Second Life are essentially modifications ("mods") that exist as a location within the Second Life world. After Unity, Adobe Flash, Second Life, Unreal, and XNA Game Studio were referenced most often in academic publications. The second survey (Survey 2) attempted to discover which game engines and frameworks were used most often to develop serious games. A search was performed using the search term "serious game." Papers detailing the creation of a serious game were downloaded and read in order to determine which engine was used. Four of the top ten engines were full-featured game creation frameworks and six of the top ten were commercial products. However, three of these commercial game engines were free for non-commercial use. Out of the top ten most utilized frameworks, eAdventure was the only one designed specifically for educational purposes.

Table 2.1 summarizes some of the most popular game engines and frameworks (as determined by the survey results detailed in Sect. 2.1) along with common features that are important to developers of serious games.

Table 2.1 Popular engines/frameworks and their features (features that are available in the form of an add-ons created by a third party or with additional licensing fees are not included)

	Unity	Flash	Second life	e-Adventure	Unreal	Torque	Ogre	Cocos	XNA	RPG maker
Level editor	■	■	■	■	■	■		■		■
Scripting	■	■	■	■	■	■		■		■
C++					■	■	■	■		
Networking	■	■	■		■	■		■	■	
3D Graphics	■		■		■	■	■	■	■	
Shader effects	■				■	■	■	■	■	
Dynamic shadows	■				■	■	■		■	
Physics	■		■		■	■		■	■	
Artificial intelligence	■				■	■				
Free non-commercial	■		■	■	■	■	■	■	■	
Free for commercial				■			■	■	■	
Mobile devices	■	■			■			■	■	■
Web player	■	■		■				■	■	■

Companies such as Unity Technologies and Epic refer to their products as "game engines" even though they provide a complete set of visual editing tools. Adding to the confusion, frameworks such as Unreal include the visual editors with shipped games. This allows the modding community to modify and extend the game by adding new content. Free content created and shared by modders helps to extend the life of commercial games, which in turn leads to increased sales [27]. Many serious games are really game mods which can be distributed freely, although they do require the player to own a copy of the original game in order to play them. Graphics engines (APIs dedicated to rendering graphics only) are often referred to as game engines even though they lack much of the functionality required to develop a complete game. Part of the confusion stems from the fact that popular graphics engines are often combined with physics and audio engines by the user base, and distributed via personal webpages or as official tutorials. Academic publications to date rarely make the distinction between game development frameworks that supply a complete set of editing tools and game engines which supply a code base of core functionality.

The results presented here suggest that serious game developers are primarily using game engines and frameworks that were designed for the creation of leisure-based or entertainment-based games. Given the disparity in available resources to serious game developers when compared to commercial (entertainment) game companies, it is peculiar that they chose the same tools. This suggests that the currently available serious game engines may be lacking many of the features found in commercial engines. The results may also indicate that game engines designed specifically for serious games do not simplify the process beyond that of commercial entertainment focused engines and frameworks. Each game engine/framework has a number of features that may lend itself to a specific serious game development project.

It should be noted that although a significant amount of time and effort was placed to conduct the searches and review the hundreds of resulting papers, there are limitations to our results. More specifically, our results are specific to papers within three academic databases (Google Scholar, IEEE Xplore Digital Library, and ACM Digital Library), and therefore, game engines and frameworks that are not described in any research-based publications may have been missed altogether. The results of Surveys 1 and 2 are also subject to any bias Google Scholar, IEEE Xplore Digital Library, or the ACM Digital Library may have relating to the ordering of the search results. Furthermore, although access to the ACM and IEEE Digital Libraries was available, 46 % of the articles revealed in the Google Scholar search were not freely accessible and thus not considered.

Therefore, significant limitation was introduced by the fact that papers were selected only if they could be freely accessed. Despite these limitations, the Surveys 1 and 2 do help to remove any personal bias on the part of the authors in selecting the frameworks that warrant further investigation. Finally, although no claim is made that the conducted search is completely exhaustive of all serious games developed and all available development tools, a number of popular and commonly used game engines and frameworks, in addition to some of their features, have been outlined.

2.3.1 Content Experts and Educators as Game Developers

Most of the frameworks revealed by our surveys are complex game development tools requiring extensive training or programming knowledge before they can be utilized effectively. Content experts and educators often need to hire skilled developers to create games on their behalf. The eAdventure framework is unique in that it is targeted to educators who are not expected to have a game development background or any programming knowledge. Many educators are prolific in the amount of educational content they author for their students. This content takes the form of PowerPoint presentations, instructional documents, quizzes, videos, and websites. However, the educational content currently being created by educators is rarely interactive.

Educators could take on a greater role in the design and development of educational software and serious games if tools were made available to them that greatly simplified the game development process. The availability of such tools coupled with an educator's educational background (instructional design in particular), may lead to more effective serious games as well as the proliferation of interactive educational content. Although it is unrealistic to expect educators to be highly skilled in game development and programming, one can expect that some educators would be willing to take the time to learn some basic skills in order to provide a better learning environment for their students. If a game creation framework could be made "simple enough", a layperson could become competent enough to develop their own games, or at least modify existing games developed by others using the framework. eAdventure may be a step in the right direction, but currently it is limited to the creation of one specific genre of game, and more specifically, 2D point-and-click adventures. Although 2D point-and-click adventures can be applied to a large number of learning applications, we believe that there is need for a simple general purpose framework for educators, or perhaps a series of frameworks each simplified for the development of one specific type of game, to ultimately provide educators (and serious games developers alike), greater freedom and flexibility.

Empowering content experts and educators to build their own interactive software and serious games may allow them to better meet the needs of their students. It could be argued that educators who do not have a background in game development may create games that do not offer an enjoyable experience for the players (students). However, it could also be argued that game developers who do not understand the educational content of the game or the needs of the students, may create games that do not meet the educational needs of the students.

To empower educators and provide them the opportunity to have a greater role within the serious game development process, we have assembled an interdisciplinary group of researchers, educators, computer scientists/engineers, and game developers, and have begun development of a novel game development framework that will greatly simplify the three primary areas of game development (art and animation, sound and music, and programming) [14]. The framework (known as the Serious Game Surgical Cognitive Education and Training Framework (SCETF)), allows for the creation of serious games by those who are not necessarily game developers or expert programmers (e.g., many educators). Although the framework may be of value to educators in general, we are currently targeting medical educators specifically. Serious gaming is growing in popularity within the medical education and training realm for a number of reasons including their ability to motivate and engage the trainees, and their cost effectiveness. However, medical procedures and practices are constantly changing in response to new technologies, new treatments, and a shifting patient demographic. By providing medical educators with the tools necessary to easily develop and/or easily modify their own interactive content (i.e., serious games), we are aiming for a proliferation of serious games for medical education and training with a current focus on cognitive-based surgical skills.

As part of the development of the SCETF, a needs analysis is being conducted to ensure that the design of the tool will meet the needs of the end user. This includes conducting a survey to gauge the technical background and computer literacy of clinical educators in order to gauge whether they possess the appropriate skillset and willingness to develop interactive software and games (given appropriate development tools). The results of this survey will then be used to guide the development process and ultimately allow us to strike a balance between ease of use and functionality, while providing the user with the necessary options and tools to ideally develop effective serious games. The survey itself has been evaluated by a panel of experts using the Delphi method to determine whether the survey questions are appropriate given the demographic and research goals. The Delphi method seeks to obtain a consensus on the opinions of experts, termed panel members, through a series of structured questionnaires [16]. As part of the process, the responses from each round are fed back in summarised form to the participants who are then given an opportunity to respond again to the emerging data. The Delphi method is therefore an iterative multi-stage process designed to combine opinion into group consensus. Experts from a variety of fields; medicine, education, and computer science, have taken part in the panel.

Although participants for this survey are still being sought, preliminary results suggest that clinical educators generally do possess a high degree of computer literacy and are interested in developing their own interactive content. Participants have also indicated that the time they are able to devote to learning to use such tools is limited. The framework should therefore be made simple enough that an educator can learn to use it by following the provided examples in their spare time over a one to two week period.

In summary, serious games provide a viable education and training option, and are becoming more widespread in use for a variety of applications. However, in contrast to traditional entertainment game design and development, serious games and serious game designers/developers must strictly adhere to the corresponding content/knowledge base, while ensuring that their end product is not only fun and engaging, but is also an effective teaching tool. Despite the importance of proper serious game design, very few tools that are specific to serious game development and account for the unique requirements and complexities inherent in the serious game development process are currently available. In this paper we have provided insight into more popular tools currently used to develop serious games. We have also provided details regarding our current work that is seeking to bridge the game between what is currently available and what is actually needed by developers of serious games. Although plenty of work remains to be done, we are confident that these first steps will at the very least bring light to this important topic and ideally lead to greater work in the area of serious game framework and engine design.

Acknowledgements This work was supported by the Natural Sciences and Engineering Research Council of Canada (NSERC), and the Social Sciences and Humanities Research Council of Canada (SSHRC), Interactive and Multi-Modal Experience Research Syndicate (IMMERSe) initiative.

References

1. Abt CC (1975) Serious games. Viking Compass, New York
2. Adobe Flash Product Information. http://www.adobe.com/ca/products/flash-builder-family. html. Accessed 27 Aug 2015
3. Arnab S (ed) (2012) Serious games for healthcare: applications and implications: applications and implications. IGI Global
4. Becker K, Parker J (2011) The guide to computer simulations and games. Wiley, Indianapolis
5. Bergeron B (2006) Developing serious games. Thomson Delmar Learning, Hingham, MA
6. Bishop L, Eberly D, Whitted T, Finch M, Shantz M (1998) Designing a PC game engine. Comput Graph Appl IEEE 18(1):46–53
7. Breuer J, Bente G (2010). Why so serious? On the relation of serious games and learning. Eludamos. J Comput Game Cult 4(1):7–24

8. Campbell J, Core M, Artstein R, Armstrong L, Hartholt A, Wilson C, Birch M (2011) Developing INOTS to support interpersonal skills practice. In: Aerospace conference, 2011 IEEE. IEEE, pp 1–14
9. Chiang O (2010) FarmVille players down 25 % since peak, now below 60 million. http://www.forbes.com/sites/oliverchiang/2010/10/15/farmville-players-down-25-since-peak-now-below-60-million/. Accessed 16 Nov 2015
10. Cocos2D game engine, Official website. http://cocos2d.org/. Accessed 13 Nov 2015
11. Cocos3D game engine, Official website. http://cocos3d.org/. Accessed 13 Nov 2015
12. Complutense University of Madrid, Home page. https://www.ucm.es/english. Accessed 23 Aug 2015
13. Corti K (2006) Game-based learning; a serious business application. PIXELearning, Coventry, UK
14. Cowan B, Kapralos B, Moussa F, Dubrowski A (2015) The serious gaming surgical cognitive education and training framework and SKY script scripting language. In: Proceedings of the 8th international conference on simulation tools and techniques. ICST (Institute for Computer Sciences, Social-Informatics and Telecommunications Engineering), pp 308–310
15. Craighead J, Burke J, Murphy R (2007) Using the unity game engine to develop sarge: a case study. Computer 4552:366–372
16. Dalkey N, Helmer O (1963) An experimental application of the Delphi method to the use of experts. Manage Sci 9(3):458–467
17. eAdventure, Home page. http://e-adventure.e-ucm.es/. Accessed 22 Aug 2015
18. Epic games, unreal engine, Official website. http://www.unrealengine.com. Accessed 8 Nov 2015
19. Feronato E (2011) Flash game development by example: build 9 classic flash games and learn game development along the way. Packt Publishing Ltd
20. Fraser M, Glover T, Vaghi I, Benford S, Hindmarsh CGJ, Heath C (2000) Revealing the realities of collaborative virtual reality. In: Proceedings of the third international conference on collaborative virtual environments, New York, NY, USA, pp 29–37
21. Gans JS, Halaburda H (2013) Some economics of private digital currency. In: economics of digitization. University of Chicago Press
22. GarageGames, Official website for the Torque 2D and Torque 3D game engines. https://www.garagegames.com/. Accessed 8 Nov 2015
23. Goldstone W (2009) Unity game development essentials. Packt Publishing
24. Hill V, Meister M (2013) Virtual worlds and libraries Gridhopping to new worlds. Coll Res Libr News 74(1):43–47
25. Hussain F, Gurung A, Jones G (2014) Cocos2d-x game development essentials. Packt Publishing Ltd
26. Iuppa N, Borst T (2010) End-to-end game development: creating independent serious games and simulations from start to finish. Focal Press, Oxford, UK
27. Jeppesen LB (2004) Profiting from innovative user communities. In: Working paper, Department of Industrial Economics and Strategy, Copenhagen Business School
28. Keller B (2006) XNA studio: introduction to XNA. In: Game developer conference
29. Kemp J, Livingstone D (2006) Putting a second life "metaverse" skin on learning management systems. In: Proceedings of the second life education workshop at the second life community convention. The University Of Paisley, CA, San Francisco, pp 13–18
30. Konstantinidis A, Tsiatsos T, Demetriadis S, Pomportsis A (2010) Collaborative learning in opensim by utilizing sloodle. In: The sixth advanced international conference on telecommunications, 9–15 May, 2010, Barcelona, Spain, pp 91–94
31. Lewis M, Jacobson J (2002) Game engines. Commun ACM 45(1):27
32. Linden BK. Q1 2011 Linden dollar economy metrics up, Users and usage unchanged, June 5th 2011. Accessed 2013
33. Linden Labs, About Linden lab. http://www.lindenlab.com/about. Accessed 8 Nov 2015
34. Michael D, Chen S (2006) Serious games: games that educate, train and inform. Thomson Course Technology, Boston, MA

35. Miller T, Johnson D (2010) XNA game studio 4.0 programming: developing for windows phone 7 and xbox 360. Pearson Education
36. Mitchell A, Savill-Smith (2004) The use of computer and video games for learning: a review of the literature. London, UK. www.LSDA.org.uk
37. MonoGame, Official website. http://www.monogame.net/. Accessed 10 Nov 2015
38. NASA, Moonbase Alpha. http://www.nasa.gov/offices/education/programs/national/ltp/games/moonbasealpha/index.html. Accessed 1 Mar 2014
39. Ogre, Official website for Ogre, An open-source 3D graphics engine. http://www.ogre3d.org/. Accessed 12 Nov 2015
40. Open simulator, Information page. http://opensimulator.org/wiki/Main_Page. Accessed 14 Oct 2015
41. Petridis P, Dunwell I, Panzoli D, Arnab S, Protopsaltis A, Hendrix M, de Freitas S (2012) Game engines selection framework for high-fidelity serious applications. Int J Interact Worlds
42. Ratan R, Ritterfeld U (2009) Classifying serious games. In: Ritterfeld U, Cody M, Vorderer P (eds) Serious games: mechanisms and effects. Routledge, New York/London
43. Roman PA, Brown D (2008) Games–just how serious are they. In: The interservice/industry training, simulation & education conference (I/ITSEC), vol 2008, no 1
44. Rose M (2013) It's official: XNA is dead. http://www.gamasutra.com/view/news/185894/Its_official_XNA_is_dead.php. Accessed 25 Sep 2015
45. RPG Maker, Official website. http://www.rpgmakerweb.com/. Accessed 9 Nov 2015
46. Rymaszewski M (2007) Second life: the official guide. Wiley
47. Sawyer B (2009) Foreword: from virtual U to serious games to something bigger. In: Ritterfeld U, Cody MJ, Vorderer P (eds) Serious games. Mechanisms and effects, pp xi–xvi
48. Second Life, Official website. http://secondlife.com/. Accessed 20 Oct 2015
49. Sherrod A (2007) Ultimate 3D game engine design & architecture. Charles River Media
50. Shute V, Ventura M, Bauer M, Zapata-Rivera D (2009) Melding the power of serious games and embedded assessment to monitor and foster learning. In: Ritterfeld U, Cody MJ, Vorderer P (eds) Serious games. Mechanisms and effects, pp 295–321
51. SLOODLE (simulation linked object oriented dynamic learning environment). http://www.sloodle.org/. Accessed 22 Oct 2013
52. Space today online. http://www.spacetoday.org/Rockets/SecondLife/SL_Spaceport_Alpha.html. Accessed 23 Oct 2013
53. Squabble studios, Sort 'Em Up project page. http://www.squabblestudios.ca/sort.html. Accessed 10 Oct 2013
54. Squire K, Jenkins H (2003) Harnessing the power of games in education. Insight 3(1):5–33
55. Susi T, Johannesson M, Backlund P (2007) Serious games—an overview (Technical report no HS-IKI-TR-07-001). School of Humanities and Informatics, University of Skovde, Sweden
56. The World Wide Web Consortium (w3C) HTML standard. http://www.w3.org/standards/webdesign/htmlcss. Accessed 26 Aug 2015
57. Torrente J, Moreno-Ger P, Fernández-Manjón B, Sierra JL (2008) Instructor-oriented authoring tools for educational videogames. In: 8th international conference on advanced learning technologies (ICALT 2008), Santander, Spain, 2008, pp 516–518
58. Unity, Official website. http://unity3d.com/. Accessed 10 Oct 2015
59. US Army, America's Army serious game. http://www.americasarmy.com/. Accessed 9 Oct 2013
60. Warburton S (2009) second life in higher education: assessing the potential for and the barriers to deploying virtual worlds in learning and teaching. Brit J Educ Technol 40(3):414–426
61. Zielke MA (2010) The game engine space for virtual cultural training. The University of Texas at Dallas Arts and Technology

Chapter 3
A Review of and Taxonomy for Computer Supported Neuro-Motor Rehabilitation Systems

Lucas Stephenson and Anthony Whitehead

Abstract Stroke and other acquired brain injuries leave a staggering number of people worldwide with impaired motor abilities. Repetitive motion exercises can, thanks to brain plasticity, allow a degree of recovery, help adaptation and ultimately improve quality of life for survivors. The motivation for survivors to complete these exercises typically wanes over time as boredom sets in. To ease the effect of boredom for patients, research efforts have tied the rehabilitation exercises to computer games. Review of recent works found through Google scholar and Carleton's summon service which indexes most of Carleton's aggregate collection, using the key terms: *stroke, acquired brain injury* and *video/computer games* revealed a number of research efforts aimed primarily at proving the viability of these systems. There were two main results; (1) A classification scheme for computer neurological motor rehabilitation systems (CNMRS) was created based on the researched systems. (2) The systems reviewed all reported some degree of positive results—small sample sizes, large range of neuro-impairments, varied motion recording technology and different game designs make it problematic to formally quantify results, beyond a general net positive trend. The taxonomy presented here can be used to classify further works, to form the basis for meta-studies or larger long term longitudinal study and by neurological rehabilitation practitioners to help select and deploy systems to match client specific needs.

L. Stephenson (✉) · A. Whitehead
Carleton School of Information Technology, Carleton University,
230 Azrieli Pavilion, 1125 Colonel By Drive, Ottawa, ON K1S 5B6, Canada
e-mail: lucas.stephenson@carleton.ca

A. Whitehead
e-mail: Anthony.Whitehead@carleton.ca

© Springer International Publishing AG 2017
A.L. Brooks et al. (eds.), *Recent Advances in Technologies
for Inclusive Well-Being*, Intelligent Systems Reference Library 119,
DOI 10.1007/978-3-319-49879-9_3

3.1 Introduction

Stroke and other traumatic brain injuries leave many with residual motor impairment. Stroke is the leading cause of adult disability in the United States [1], and along with other types of acquired brain injuries, results in a variety of neurological impairments, including impaired motor ability. A primary method for aiding recovery are repetitive motion exercises. The goal of these exercises is to utilize the brain's plasticity to (re)build and strengthen neural pathways to affected motor systems which allows a degree of recovery [2]. This type of rehabilitation differs from typical physical rehabilitation, in that the *focus* is on rebuilding neurological control, and not directly on muscle strength and flexibility.

The effectiveness of neurological motor rehabilitation is dependent on the volume of repetitions of prescribed exercises, in other words: a high volume of repetitions is prescribed by clinicians [3]. Patient motivation, especially when away from a clinician's watch, often wanes over time [4]. The primary causes of declining interest and motivation are the declining benefit to exercise ratio and mental fatigue (boredom). The mental boredom of repeatedly performing "tedious" movements can be mitigated by attaching the exercises to additional stimulation, such as interactive games. This connection enables the client to potentially sustain interest and maximise the possible movement-based therapy recovery.

By digitally capturing human movement, computer system processing is possible; passing this information through a computer can provide verification to help ensure that movements comply with prescribed therapeutic exercises. Supporting compliance provides confidence to stakeholders; therapists can verify that their clients are performing required therapeutic movements and clients can be sure that they are optimising their recovery whilst performing their exercises. A system that helps verify that exercises are being executed correctly allows prescribed rehabilitation to be most effective [5]. Further, a system that allows the capture of motion data allows a trained professional to review and enforce proper form without their physical presence during the exercises. Indirect monitoring of clients allows greater coverage of patients at lower cost and allows for remote (outpatient) rehabilitation and facilitates long term tracking by therapists, which has been shown to be instrumental in maximising recovery [6].

The purpose of a computer neurological motor rehabilitation system (CNMRS) is to facilitate the provisioning of effective and long-term rehabilitation services to patients. A typical CNMRS is composed of 4 main logical parts: capture, process, interact and report, illustrated in Fig. 3.1.

A therapist is a primary stakeholder; deciding which CNMRS best fit with a client's current needs, monitoring progress and re-assessing requirements. In practise, a system that makes use of a CNMRS will have at least one software application configured; that users will interact with, the software would provide the user feedback including the cognitive and motivational aspects of the therapy. When choosing applications to be used in conjunction with the CNMRS, the goals of the client must be considered. For example, if the application is to be used as pure motivation/entertainment, then a commercially available game might be

Fig. 3.1 Logical structure of
a CNMRS

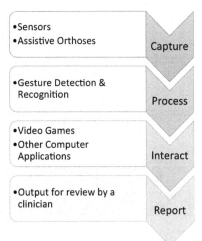

applicable by mapping gestures and movement input from the CNMRS to in-game controls. Alternatively, if specific learning objectives are desired, custom or specialized software could be provided, optimally a mix of these two ends of the spectrum could be used to meet a specific client's needs.

3.2   Taxonomy Overview

Identifying and classifying the characteristics of computer systems intended to provide support for neurological motor-rehabilitation enables clarification of the available systems and technologies, through categorization. This categorization allows efficient evaluation of current systems based on case specific need. Further, it allows a simple path to develop goals for new and evolving systems.

The cognitive aspect of these systems is either motivation based or client-specific-goal based, which are considered aspects of application design. It is assumed that a CNMRS provides the user with control of computer applications; ranging from commercially available games to custom designed software for cognitive rehabilitation and development. The software user interface, or game design aspect of these systems is related, but distinct and separable from the sensor systems that support these systems. This taxonomy is a classification system for CNMR hardware and software systems used to acquire and process raw input and provide directly usable data, for use as input into a computer system. It does not include terms to describe games or other software used with these systems.

To develop a taxonomy for CNMRSs an existing related taxonomy [7] was used as a starting point. The proposed taxonomy in [7] was deemed insufficient primarily because it couples application design and control aspects, a coupling that is felt artificially limits the applicability of the control apparatus. Further, the classification scheme assumed systems were pre-configured; we wanted to include basic setup and deployment information, in order to give stakeholders an idea of the costs involved with providing a system to the end users.

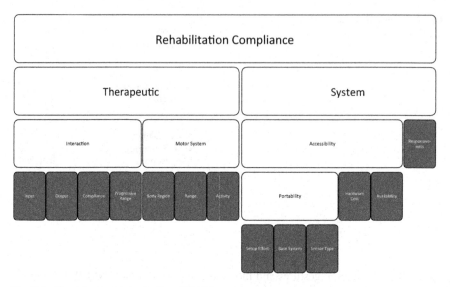

Fig. 3.2 Classification hierarchy for CNMRS

A review of the available literature was performed to identify key features of these systems. After features were identified, terms that allow classifying a system's implementation details were added so that review of the potential barriers to acquisition, setup and use of a particular CNMRS is possible.

The resultant taxonomy presented in Fig. 3.2, is presented as a hierarchy, intended to categorize the taxonomy terms, and thus classified systems into digestible chunks. The classification terms themselves are presented in a darker shade.

3.3 Taxonomy Development

There were 3 stages to the survey that informed the development of the taxonomy. Stage 1 included reviewing the works cited in Serious Games for Rehabilitation [7] and focusing on the research that was classified as having a "Motor" Application Area in the presented taxonomy. From these works, we were able to identify a variety of input mechanisms that used a number of sensor input types further described in Sect. 3.3.1.1. We also found that these systems are usually designed specifically for a particular body region/joint(s) [8–11]. While all systems surveyed provided video (and assumed audio ability) output [8–11], Ma and Bechkoum [9] made use of a haptic glove to provide an additional output modality, namely, haptic. Also noted was the granularity of movements supported. For example, Burke et al. [10] required explicit positions be held by the user, while others [8, 9, 11] proposed systems with a dynamic movement range that could be adjusted in software. This enabled us to craft a draft intermediary CNMRS taxonomy.

The 2nd stage of the review involved querying research databases, primarily Google scholar to seek out recent (since 2011) projects that involved stroke or acquired brain injury (ABI) computer aided motor rehabilitation. Those works found were filtered for relevance (many of the terms are common or have homonyms) and the references in found works were inspected for additional projects. Subsequently, works were classified using the intermediary taxonomy, and minimal refinement and expansion of terms occurred. The individual terms of the taxonomy are discussed in Sect. 3.3.1 along with the projects that effected their inclusion within the final proposed taxonomy.

Stage 3 was a validation stage, a thought exercise, cognitively inspecting novel input and output mechanisms such as Google Soli [12] and Microsoft HoloLens [13], verifying that these could be represented within the proposed taxonomy. This stage resulted in no changes to the taxonomy.

3.3.1 Taxonomy Details

To provide clarity, subgroups of related terms were implemented. There are two main groups that delineate the therapeutic and system classifiers. The therapeutic side includes all the portions of a system that would be relevant to treatment and the therapy capabilities of the system. The system stem includes classifiers related to the setup, cost and mobility and hardware of the system.

3.3.1.1 Therapeutic

The therapeutic aspect of the systems reviewed was further divided into 2 subgroups: (1) Interaction: to identify a system's input and output modalities and how and if the system can support progressive range, and (2) Motor System: provides terms for the granularity (fine vs. gross) and type(s) of motor system that is intended to be supported by a system.

Interaction

The interaction group holds classifiers that identify the ways in which the client interacts with the rehabilitation system; how movement data is provided to the system, any output mechanisms inherent in the system and whether the system provides progressive range adjustment (e.g. range of motion can increase over time).

Input
The *Input* classifier indicates the type of input the system tracks. This can be motion or tangible. A tangible system tracks the user through their interaction with explicit objects, potentially providing an augmentation path for existing therapeutic techniques, such as in [14] where a box and blocks therapeutic game was augmented with sensors. A motion system uses sensors that track the client's movements directly, capturing motion, confirming the client's compliance with the prescribed gestures

and poses, either from external sensors, sensors attached to the body or explicit actions. The majority of the reviewed works included: basic video camera setups that do *blob detection* [10, 15, 16], systems that provide inverse kinematic skeletal reconstruction simulations using external (unattached sensors) [17–19], systems that use inertial measurement units (IMU) that are attached to the limbs [11], EMG sensors that measure muscle flexion [20], and pressure sensor arrays that can be used to measure weight distribution [8, 11]. Additionally there were a few examples of systems that used more direct interaction, such as touch [21], or provided classical options for input such as mouse and keyboard to record motion data [9].

Output
Output indicates the methods the client receives live feedback from the system. Most of the reviewed systems output standard video, and sometimes audio. There was an example of specialized video; a head mounted display, allowing for a 3D-immersive environment [9]. An additional form of output that was recorded were haptic gloves that can provide some tactile based feedback to clients when worn [9, 22].

Compliance
The Compliance term is used to indicate if the system helps the client complete the required rehabilitative motions correctly, either through software analyzed feedback or (inclusively) physical orthosis. Initial construction of the taxonomy did not include *compliance,* primarily because it was seen as something purely provided by analyzing sensor input and providing feedback through the software interface. There was, however a number of works that included physical feedback by way of orthosis [23], control of objects (tangible) [14] or haptic feedback [9] and the term was added.

Progressive range
Progressive range indicates that the system provides the ability for incremental changes to required motions, for example to extend (or contract), the required range of motion. This can be therapeutically beneficial allowing the client to progress over time with the use of the system [24]. The reviewed works were implicitly able to support progressive range, and this data would be interpreted by the target game or application, however, in most configurations there is a logical maximum range. For example, a camera-based motion capture sensor would be restricted to the camera's viewable area.

There is a variety of options for how progressive range is implemented:

1. The CNMRS provides the progressive range classification, providing the underlying system with the corresponding executed movement input only when the range is sufficient to overcome the current target extent.
2. The CNMRS passes sufficient motion data for applications to interpret range along with recognized motions, allowing in-game feedback.

The CNMRS could be developed to be flexible enough to provide both. Option 1 would be ideal for existing software (where the CNMRS emulates standard input mechanisms) while 2 is potentially more interactive, as it allows the target software to provide a varying experience and direct feedback.

Motor System

The motor system subgroup holds terms to classify a system's target body region, and target motor range maximum and minimum, which can be used to determine what range of the fine/gross motor spectrum a system can analyze.

Body Region
The term *Body Region* is used to specify what regions of the body can be tracked and in what dimensional order (1D, 2D, 3D). This is important for therapists selecting a CNMRS so that they can provide targeted rehabilitation. It was described in every work surveyed; explicitly classifying supported body region(s) allows clinicians to select relevant systems more efficiently. Appropriate values would include; whole body/balance, major limb (arm/leg/head), minor joints (hands/feet/fingers/toes).

Range
Range is used to specify where on the gross versus fine motor spectrum the system lies. It specifies the minimum and maximum recordable movements in terms of distance and speed. Underlying sensor systems generally have limitations—a touch panel has specific dimensions, a video capture system can only record in a well-defined area, and IMU systems have maximum speeds.

Activity
The *Activity* classifier allows a brief description of the typical rehabilitative motions supported by the system; that is, the motions the researchers designed the system to support. This classifier, in combination with the body region classifier, enables more comprehensive coverage of the system's designed rehabilitation. Researchers looking to perform further work in supporting a specific motion type can use this term to focus their efforts on similar projects.

3.3.1.2 System

The system group includes classifiers for how accessible and responsive the system is. Table 3.2 shows the system subgroup classification results for the reviewed works.

Accessibility

The accessibility subtree was included to provide classification of the system within potential time, monetary and mobility restrictions. It is used to identify *where, when and at what cost can a CNMRS be used?*

Portability

The portability group explicitly states if a system is movable during use (base system). The amount and type of setup were also viewed to be relevant to clinicians (setup effort). All the systems reviewed were attached to stationary systems, and thus were classified as having a fixed range system value for portability. Subsequently, an additional term (sensor type) was added, to allow identification of sensor mobility. The intention of this field is to allow researchers looking to extend and create mobile rehabilitation frameworks to be able to identify and leverage existing works.

Hardware Cost

Hardware cost directly affects the level of access a client will have to the system. However, none of the systems reviewed specified this information, in addition, the price of these systems is constantly in flux. Therefore, an estimation is provided with a relative rating of *Low*, *Medium* or *High* based on estimates of hardware cost.

Availability

The *Availability* term is used to indicate the availability of the hardware involved, the values *Off the Shelf (OTS)*, *Specialized* and *Multiple* can be specified. *OTS* indicates the sensors are generally available commercially, and easy to access, service and replace. *Specialized* means it is not mainstream hardware, and is likely more difficult to acquire, and repair/replace. *Multiple* is intended to represent systems that require more than one piece of sensor hardware, with at least one being *OTS* and another *Specialized.*

Responsiveness

In order to be considered valid for use in a game, a sensor system, must be able to provide an interactive experience for clients. However, real and perceived system reaction time is a factor in a system's appropriateness for use. For example, if the system requires that a pose be held, or has a client-noticeable delay in recognizing a movement, then the system is unlikely suitable for use with many real-time applications. For this reason, the responsiveness term was added. Systems that indicated perceptively real-time responsiveness were indicated as "real-time", systems that required a position be held for any length time for technical reasons were indicated as "delayed".

3.4 Classifying CNMRS Systems

The crafted taxonomy detailed in Sects. 3.2 and 3.3 was applied to 55 published research papers. The *System* and *Therapeutic* areas, as seen in Fig. 3.2, were classified independently, and are presented below in Tables 3.1 and 3.2.

Table 3.1 Therapeutic terms: classification of reviewed CNMRSs

Research	Input	Output	Compliance	Progressive range	Body region	Range	Activity
Betker et al. [8, 27]	Motion	Video/audio	Virtual	Software	Sitting balance	Gross	Leaning
Ma and Bechkoum [9]	Motion	3D Video/audio/ haptic	Virtual/physical	Software	Hand	Fine	Pointing and grasping
Burke et al. [10], Crosbie et al. [28]	Motion	Video/audio	Virtual	Manual	Arms	Gross	Arm pointing and simple gestures
Ryan et al. [11]	Motion	Video/audio	Virtual	Software	Balance	Gross	Standing balance/in place navigation
Zhao et al. [14]	Tangible	Video/audio	Virtual/physical	Manual	Arms/hands	Fine	Grasping
Adamovich et al. [22]	Motion	Video/audio/haptic	Virtual/physical	Software	Hand	Fine	Pointing/hand dexterity
Chang et al. [17]	Motion	Video/audio	Virtual	Software	Arms and legs	Gross	Shoulder flexion and rotation
Fraiwan et al. [18]	Motion	Video/audio	Virtual	Software	Arms and legs	Gross	Various: gross full body
Cameirão et al. [15, 29, 30]	Motion	Video/audio	Virtual	Software	Arms	Gross	Pointing/grasping
Harley et al. [31]	Motion	Video/audio	Virtual	Software	Arms	Gross	Arm tilt
Okošanovi et al. [21]	Motion	Video/audio	Virtual	Software	Arms	Gross	Mouse gestures
Rahman et al. [19, 32, 33]	Motion	Video/audio	Virtual	Software	Arms, legs and hands	Gross and Fine	Various: full body
Tan et al. [16]	Motion	Video/audio	Virtual	Software	Hand	Fine	Hand gestures
Fukamoto [20]	Motion	Video/audio	Virtual	Manual	Foot/lower leg	Gross	Foot tilt
De et al. [23]	Motion	Video/audio	Virtual/physical	Manual	Arm	Gross	Assisted elbow flexion
Bhattacharya et al. [34]	Explicit	Video/audio	Virtual	None	Arm	Gross	Tilting arm/jostiq
Brokaw and Brewer [35]	Motion	Video/audio	Virtual	Software	Arms/shoulders	Gross	Arm gestures

(continued)

Table 3.1 (continued)

Research	Input	Output	Compliance	Progressive range	Body region	Range	Activity
Crocher et al. [36]	Motion	Video/audio	Virtual	Software	Hand	Fine	Hand movement/grasping/selecting
Dukes et al. [37]	Motion	Video/audio	Virtual	Software	Upper arm	Gross	Hand movement/grasping/selecting, arm movement
Erazo et al. [38]	Motion	Video/audio	Virtual	Software	Arm	Gross	Arm gestures
Gil-Gómez et al. [39]	Motion/speech	Video/audio	Virtual	Software	Upper arm	Gross	Arm gestures
Gonçalves et al. [40]	Motion	Video/audio	Virtual	Software	Standing balance	Gross	Standing or sitting balance
Kafri et al. [41]	Motion	Video/audio	Virtual/physical	Software	Ankle/leg	Gross	Ankle rotation
Kim et al. [42]	Motion	Video/audio	Virtual	Software	Arm/balance	Gross	Tilting/jogging in place/boxing
Kizony et al. [43]	Motion	Video/audio	Virtual	Software	Arm	Gross	Various
Labruyère et al. [44]	Motion	Video/audio	Virtual	Software	Arm	Gross	Arm gestures
Maier et al. [45]	Motion	Video/audio	Virtual	Software	Gait/balance	Gross	Gait/ambulating
Mainetti et al. [46]	Motion	Video/audio	Virtual	Software	Upper body	Gross	Arm gestures
Parafita et al. [47]	BCI*	Video/audio	Virtual	Software	General	Gross	Navigation/w BCI
Siqueira et al. [48]	Motion	Video/audio	Virtual/physical	Software	Ankle/leg	Gross	Ankle rotation
Saposniket al. [49]	Motion	Video/audio	Virtual	Software	Upper body	Gross	Arm gestures
Sucar et al. [50]	Motion/explicit	Video/audio	Virtual	Software	Arm	Gross	Hand grip and arm pointing
Vandermaesen et al. [51]	Tangible	Physical/audio	Physical	Manual	Arms	Gross	Griping and placing object
Méndez [52]	Motion	Video/audio	Virtual	Software	Balance	Gross	Balance activities
Vourvopoulos et al. [53]	Motion/explicit/BCI	Video/audio	Virtual/physical	Software	Sensor dependent	Sensor Dependant	Various, sensor dependant
Yavuzer [54]	Motion	Video/audio	Virtual	Software	Upper body	Gross	Arm gestures
Caglio et al. [55, 56]	Motion	Video/audio	Virtual	Software	Upper body	Gross	Driving/steering wheel

(continued)

Table 3.1 (continued)

Research	Input	Output	Compliance	Progressive range	Body region	Range	Activity
Rábago al. [57]	Motion	Video/audio	Physical/virtual	Software	Gait/balance/upper body	Gross	Gait/balance/game dependent
Holden [58]	Motion	Video/audio	Virtual	Software	Arms	Gross	Pouring
Mumford et al. [59, 60]	Tangible	Video/audio	Physical/virtual	Software	Upper body	Gross	Grasping/griping/moving objects
Ustinova et al. [61, 62]	Motion	Video/audio	Virtual	Software	Upper body	Gross	Arm gestures/extensions
Grealy et al. [63]	Motion	Video/audio	Physical/virtual	Software	Lower body	Gross	Cycling
Housman et al. [64]	Motion	Video/audio	Physical/virtual	Software	Arm	Gross	Orthosis supported arm movements
Subramanian et al. [65]	Motion	Video/audio	Virtual	Manual	Upper body	Gross	Pointing
Jinhwa et al. [66]	Motion	Video/audio	Physical	Manual	Gait/balance	Gross	Gait/ambulation
Yang et al. [67]	Motion	Video/audio	Physical	Manual	Gait/balance	Gross	Ambulation
Broeren, et al. [68]	Tangible	3D video/audio	Virtual/physical	Software	Upper extremities	Fine	Tangible hand/grasping light force feedback
Kim et al. [69]	Motion	Video/audio	Physical	Manual	Gait/balance	Gross	Ambulation
Jo et al. [70]	Motion	Video/audio	Virtual	Software	General	Gross	Various standing upper body
Kwon et al. [71]	Motion	Video/audio	Virtual	Software	General	Gross	Various standing upper body
Cikajlo et al. [72]	Motion	Video/audio	Virtual	Software	Balance	Gross	Supported standing and leaning
Kiper et al. [73–75]	Motion	Video/audio	Virtual	Software	Upper extremities	Gross and Fine	Grasping objects
Mirelman et al. [76, 77]	Motion	Video/audio	Virtual/physical	Software	Ankle/leg	Gross	Foot tilt
You et al. [78]	Motion	Video/audio	Virtual	Software	General	Gross	Various standing upper body
Agmon et al. [79]	Motion	Video/audio	Virtual	Software	Balance	Gross	Leaning/stepping

Table 3.2 System terms: classification of reviewed CNMRSs

Research	Portability: setup effort	Portability: base system	Portability: sensor	System availability	Hardware cost	Responsiveness
Betker et al. [8, 27]	Initial configuration assist per use	Fixed range	Wheelchair	Specialized	Medium	Real-time
Ma and Bechkoum [9]	Configuration per use	Fixed range	Free range	Specialized	High	Real-time
Burke et al. [10], Crosbie et al. [28]	Configuration per use	Fixed range	Fixed area	Specialized	Low	Delayed
Ryan et al. [11]	Initial configuration	Fixed range	Fixed area	OTS	Low	Real-time
Zhao et al. [14]	Initial configuration	Fixed range	Fixed area	Multiple	Medium	Real-time
Adamovich et al. [22]	Initial configuration	Fixed range	Free range	Multiple	High	Real-time
Chang et al. [17]	Initial configuration	Fixed range	Fixed area	OTS	Low	Real-time
Fraiwan et al. [18]	Initial configuration	Fixed range	Fixed area	OTS	Low	Real-time
Cameirão et al. [15, 29, 30]	Configuration per use	Fixed range	Fixed area	Specialized	Low	Delayed
Harley et al. [31]	Initial configuration	Fixed range	Fixed area	OTS	Low	Real-time
Okošanovi et al. [21]	Monitoring only	Fixed range	Both mobile and fixed area sensor	OTS	Medium	Real-time
Rahman et al. [19, 32, 33]	Initial configuration assist per use	Fixed range	Fixed area	Multiple	Medium	Real-time
Tan et al. [16]	Configuration per use	Fixed range	Fixed area	OTS	Low	Real-time
Fukamoto [20]	Configuration per use	Fixed range	Free range	Specialized	Medium	Real-time
De et al. [23]	Initial configuration	Fixed range	Free range	Specialized	Medium	Real-time
Bhattacharya et al. [34]	Demonstration	Fixed range	Free range	OTS	Low	Real-time
Brokaw and Brewer [35]	Demonstration	Fixed range	Fixed area	OTS	Low	Real-time
Crocher et al. [36]	Demonstration	Fixed range	Free range	OTS	Medium	Real-time

(continued)

Table 3.2 (continued)

Research	Portability: setup effort	Portability: base system	Portability: sensor	System availability	Hardware cost	Responsiveness
Dukes et al. [37]	Demonstration	Fixed range	Fixeda	Multiple	Medium	Real-time
Erazo et al. [38]	Initial configuration/calibration per use	Fixed range	Fixed area	OTS	Low	Real-time
Gil-Gómez et al. [39]	Initial Configuration	Fixed range	Fixed area	OTS	Low	Real-time
Gonçalves et al. [40]	Configuration per use	Fixed range	Fixed area	OTS	Low	Real-time
Kafri et al. [41]	Configuration per use	Fixed range	Fixed area	Specialized	Medium	Real-time
Kim et al. [42]	Demonstration	Fixed range	Fixed area	OTS	Low	Real-time
Kizony et al. [43]	Demonstration	Fixed range	Fixed area	OTS	Low	Real-time
Labruyère et al. [44]	Demonstration	Fixed range	Fixed area	OTS	Low	Real-time
Maier et al. [45]	Configuration per use	Fixed range	Free range	Specialized	Medium	Real-time
Mainetti et al. [46]	Configuration per use	Fixed range	Fixed area	Multiple	Low	Real-time
Parafita et al. [47]	Configuration per use	Fixed range	Fixed area	Specialized	Medium	Real-time
Siqueira et al. [48]	Configuration per use	Fixed range	Fixed area	Specialized	Medium	Real-time
Saposnik et al. [49]	Demonstration	Fixed range	Fixed area	OTS	Low	Real-time
Sucar et al. [50]	Demonstration	Fixed range	Fixed area	Specialized	Low	Real-time
Vandermaesen et al. [51]	Configuration per use	Fixed range	Fixed area	Specialized	Medium	Real-time
Méndez [52]	Demonstration	Fixed range	Fixed area	OTS	Low	Real-time
Vourvopoulos et al. [53]	Configuration per use	Fixed range	Fixed area	Multiple	High	Real-time
Yavuzer [54]	Demonstration	Fixed range	Fixed area	OTS	Low	Real-time
Caglio et al. [55, 56]	Demonstration	Fixed range	Fixed area	OTS	Low	Real-time
Rábago et al. [57]	Configuration per use	Fixed range	Fixed area	Specialized	High	Real-time

(continued)

Table 3.2 (continued)

Research	Portability: setup effort	Portability: base system	Portability: sensor	System availability	Hardware cost	Responsiveness
Holden [58]	Configuration per use	Fixed range	Free range	Specialized	Low	Real-time
Mumford et al. [59, 60]	Demonstration	Fixed range	Fixed area	Specialized	Medium	Real-time
Ustinova et al. [61, 62]	Configuration per use	Fixed range	Fixed area	Specialized	High	Real-time
Grealy et al. [63]	Demonstration	Fixed range	Fixed area	Specialized	Medium	Real-time
Housman et al. [64]	*	Fixed range	Fixed area	Specialized	*	Real-time
Subramanian et al. [65]	Configuration per use	Fixed range	Fixed area	Specialized	*	Real-time
Jinhwa et al. [66]	Configuration per use	Fixed range	Fixed area	Multiple	High	Real-time
Yang et al. [67]	Demonstration	Fixed range	Fixed area	Multiple	Medium	Real-time
Broeren et al. [68]	Demonstration	Fixed range	Fixed area	Specialized	High	Real-time
Kim et al. [69]	Demonstration	Fixed range	Fixed area	Multiple	Medium	Real-time
Jo et al. [70]	Demonstration	Fixed range	Fixed area	Specialized	High	Real-time
Kwon et al. [71]	Demonstration	Fixed range	Fixed area	Specialized	High	Real-time
Cikajlo et al. [72]	Configuration per use	Fixed range	Fixed area	OTS	Low	Real-time
Kiper et al. [73–75]	Assist per use	Fixed range	Fixed area	Specialized	Low	Real-time
Mirelman et al. [76, 77]	Assist per use	Fixed range	Fixed area	Specialized	Medium	Real-time
You et al. [78]	Demonstration	Fixed range	Fixed area	Specialized	High	Real-time
Agmon et al. [79]	Demonstration	Fixed range	Fixed area	OTS	Low	Real-time

3.5 Conclusion

A review of a number of computer based rehabilitation systems revealed that these were created as singletons; novel input mechanisms are used as input to custom designed software. This pattern works well for research efforts and proof of concept prototypes. However, it tightly couples the entire logical system stack (as outlined in Fig. 3.1: Logical Structure of a CNMRS) and limits re-use of both components and software. This artificially limits the availability of such systems, as the time and monetary cost of developing them is higher than a system made primarily of reused components, such as off the shelf hardware and community developed software modules [25].

This taxonomy was created in response to a perceived need—there is significant evidence [8–23] that CNMRSs are effective in boosting patient outcomes, however these systems are not widely available to patients. The primary reason is that the existing solutions are developed and tested for research, not necessarily driven by therapists and thus the cost to deploy such systems widely is prohibitive.

The taxonomy described above, in Sects. 3.2 and 3.3 provides a categorization scheme that can be applied as a tool to identify input apparatus in support of neuro-motor rehabilitation clients. Specifically, systems that can monitor specific motor actions for compliance, record progress, provide feedback and provide digitized input signals. This taxonomy is purposefully external to application design, focusing on the neuro-motor rehabilitation capabilities of the systems. Future work would ideally integrate this taxonomy into a useable interactive database tool that would allow systems to be discovered through taxonomy-based query parameters. This would allow stakeholders to both locate the most appropriate system for a situation, and identify the need for CNMRS with specific parameters.

The collected works used to craft the included taxonomy, feature few works that are built around highly portable mobile systems such as phones or portable tablets. The difficulty in locating work that covered these portable systems might indicate a gap that could be explored further in future works. Further, an analysis of the relative (taxonomy) feature sets could identify over- and underexplored feature sets. To analyze the available research feature sets, an online user contributed, peer-reviewed database of taxonomy classified CNMRS research projects is proposed. This database could be used as a basis for performing multidimensional feature-space analysis.

Some combinations of feature sets could be beneficial to clients. Rehabilitation programs usually indicate to patients that they should make an effort to continue their rehabilitative exercises long term [5, 26]. Providing more accessible feature sets, including portable or mobile options to clients could increase the longevity and compliance with movement therapy programs. With the array of sensors available on consumer mobile phones a project evaluating the long term effectiveness of a mobile phone based CNMRS is needed.

To support the availability of CNMRSs, a tool to process and aggregate movement data, from a variety of sensors is proposed. Such a system would allow the free selection of supported sensors, by the clinician, allowing them to select the most appropriate system for their client's situation. The signals provided as output from the tool would then be mapped to discrete or ranged computer input. Researchers and developers would be able to improve upon the system by developing modules for the system.

References

1. American Stroke Association About Stroke. In: Am. Stroke Assoc. Build. Heal. lives, Free stroke Cardiovasc. Dis. http://www.strokeassociation.org/STROKEORG/AboutStroke/About-Stroke_UCM_308529_SubHomePage.jsp. Accessed 19 May 2015
2. Murphy TH, Corbett D (2009) Plasticity during stroke recovery: from synapse to behaviour. Nat Rev Neurosci 10:861–872. doi:10.1038/nrn2735
3. Bonita R, Beaglehole R (1988) Recovery of motor function after stroke. Stroke 19:1497–1500. doi:10.1002/dev.20508
4. Poltawski L, Boddy K, Forster A et al (2015) Motivators for uptake and maintenance of exercise: perceptions of long-term stroke survivors and implications for design of exercise programmes. Disabil Rehabil 37:795–801. doi:10.3109/09638288.2014.946154
5. Werner RA, Kessler S (1996) Effectiveness of an intensive outpatient rehabilitation program for postacute stroke patients. Am J Phys Med Rehabil. doi:10.1097/00002060-199603000-00006
6. Alankus G, Proffitt R, Kelleher C, Engsberg J (2011) Stroke therapy through motion-based games. ACM Trans Access Comput 4:1–35. doi:10.1145/2039339.2039342
7. Rego P, Moreira PM, Reis LP (2010) Serious games for rehabilitation: a survey and a classification towards a taxonomy. In: 2010 5th Iberian conference on information systems and technologies (CISTI). ISBN: 978-1-4244-7227-7
8. Betker AL, Desai A, Nett C et al (2007) Game-based exercises for dynamic short-sitting balance rehabilitation of people with chronic spinal cord and traumatic brain injuries. Phys Ther 87:1389–1398. doi:10.2522/ptj.20060229
9. Ma M, Bechkoum K (2008) Serious games for movement therapy after stroke. IEEE Int Conf Syst Man Cybern 1872–1877. doi:10.1109/ICSMC.2008.4811562
10. Burke JW, McNeill MDJ, Charles DK et al (2009) Optimising engagement for stroke rehabilitation using serious games. Vis Comput 25:1085–1099. doi:10.1007/s00371-009-0387-4
11. Ryan M, Smith S, Chung B, Cossell S (2012) Rehabilitation games: designing computer games for balance rehabilitation in the elderly
12. Google ATAP (2015) Welcome to Project Soli. https://www.youtube.com/watch?v=0QNiZfSsPc0. Accessed 11 June 2015
13. Microsoft (2015) Microsoft HoloLens| Official Site. https://www.microsoft.com/microsoft-hololens/en-us. Accessed 11 June 2015
14. Zhao C, Hsiao C-P, Davis NM, Yi-Leun Do E (2013) Tangible games for stroke rehabilitation with digital box and blocks test. CHI'13 extended abstracts on human factors in computing systems—CHI EA'13, pp 523–528. doi:10.1145/2468356.2468448
15. Cameirão MS, Bermúdez i Badia S, Zimmerli L et al (2007) The rehabilitation gaming system: a virtual reality based system for the evaluation and rehabilitation of motor deficits. In: 2007 virtual rehabilitation. IWVR, pp 29–33

16. Tan CW, Chin SW, Lim WX (2013) Game-based human computer interaction using gesture recognition for rehabilitation. In: Proceedings of 2013 IEEE international conference on control system, computing and engineering, ICCSCE 2013, pp 344–349. doi:10.1109/ICCSCE.2013.6719987

17. Chang C-Y, Lange B, Zhang M, et al. (2012) Towards pervasive physical rehabilitation using microsoft kinect. In: 6th international conference on pervasive computing technologies for healthcare, San Diego, USA, pp 159–162

18. Fraiwan MA, Khasawneh N, Malkawi A et al (2013) Therapy central: on the development of computer games for physiotherapy. In: 2013 9th international conference on innovation information technologies IIT, pp 24–29. doi:10.1109/Innovations.2013.6544388

19. Rahman MA, Hossain D, Qamar AM, et al. (2014) A low-cost serious game therapy environment with inverse kinematic feedback for children having physical disability. In: Proceedings of international conference on multimedia retrieval—ICMR'14 529–531. doi:10.1145/2578726.2582619

20. Fukamoto T (2010) NeuroRehab + the "Fun" factor. In: Proceedings of 5th ACM SIGGRAPH symposium on video games—sandbox'10 1:69–78. doi:10.1145/1836135.1836146

21. Okošanovi MT, Kljaji J, Kosti MD (2014) Platform for integration of internet games for the training of upper extremities after stroke. In: 2014 12th symposium on neural network applications in electrical engineering (NEUREL). IEEE, Belgrade, pp 167–172

22. Adamovich SV, Merians AS, Lewis JA (2005) A virtual reality—based exercise system for hand rehabilitation. Presence 161–175

23. De O, Andrade K, Fernandes G, Martins J et al (2013) Rehabilitation robotics and serious games: an initial architecture for simultaneous players. ISSNIP Biosignals Biorobotics Conf BRC. doi:10.1109/BRC.2013.6487455

24. Perry JC, Zabaleta H, Belloso A et al (2013) ArmAssist: an integrated solution for telerehabilitation of post-stroke arm impairment. Converging Clin Eng Res NR 1:255–258. doi:10.1007/978-3-642-34546-3

25. Martin MV, Ishii K (2002) Design for variety: developing standardized and modularized product platform architectures. Res Eng Des 13:213–235. doi:10.1007/s00163-002-0020-2

26. Calautti C, Baron J-C (2003) Functional neuroimaging studies of motor recovery after stroke in adults: a review. Stroke 34:1553–1566. doi:10.1161/01.STR.0000071761.36075.A6

27. Betker AL, Szturm T, Moussavi ZK, Nett C (2006) Video game-based exercises for balance rehabilitation: a single-subject design. Arch Phys Med Rehabil 87:1141–1149. doi:10.1016/j.apmr.2006.04.010

28. Crosbie J, Lennon S, McGoldrick M et al (2012) Virtual reality in the rehabilitation of the arm after hemiplegic stroke: a randomized controlled pilot study. Clin Rehabil 26:798–806. doi:10.1177/0269215511434575

29. Da Silva Cameiro M, Bermúdez I Badia S, Duarte E et al (2011) Virtual reality based rehabilitation speeds up functional recovery of the upper extremities after stroke: a randomized controlled pilot study in the acute phase of stroke using the rehabilitation gaming system. Restor Neurol Neurosci 29:287–98. doi:10.3233/RNN-2011-0599

30. Cameirão M, Bermúdez I (2009) The rehabilitation gaming system: a review. Stud Health Technol

31. Harley L, Robertson S, Gandy M (2011) The design of an interactive stroke rehabilitation gaming system. Interact Users 167–173

32. Rahman A, Ahmed M, Qamar A et al (2014) Modeling therapy rehabilitation sessions using non-invasive serious games 1–4

33. Qamar A, Rahman MA, Basalamah S (2014) Adding inverse kinematics for providing live feedback in a serious game-based rehabilitation system, pp 215–220. doi:10.1109/ISMS.2014.43

34. Bhattacharya S, Joshi C, Lahiri U, Chauhan A (2013) A step towards developing a virtual reality based rehabilitation system for individuals with post-stroke forearm movement

disorders. In: CARE 2013—2013 IEEE international conference on robotics and automation, robotics and embedded systems. doi:10.1109/CARE.2013.6733743

35. Brokaw EB, Brewer BR (2013) Development of the home arm movement stroke training environment for rehabilitation (HAMSTER) and evaluation by clinicians. In: Lecture notes in computer sciences (including subseries lecture notes in artificial intelligence and lecture notes in bioinformatics), pp 22–31

36. Crocher V, Hur P, Seo NJ (2013) Low-cost virtual rehabilitation games: house of quality to meet patient expectations. In: 2013 international conference on virtual rehabilitation, ICVR 2013, pp 94–100

37. Dukes PS, Hayes A, Hodges LF, Woodbury M (2013) Punching ducks for post-stroke neurorehabilitation: system design and initial exploratory feasibility study. In: IEEE symposium on 3D user interface 2013, 3DUI 2013—Proc 47–54. doi:10.1109/3DUI.2013.6550196

38. Erazo O, Pino JA, Pino R, Fernández C (2014) Magic mirror for neurorehabilitation of people with upper limb dysfunction using kinect. In: Proceedings of annual Hawaii international conference on system sciences, pp 2607–2615. doi:10.1109/HICSS.2014.329

39. Gil-Gómez J-A, Lloréns R, Alcañiz M, Colomer C (2011) Effectiveness of a Wii balance board-based system (eBaViR) for balance rehabilitation: a pilot randomized clinical trial in patients with acquired brain injury. J Neuroeng Rehabil 8:30. doi:10.1186/1743-0003-8-30

40. Gonçalves ACBF, Consoni LJ, Amaral LMS (2013) Development and evaluation of a robotic platform for rehabilitation of ankle movements 8291–8298

41. Kafri M, Myslinski MJ, Gade VK, Deutsch JE (2013) Energy expenditure and exercise intensity of interactive video gaming in individuals poststroke. Neurorehabil Neural Repair 28:56–65. doi:10.1177/1545968313497100

42. Kim EK, Kang JH, Park JS, Jung BH (2012) Clinical feasibility of interactive commercial nintendo gaming for chronic stroke rehabilitation. J Phys Ther Sci 24:901–903. doi:10.1589/jpts.24.901

43. Kizony R, Weiss PL, Feldman Y et al (2013) Evaluation of a tele-health system for upper extremity stroke rehabilitation. In: 2013 international conference on virtual rehabilitation ICVR, pp 80–86. doi:10.1109/ICVR.2013.6662096

44. Labruyère R, Gerber CN, Birrer-Brütsch K et al (2013) Requirements for and impact of a serious game for neuro-pediatric robot-assisted gait training. Res Dev Disabil 34:3906–3915. doi:10.1016/j.ridd.2013.07.031

45. Maier M, Rubio Ballester B, Duarte E, et al (2014) Social integration of stroke patients through the multiplayer rehabilitation gaming system. Lecture notes in computer science (including Subser Lect Notes Artif Intell Lect Notes Bioinformatics) 8395 LNCS:100–114. doi:10.1007/978-3-319-05972-3_12

46. Mainetti R, Sedda A, Ronchetti M et al (2013) Duckneglect: video-games based neglect rehabilitation. Technol Heal Care 21:97–111. doi:10.3233/THC-120712

47. Parafita R, Pires G, Nunes U, Castelo-Branco M (2013) A spacecraft game controlled with a brain-computer interface using SSVEP with phase tagging. In: SeGAH 2013—IEEE 2nd international conference on serious games and application for health. doi:10.1109/SeGAH.2013.6665309

48. Siqueira AAG, Michmizos KP, Krebs HI (2013) Development of a robotic system for bilateral telerehabilitation. Ribeirão Preto, Brazil, pp 8427–8436

49. Saposnik G, Teasell R, Mamdani M et al (2010) Effectiveness of virtual reality using wii gaming technology in stroke rehabilitation: a pilot randomized clinical trial and proof of principle. Stroke 41:1477–1484. doi:10.1161/STROKEAHA.110.584979

50. Sucar LE, Orihuela-Espina F, Velazquez RL et al (2014) Gesture therapy: an upper limb virtual reality-based motor rehabilitation platform. IEEE Trans Neural Syst Rehabil Eng 22:634–643. doi:10.1109/TNSRE.2013.2293673

51. Vandermaesen M, Weyer T De, Coninx K et al (2013) Liftacube: a prototype for pervasive rehabilitation in a residential setting categories and subject descriptors. doi:10.1145/2504335.2504354

52. Méndez AV (2013) The effects of Nintendo Wii®on the postural control of patients affected by acquired brain injury: a pilot study 3:76–94

53. Vourvopoulos A, Faria AL, Cameirao MS, Bermudez I Badia S (2013) RehabNet: a distributed architecture for motor and cognitive neuro-rehabilitation. In: 2013 IEEE 15th international conference on e-health networking, application and services (Healthcom), pp 454–459. doi:10.1109/HealthCom.2013.6720719

54. Yavuzer G, Senel A, Atay MB, Stam HJ (2008) "Playstation eyetoy games" improve upper extremity-related motor functioning in subacute stroke: a randomized controlled clinical trial. Eur J Phys Rehabil Med 44:237–244

55. Caglio M, Latini-Corazzini L, D'Agata F et al (2012) Virtual navigation for memory rehabilitation in a traumatic brain injured patient. Neurocase 18:123–131. doi:10.1080/13554794.2011.568499

56. Caglio M, Latini-Corazzini L, D'Agata F et al (2009) Video game play changes spatial and verbal memory: rehabilitation of a single case with traumatic brain injury. Cogn Process. doi:10.1007/s10339-009-0295-6

57. Rábago CA, Wilken JM (2011) Application of a mild traumatic brain injury rehabilitation program in a virtual realty environment: a case study. J Neurol Phys Ther 35:185–93. doi:10.1097/NPT.0b013e318235d7e6

58. Holden MK, Dettwiler A, Dyar T et al (2001) Retraining movement in patients with acquired brain injury using a virtual environment. Stud Health Technol Inform 81:192–198

59. Mumford N, Duckworth J, Thomas PR et al (2012) Upper-limb virtual rehabilitation for traumatic brain injury: a preliminary within-group evaluation of the elements system. Brain Inj 26:166–176. doi:10.3109/02699052.2011.648706

60. Mumford N, Duckworth J, Thomas PR et al (2010) Upper limb virtual rehabilitation for traumatic brain injury: initial evaluation of the elements system. Brain Inj 24:780–791. doi:10.3109/02699051003652807

61. Ustinova KI, Leonard WA, Cassavaugh ND, Ingersoll CD (2011) Development of a 3D immersive videogame to improve arm-postural coordination in patients with TBI. J Neuroeng Rehabil 8:61. doi:10.1186/1743-0003-8-61

62. Ustinova KI, Ingersoll CD, Cassavaugh N (2011) Short-term practice with customized 3D immersive videogame improves arm-postural coordination in patients with TBI. In: 2011 international conference virtual rehabilitation, ICVR 2011. doi:10.1109/ICVR.2011.5971864

63. Grealy MA, Johnson DA, Rushton SK (1999) Improving cognitive function after brain injury: The use of exercise and virtual reality. Arch Phys Med Rehabil 80:661–667. doi:10.1016/S0003-9993(99)90169-7

64. Housman SJ, Scott KM, Reinkensmeyer DJ (2009) A randomized controlled trial of gravity-supported, computer-enhanced arm exercise for individuals with severe hemiparesis. Neurorehabil Neural Repair 23:505–514. doi:10.1177/1545968308331148

65. Subramanian SK, Lourenco CB, Chilingaryan G et al (2012) Arm motor recovery using a virtual reality intervention in chronic stroke: randomized control trial. Neurorehabil Neural Repair. doi:10.1177/1545968312449695

66. Jinhwa J, Jaeho Y, Hyungkyu K (2012) Effects of virtual reality treadmill training on balance and balance self-efficacy in stroke patients with a history of falling. J Phys Ther Sci 24:1133–1136. doi:10.1589/jpts.24.1133

67. Yang YR, Tsai MP, Chuang TY et al (2008) Virtual reality-based training improves community ambulation in individuals with stroke: a randomized controlled trial. Gait Posture 28:201–206. doi:10.1016/j.gaitpost.2007.11.007

68. Broeren J, Claesson L, Goude D et al (2008) Virtual rehabilitation in an activity centre for community-dwelling persons with stroke: the possibilities of 3-dimensional computer games. Cerebrovasc Dis 26:289–296. doi:10.1159/000149576

69. Kim JH, Jang SH, Kim CS et al (2009) Use of virtual reality to enhance balance and ambulation in chronic stroke: a double-blind, randomized controlled study. Am J Phys Med Rehabil 88:693–701. doi:10.1097/PHM.0b013e3181b33350

70. Jo K, Jung J, Yu J (2012) Effects of virtual reality-based rehabilitation on upper extremity function and visual perception in stroke patients: a randomized control trial. J Phys Ther Sci 24:1205–1208. doi:10.1589/jpts.24.1205

71. Kwon J-S, Park M-J, Yoon I-J, Park S-H (2012) Effects of virtual reality on upper extremity function and activities of daily living performance in acute stroke: a double-blind randomized clinical trial. Neurorehabilitation 31:379–385. doi:10.3233/NRE-2012-00807

72. Cikajlo I, Rudolf M, Goljar N et al (2012) Telerehabilitation using virtual reality task can improve balance in patients with stroke. Disabil Rehabil 34:13–18. doi:10.3109/09638288. 2011.583308

73. Kiper P, Piron L, Turolla A et al (2011) The effectiveness of reinforced feedback in virtual environment in the first 12 months after stroke (Skutecznoœæ terapii w œrodowisku wirtualnym w pierwszych 12 miesi[1]cach po udarze mózgu). Neurol Neurochir Pol 45:436–444. doi:10.1016/S0028-3843(14)60311-X

74. Piron L, Turolla A, Agostini M et al (2010) Motor learning principles for rehabilitation: a pilot randomized controlled study in poststroke patients. Neurorehabil Neural Repair 24:501–508. doi:10.1177/1545968310362672

75. Piron L, Tombolini P, Turolla A et al (2007) Reinforced feedback in virtual environment facilitates the arm motor recovery in patients after a recent stroke. Virtual Rehabil IWVR 2007:121–123. doi:10.1109/ICVR.2007.4362151

76. Mirelman A, Bonato P, Deutsch JE (2009) Effects of training with a robot-virtual reality system compared with a robot alone on the gait of individuals after stroke. Stroke 40:169–174. doi:10.1161/STROKEAHA.108.516328

77. Mirelman A, Patritti BL, Bonato P, Deutsch JE (2010) Effects of virtual reality training on gait biomechanics of individuals post-stroke. Gait Posture 31:433–437. doi:10.1016/j.gaitpost. 2010.01.016

78. You SH, Jang SH, Kim YH et al (2005) Virtual reality-induced cortical reorganization and associated locomotor recovery in chronic stroke: an experimenter-blind randomized study. Stroke 36:1166–1171. doi:10.1161/01.STR.0000162715.43417.91

79. Agmon M, Perry CK, Phelan E, et al (2011) A pilot study of wii fit exergames to improve balance in older adults. J Geriatr Phys Ther 1. doi:10.1519/JPT.0b013e3182191d98

Part II
Physical Therapy

Chapter 4
A Customizable Virtual Reality Framework for the Rehabilitation of Cognitive Functions

Gianluca Paravati, Valeria Maria Spataro, Fabrizio Lamberti, Andrea Sanna and Claudio Giovanni Demartini

Abstract Brain injury can cause a variety of physical effects and cognitive deficits. Although it has not yet been systematically adopted in clinical settings, virtual reality promises to be an excellent therapeutic tool for regaining both locomotor and neurological capacities. This work presents the design and implementation of VR^2 (Virtual Reality Rehabilitation), a customizable rehabilitation framework intended to enable the creation of motivating rehabilitation scenarios based on an ecologically valid semi-immersive system. Following the implementation phase, a study to test the acceptability of VR^2 in a group of subjects with cerebral lesions was conducted to investigate the usability of the framework. The group consisted of 11 people from 22 to 70 years of age, who were divided into two groups depending on the chronicity of disorder. The adequacy of the interface between patient and system was verified through questionnaires containing subjective questions, which revealed good overall acceptance and enjoyment of the tool. Moreover, to obtain early results useful for tuning the overall system in preparation for rigorous clinical trials, a set of preliminary cognitive tests concerning the rehabilitation protocol was conducted within the same group. Although the preliminary findings are promising and reveal a positive trend in neurocognitive investigations, the system should undergo clinical trials before being used in real clinical settings.

G. Paravati (✉) · F. Lamberti · A. Sanna · C.G. Demartini
Dipartimento di Automatica e Informatica, Politecnico di Torino,
Corso Duca Degli Abruzzi 24, 10129 Turin, Italy
e-mail: gianluca.paravati@polito.it

F. Lamberti
e-mail: fabrizio.lamberti@polito.it

A. Sanna
e-mail: andrea.sanna@polito.it

C.G. Demartini
e-mail: claudio.demartini@polito.it

V.M. Spataro
Dipartimento di Psicologia, Università Degli Studi di Torino,
Via Verdi 10, 10124 Turin, Italy
e-mail: valeriamspataro@gmail.com

© Springer International Publishing AG 2017   61
A.L. Brooks et al. (eds.), *Recent Advances in Technologies
for Inclusive Well-Being*, Intelligent Systems Reference Library 119,
DOI 10.1007/978-3-319-49879-9_4

4.1 Introduction

Brain injury can cause a large variety of physical effects and cognitive deficits. Depending on the extent of injury, people with cognitive deficits can experience serious difficulties in performing even basic actions during everyday life. Thus, rehabilitation plays a key role in helping patients recover as much of their independence as possible.

In this context, it is well known that traditional rehabilitation methods (e.g., those based on paper-and-pencil psychometric tests) can be boring and demotivating in the long term for people undergoing treatment, especially because of the repetitiveness of the tasks [39]. Preliminary investigations considering the use of virtual environments for both motor [20, 36] and cognitive rehabilitation [37, 41] started by the late 1990s. Virtual Reality (VR) has become an effective basis for rehabilitation tools by virtue of its high ecological validity (i.e., the degree of relevance or similarity that a test or training system possesses with respect to the real world). A typical VR rehabilitation tool consists of a three-dimensional scenario that can be navigated by the user, allowing he or she to feel part of the scene by interacting with it [44].

The primary benefit of using VR for rehabilitation with respect to non-VR cognitive retraining treatments concerns the motivation of the participant [39]. Indeed, performing a task within an ecological interactive simulated environment requires active participation and learning by the patient. From the point of view of the therapist, if properly designed, a VR system provides the ability to easily control the complexity of the environment, thus enabling the creation of individualized treatments depending on each patient's capabilities. For instance, the therapist can control the quantity and type of stimuli presented to the patient as well as their features (such as their visual appearance). Moreover, compared with the real world, VR offers the therapist a greater ability to measure the progress of the user in safe but challenging environments, such as crossing a street or confronting phobias [5].

However, VR has not yet been extensively adopted as an assistive technology in clinical environments, primarily because it has not yet reached an adequate level of maturity [50]. Indeed, the setup and operation of current VR systems may require the intervention of specialists. Moreover, if not well and carefully designed, activities performed within virtual environments may be difficult to enjoy.

From a technological perspective, the design and implementation of a targeted and specific rehabilitation tool based on an interactive simulated environment often requires that a new project be started from scratch. In fact, each scenario and its associated tasks are, in general, specifically tailored to a specific rehabilitation goal. It is therefore anticipated that a beneficial characteristic of a novel rehabilitation tool would be the ability to modify each environment to address the needs of different patients or groups of patients. By modifying the complexity of the scene and the tasks to be performed, each environment could be customized to achieve precise control over a large number of physical variables that influence human behavior and perceptions. For instance, the therapist could control the exact delivery of the presented stimuli (which one, how many, etc.).

The contributions of this work are twofold. From a neuropsychological perspective, the focus is placed on selective attention. Most existing VR rehabilitation tools lack the possibility to introduce variability in the administration of tests to adapt the presented task to the cognitive abilities of the user (e.g., by increasing or decreasing the number of stimuli). Variables and factors that influence the selective ability of individuals have been extracted from a thorough analysis of the literature and practice. This theoretical approach laid the groundwork for the formulation of a series of principles to drive the design of the VR platform and the evolution of its scenarios. From a technological perspective, the system discussed in this work adapts the above principles into requirements for the VR platform such that the devised tool can adapt to the abilities of individuals by presenting personalized training trials. In fact, the scenario as well as individual trials and their parameters can be tuned according to the level of cognitive deficit of each patient or group of patients. In this sense, from a technological perspective, the framework is generic. Although it is presented in this work with a focus on the rehabilitation of cognitive functions, it could also be used for various other kinds of rehabilitation. Moreover, the system is easily extendable and amenable to further developments by virtue of its ability to integrate scenarios developed using different tools and requiring heterogeneous I/O devices.

The objective of the present work was to investigate the usability of the developed framework for the rehabilitation of cognitive functioning through direct involvement with people with attentional disorders. In this stage of implementation, the investigation was concerned with the users' level of acceptance of the required means of interaction with the VR system and their motivation. Moreover, a pilot experimental study was conducted to gather preliminary observations regarding the neurocognitive evaluation of patients using the proposed system through neuropsychological pre-tests and post-tests. These observations will serve as a guide both for clinical tests, which will be needed for an effective assessment of the validity of the devised tool, and for future developments of the platform.

The framework was implemented and tests were performed in a clinical scenario under the supervision and guidance of a specialist. Nonetheless, by virtue of its affordable implementation based on mass-market components and open-source software, the system can also be used in the home or in other low-resource scenarios.

This chapter is organized as follows. Section 4.2 reviews state-of-the-art VR systems for rehabilitation, in the fields of both motor and cognitive training. The principles upon which the current project was built, the architecture of the designed framework, and the implemented scenario and tasks are described in Sect. 4.3. A usability study and a pilot study conducted based on pre-tests and post-tests are discussed in Sect. 4.4. Finally, conclusions are drawn in Sect. 4.5.

4.2 Virtual Reality and Rehabilitation

VR technology is widely considered to be an innovative tool for rehabilitation assessment and treatment in a number of application areas, by virtue of its ecological validity and its value for predicting or improving everyday functioning [31]. VR can be used to develop a wide range of applications for the rehabilitation and treatment of patients with a very large spectrum of problems and special needs. To date, various stand-alone and targeted applications have been successfully proposed (e.g., for the rehabilitation of driving skills [43], street-crossing skills for people with unilateral spatial neglect [57], and daily living activities such as hospital and university way-finding [11, 15] as well as for the treatment of phobias [18], among other purposes). VR has also been used in pain management therapies such as burn care, in venipuncture, and for treating the pain experienced by wheelchair users and spinal cord injury patients, among others [1, 48, 56]. Most studies concerning VR for rehabilitation have targeted individuals with motor disabilities. However, a significant amount of research has also focused on the the rehabilitation of individuals with cognitive impairments. The amount of recent research activity that has been conducted on this topic confirms that the use of VR in rehabilitation is a subject of extremely high interest. In the following, a comprehensive overview of the most recent applications of VR in the fields mentioned above is provided.

4.2.1 Sensory-Motor Investigation and Rehabilitation

Considerable research has focused on the design of devices to support sensory-motor rehabilitation tasks. In this context, ad hoc VR applications are developed to be used in combination with such devices.

A platform for evaluating sensory-motor skills was designed in [30], with the purpose of quantifying the movements and related problems (such as tremors) of patients with Parkinson's disease. From a technological point of view, the system must share its computational load among several networked peers. Various input and output devices can be used, and stimuli can be modified using XML-based descriptions. Several prototype applications have been developed, but no evaluation of their rehabilitation capabilities has yet been reported. Preliminary evidence of the feasibility of VR-based motor rehabilitation was presented in [50]. The main features of the proposed platform include adherence to the design principles for motor rehabilitation, the use of specific controllers for arm rehabilitation, and the implementation of therapies based on artificial intelligence decision models. Similarly, in [56], a pilot study was presented in which VR applications were used to improve motor function in spinal cord injury patients and to reduce the associated neuropathic pain. Size-adjustable shoes equipped with motion sensors provided data for interacting with the developed training scenarios. As an advancement with respect to previously developed systems, this work combined the observation and execution of actions.

VR is also used as a means for providing therapists in remote locations with the ability to monitor and inspect patient's rehabilitation sessions. The authors of [14] reported a feasibility study of the integration of a robot into such a tele-rehabilitation system to provide support and present resistance to patients' movements while they are navigating in virtual environments. Similarly, [8] presented a tool for monitoring patients' motion that can also be used by therapists to create exercises and teach them through direct demonstration. In particular, [8] focused on the possibility of creating exercise libraries for physical therapy and collaborative therapy sessions. Tele-rehabilitation systems present additional challenges because they require a communication infrastructure between the sites at which the patient and therapist are located. In particular, bandwidth and latency issues must be carefully considered during the design phase [33–35]. However, suitable network technologies already exist that are able to guarantee real-time data delivery over packet-switched networks. For example, it has been shown that pipeline forwarding technology can offer deterministic delivery over wireless [3], wired [4], and even all-optical [2] networks.

An important aspect that is especially necessary to consider in VR systems for rehabilitation is human-computer interaction. In the context of sensory-motor rehabilitation, several works have addressed the conditions created by weakness of the upper limbs, such as arm and hand hemiparesis. Often, the measurement of arm and hand kinematics is performed using external devices such as gloves [6, 45, 58] or arm training systems [7] designed for use with VR systems. Other studies have addressed the possibility of supporting movements with ad hoc devices. For instance, in [10], hand rehabilitation for stroke survivors was accomplished by means of grasp-and-release movements using a pneumatic glove that was able to assist in finger extension within a virtual training environment. In [55], the same task was accomplished using a device that replicated the motion of a healthy hand on the impaired hand. In these cases, pilot studies have indicated the feasibility of applying the proposed approaches to enhance efficacy in patients with chronic hand hemiparesis. Under the assumption that introducing discordance between a real hand movement and the corresponding movement that is observed in a VR hand could facilitate functional recovery, experimental data obtained via functional magnetic resonance imaging as reported in [54] indicate that it is possible to enhance brain reorganization after injury. In all these scenarios, the relevant VR applications do not require complex management of the variables involved variables because the VR system is fundamentally used to perform quite simple tasks.

The integration of Natural User Interfaces (NUIs) based on motion-sensing technologies is considered to be a promising means of improving interaction with virtual environments in motor rehabilitation [42] because such interfaces need not be attached to patient sensors that would restrict their movements. Moreover, the widespread availability of NUI devices currently permits the development of low-cost systems, which could facilitate their diffusion and adoption in clinical rehabilitation scenarios. Specifically, in [42] and [28], the Microsoft Kinect was used to develop applications to enhance patients' movement capabilities, stability and coordination. A preliminary evaluation of the kinematic validity of the Microsoft Kinect for use in rehabilitation systems for upper limb hemiparesis was presented in [52].

Following the shift toward new types of interaction techniques, gaze-based tracking technologies [25] and brain-computer interfaces (BCI) [17, 19] have recently been investigated. In the usability study presented in [25], eye-gaze patterns were monitored in real time during VR-based social interactions to present individualized intervention strategies to children with autism spectrum disorders. In [29, 32], brain signals sensed by electroencephalographic (EEG) devices were mapped to the movements of avatars or first-person viewpoints in virtual environments for rehabilitation tasks. The basic idea was to determine the intended hand movements by analyzing the frequency responses with respect to rest poses. Relations between brain and behavior were investigated by [46]. In that study, while a user was moving within a virtual environment, the experimental system monitored neural data by means of EEG recordings. Synchronized brain imaging and body motion capture data were then combined and analyzed. Mental activity can represent an interesting alternative to other conventional human-computer interaction techniques for enhancing the participation of individuals with cognitive impairments. However, interfaces of this kind and, in particular, the necessary means of interpreting brain signals, still require further developments prior to use in real clinical scenarios to reach an accuracy comparable to that of other established interaction techniques.

4.2.2   Cognitive Rehabilitation

Several studies reported in the literature have been dedicated to the use of VR for cognitive rehabilitation tasks, although most have presented only preliminary research results. For instance, the objective of Cyber Care Clinique, as reported in [51], is to develop a unified mixed VR system for cognitive, psychological and physical rehabilitation. The framework is oriented toward personalization, and panoramic video cameras and microphones are used to replicate places that look familiar to patients. In this respect, the framework is particularly well suited for the treatment of phobias, as it can replicate their stimuli in safe virtual environments. The AGATHE project [23] aims to develop an assessment and therapy tool for the virtual implementation of daily living activities. The system has been built to be easy to use for both therapists and patients. To adapt to the capabilities of individual patients, customization of the objectives is possible. Unfortunately, more in-depth details regarding the functionalities of the platform are not available. The customization of virtual environments for the rehabilitation of psychological and neurological disorders is a key feature of NeuroVR [38]. An editor allows non-expert users to modify virtual scenes. However, the customizability is based on the ability to choose stimuli from among a database of objects with the aim of tailoring the scene contents based on the needs of a specific clinical problem (e.g., the treatment of phobias).

Distinct from the existing works, the objective of VR^2 is to enable customization with a focus on adaptation to the abilities of individuals or groups. Moreover, in VR^2, a model interaction strategy with 3 degrees of freedom (DOFs) is used.

Attention and executive functioning were studied in [16]. The authors found that together with risk-taking propensity, their combined contributions could predict risky driving performance in VR driving tasks. In [24], an office scenario was developed to evaluate the performances in separate tasks addressing selective and divided attention, problem-solving capabilities, working memory components and prospective memory. A certain amount of flexibility was a fundamental requirement of this study to allow the user to switch among the various tasks depending on his or her particular needs (replying to emails, printing, etc.). In [21], the effectiveness of VR for the treatment of executive functions in patients with TBI was investigated.

In several works, VR supermarkets have been used to train participants' abilities to perform instrumental activities of daily living (IADL) [21, 22, 38]. In [21], a commercial solution was adopted to navigate and operate within a supermarket through a video capture system. The task consisted of checking a shopping list, selecting the correct supermarket aisle by touching one of the figures depicting product categories (food, personal care, etc.) and choosing the correct products while standing in front of the shelves. In [22], planning ability was specifically considered. Planning is a complex cognitive function because it requires the ability to imagine and design complex actions; hence, other low-level cognitive functions (e.g., selective attention, working memory, etc.) must be intact for a person to be able to perform these more complex ones.

The system proposed in the present work differs from those discussed above because it is based on a theoretical model that allows therapists to build tests of gradually increasing difficulty. In a gradual rehabilitation environment, basic cognitive functions can be trained in the first phases of the virtual scenarios. Then, patients can be trained to perform more complex tasks (e.g., those concerning the ability to categorize and plan) only at a later stage. Moreover, particular attention has been devoted to eliminating the effect of learning. This effect arises when patients repeat the same test with the same control parameters over a long duration. Each single test presented in the proposed framework is unique by virtue of the generation of randomly distributed stimuli (in terms of location, color, etc.). Furthermore, each rehabilitation session is personalized for each individual by means of an automatic calibration method that attempts to adapt the difficulty of each single test to the ability of the current patient. For instance, the maximum time to accomplish a task is computed based on the patient's previous performances.

4.3 Proposed Framework

4.3.1 Theoretical Model

VR^2 was designed in collaboration with neuropsychology experts. The rehabilitation workflow presented in this chapter is focused on selective attention. The reasons for this choice are manifold. First, the role of selective attention in the attentional system is fundamental, so much so that for many years, the entire definition of attention was associated with the ability to select stimuli. The identification of the entire attention system with selective attention is due to the pervasiveness of selective attention in everyday life. Second, by practicing selective attention, it is possible to simultaneously train two other components, namely, the speed of execution and sustained attention. In fact, the devised VR trials involve the selection of salient stimuli among distractors (selective attention) as rapidly as possible (speed of execution) over a prolonged period of time (sustained attention). Finally, the selective ability poses a reduced demand in cognitive processing; consequently, in the case of the impairment of all attentional components, it is necessary to first rehabilitate the simplest (selective attention), followed by the more complex ones (alternating and divided attention).

Selective attention represents the ability to focus on a stimulus or relevant information while neglecting those in competition. A stimulus is considered to be relevant in comparison with irrelevant ones when it is characterized by features that are of interest to the individual. These features depend on many different factors. Thus, a number of variables involved in the selection of stimuli were identified to carefully guide the design and implementation of the rehabilitation platform. In this way, the framework is able to host multiple scenarios with various levels of difficulty, in which the stimuli are varied to control the perceptive and cognitive load required for interacting with the environment. The variables considered are the following:

- Level of selection of relevant stimuli: The patient is asked to select stimuli based on their physical or semantic features [13].
- Number of features: Relevant stimuli can be characterized by one or more features simultaneously [53].
- Cognitive load: Three cognitive mechanisms are involved. The *inhibition of information* permits the suppression of the interfering effect of conflicting information; *working memory* and *coordination among processes* seem crucial for maintaining priorities between relevant and irrelevant stimuli and serve to govern the behavior of an individual according to his or her current goals [26].
- Perceptual load: This variable is the amount of information that can be perceived within a given time [26].
- Selection by interaction: The selection of salient stimuli occurs through interaction with the environment; it is based on experience, knowledge, expectations, motivations, and the goals that are to be achieved [9].

4.3.2   Design of the System Architecture

The main objectives of the architectural design are simplicity of the interface and expandability of the framework. The first objective addresses the requirement that the therapist must be able to configure and manage, in a simple manner, a system that comprises several elements. Expandability is needed to allow new VR scenarios to be implemented in the rehabilitation framework regardless of the software tools or versions used to create the already existing trials. For these reasons, the modular approach that is shown in Fig. 4.1 was conceived. The core elements of the system are the VR Manager, which is responsible for controlling the application as a whole, and the application scenarios; for each scenario, a series of tasks can be implemented. A peculiarity of each scenario is the possibility to configure a set of parameters representing particular features of the specific environment (depending

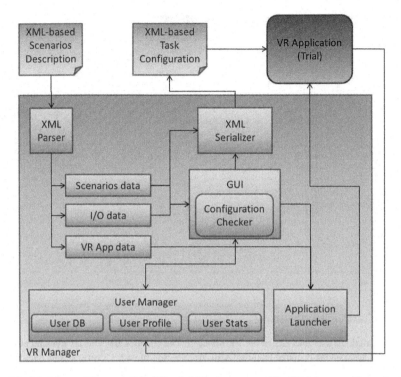

Fig. 4.1 Modular architecture of the VR rehabilitation system. The VR Manager allows the therapist to modify the configuration and the control parameters of each VR application

on the task); further details regarding the implemented scenarios and parameters are presented in Sect. 4.3.3.

The user interface of the VR Manager is built dynamically at run time. Each scenario, with its related tasks (rehabilitation trials) and customizable features, is declared in an XML-based interface description. The XML Parser module is responsible for translating such descriptions into the data structures that are required to build the elements of the interface (tree views, check boxes, sliders, menus, etc.). Three main blocks are extracted from the interface description: Scenario Data, I/O Data and Application Data. Scenario Data contains the structure of the scenario and the definition of the parameters for each trial. I/O Data contains the information that allows the user to choose the devices used for input (mouse, keyboard, joystick, data glove, depth camera, etc.) and output (flat screen, projection system, 3D glasses, autostereoscopic display, etc.). Application Data stores information about the software version to ensure the correct launching of each single trial. As shown in Fig. 4.2, the Graphical User Interface (GUI) is predominantly composed of three columns. The leftmost column contains a hierarchical tree view of the available scenarios, tasks, and sub-tasks. In the middle, for each selected element, a comprehensive list

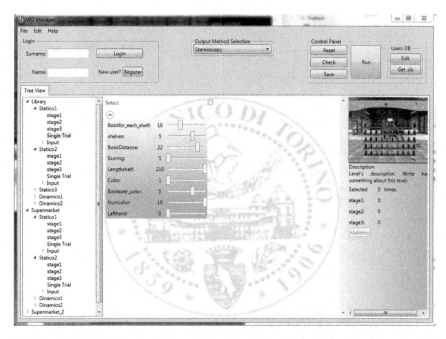

Fig. 4.2 Graphical user interface of the VR Manager. The graphics elements are dynamically built at run time based on the declared features (XML description) of each scenario and its related tasks

of customization parameters is shown. The rightmost column shows support information concerning the selected element (preview, trial description, statistics, etc.).

The Configuration Checker module is responsible for checking and verifying the changes introduced by the user and for providing support regarding the system setup (e.g., step-by-step instructions to the therapist for connecting I/O devices). Once parameters have been modified in the GUI, the XML Serializer attempts to build a formal description of the current configuration of the individual trials. Again, this configuration is XML-based and forms the configuration input file for the VR application. Because scenarios can be implemented using various software tools and versions (sometimes not completely compatible among each other, as in the case of open-source software such as Blender [40]), the Application Launcher module is included to recognize and manage the correct execution environment for each scenario.

Finally, the User Manager module manages patient profiles and related statistics. More specifically, this module allows the therapist to register a new user and store his or her personal data into a database, including the results of neuropsychological tests (attentional matrices, trail making tests, verbal MBT, working memory, etc.). Moreover, each user's history in the presented VR trials is automatically recorded to accrue statistics.

4.3.3   Scenarios and Tasks

The platform currently hosts two VR scenarios, named *library* and *supermarket*. Each scenario includes multiple tasks to be performed. The patient can move to the next task when the previous one has been completed. The library scenario consists of five tasks, each comprising three trials. Both tasks and trials are organized such that the difficulty increases incrementally. The supermarket scenario is divided into tasks performed by customers (four tasks, each comprising three trials) and tasks completed by a warehouseman (four tasks, ten trials).

The library is the first scenario presented to patients, and the stimuli involved are particularly simple to process from a cognitive point of view. All trials within the library environment require the patient to select stimuli based on a single physical feature (the color), while the variables concerning the level of selection and the number of features for the relevant stimulus are held fixed. The first three tasks can be regarded as "static" because the degrees of freedom available to the user are limited to the movement of a virtual hand to select the target objects. The last two tasks are regarded as "dynamic" because the user must navigate through the virtual room to accomplish the task.

Several sample screenshots depicting the overall library scenario are shown in Figs. 4.3 and 4.4. The objective of the first trial (Fig. 4.3a) in the first task is to collect all books of a given color (shown in the upper part of the user interface) from a shelf. As stated above, the VR Manager and the scenarios are designed to dynamically build libraries with varying parameters. In this way, the cognitive load can be easily increased in the next trial by changing the relevant parameters, e.g., the number of shelves, the number of books, etc. Specifically, the second and third trials (Fig. 4.3c and e, respectively) progressively increase the number of stimuli (two and three different colors are considered, respectively), the number of books per shelf and the number of shelves. No time limit is set during the first trial, but the time used to complete the exercise is recorded to use as a basis for computing time limits for the other trials that are proportional to the number of stimuli and account for the patient's capabilities. The object features and stimuli are randomized to avoid a possible learning effect. The second task concerns the reorganization of a series of books initially placed in a cart based on their color. Essentially, all books of a given color are to be placed into shelves/boxes of the same color. Similar to the previous case, the VR environment builds a suitable number of boxes depending on the number of stimuli (Fig. 4.3b–f). The goal of the third task is the same as the first task, but the perceptive load is increased by randomly assigning the books different sizes and randomly positioning them at different angles, while also introducing the possibility that the books may be placed on top of each other (an example is provided in Fig. 4.4a). The fourth task requires the patient to check stimuli while moving within the room (Fig. 4.4c). Specifically, during each of the three trials, the user is positioned in the center of the room and he or she can turn left and right to inspect four different shelves (one per wall). Finally, the last task (Fig. 4.4e) concerns the selection of books while moving through the room. The viewpoint of the user moves along

(a) Task I - 1st trial

(b) Task II - 1st trial

(c) Task I - 2nd trial

(d) Task II - 2nd trial

(e) Task I - 3rd trial

(f) Task II - 3rd trial

Fig. 4.3 Sample screenshots from the library (*L*) scenario. **a, c, e** *L* Task I; **b, d, f** *L* Task II

a fixed path at a pre-defined speed, thus allowing the patient to concentrate on performing the assigned task without focusing on movement commands. These trials are characterized by a high cognitive load, because the individual must remember the physical features of the stimuli, which must be collected before they disappear from the field of view.

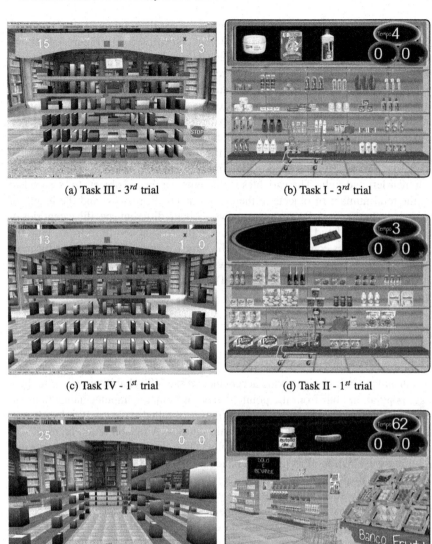

(a) Task III - 3rd trial   (b) Task I - 3rd trial

(c) Task IV - 1st trial   (d) Task II - 1st trial

(e) Task V - 3rd trial   (f) Task IV - 2nd trial

Fig. 4.4 Sample screenshots from the library (*L*) and supermarket (*S*) scenarios

The supermarket scenario follows a similar workflow. However, as indicated in Sect. 4.3.1, the difficulty is gradually increased. Similar to the supermarket scenario, the objective is to select products that are randomly presented by the system on a virtual shopping list. However, in this case, the physical features of the stimuli simultaneously include the product shape, color and name. Figure 4.4b–f show several sample screenshots from the supermarket trials. In the first task, the user must select products based on their physical features. Because the variables include both the number of simultaneous stimuli to be searched for and the total number of products, it is possible to devise more trials with different perceptive loads. Figure 4.4b shows an example of a trial in which the user needs to search for 3 products. The selection phase in the second task involves semantic reasoning, because the user must select the products to put in the cart based on their name or a representative image (i.e., the user cannot rely on the shapes and colors of the stimuli, as done before). For example, Fig. 4.4d depicts a case in which the patient is asked to pick up a chocolate bar. Also in this scenario, the last two tasks offer the user more degrees of freedom compared with the first tasks. Specifically, in the third task, products must be searched for in an entire aisle of the supermarket, meaning that patients can find objects to both their left and right. The variables include the number of objects to be searched for, the total number of objects in the environment (distractors) and the length of the shelves. In the first trial, opposite shelves contain different specific categories of products (e.g., food and cleaning), whereas in the second and third trials, both the left and right shelves contain products in the same category. The last task is the most interactive, because the user can freely navigate the supermarket to search for products in their corresponding areas (Fig. 4.4f). In this case, the objective is to train the implicit memory.

4.4 Results

The current study investigated the acceptance of the VR rehabilitation system by a target population both from the point of view of human-computer interaction and from a motivational perspective. The system was installed at Centro Puzzle in Turin, Italy, a rehabilitation center for severe brain injuries. A rehabilitation protocol was administered to patients, and neuropsychological tests were conducted to perform a preliminary evaluation of its potential benefits.

The study involved 11 individuals (eight males and three females, from 22 to 70 years of age, with an average age of 35 years and a standard deviation of 17 years) with cerebral lesions. These lesions were caused by traumatic brain injuries, except in one case, in which the lesion was caused by a cerebral hemorrhage. The participants were classified in two different groups based on the chronicity of the injury; five participants (group A) had suffered the cerebral lesion within the past 30 months, whereas the remaining six (group B) had suffered from a chronic illness for over 30 months.

The setup consisted of a semi-immersive VR installation. Each participant was seated in front of the virtual scenario, which was displayed on a projection screen. Stereoscopic visualization was provided by exploiting quad-buffered OpenGL support on a 120 Hz projector and 3D shutter glasses. A Microsoft Kinect device was used for gesture recognition with the aim of translating the user's arm gestures into virtual hand movements. In this manner, natural control of the stimuli selection process was achieved.[1]

---

[1]A video showing the interaction with the system is available at http://youtu.be/HCldmXLUz8E.

The rehabilitation protocol consisted of 15 training sessions per patient, held two times per week. Each session had a duration of 30 min. In this pilot study, the rehabilitation process encompassed all five tasks of the library scenario. Each task consisted of three trials, whose parameters were the same for all participants. Starting from the first trial (the simplest one), each participant advanced to the next trial whenever a maximum number of errors (i.e., the user selected an incorrect stimulus, such as a book of a different color than that of the reference book) or omissions (i.e., the user missed the selection of a stimulus) were committed during the selection process. Thus, depending on the number of errors and omissions, most participants repeated the entire protocol several times. In subsequent repetitions of the protocol, qualitative changes in its administration were explored by comparing tests presented with and without 3D stereoscopic glasses.

4.4.1 Usability

At the end of each session, the users were asked to complete a questionnaire concerning the usability of the system. The questions reflected subjective evaluations of characteristics that were difficult to measure. Therefore, the Visual Analog Scale (VAS) was used. In the VAS evaluation, the users were asked to specify their levels of agreement with the presented statements by indicating a position along a continuous line between two endpoints.

Table 4.1 reports the main findings from the answers provided. In particular, the average evaluation μ and the standard deviation σ for the first and last sessions are shown. The VAS scale was defined with a range of 0–10. Overall, the users enjoyed the presented exercises and found them easy to perform. This means that they were not frustrated by errors and omissions committed during the tasks. The interaction with the virtual objects became more natural during repetition of the training. The participants experienced little fatigue or boredom during the rehabilitation and did

Table 4.1 Usability questionnaire. Results after the first and last sessions

Item	Topic	First session		Last session	
		μ	σ	μ	σ
1	Ease of grasping	5.7	2.6	6.9	3.3
2	Ease of the tasks	8.6	1.7	7.7	1.4
3	Fatigue	2	2.3	2.5	3.3
4	Enjoyment	5.7	3.9	7.3	3.1
5	Boredom	1.6	2	1.2	1.8
6	Realism	5	3.7	5.9	3.8
7	VR glasses added value	5.1	3.3	5.8	3.4
8	Nausea	0.3	0.3	0.4	0.4

not report any nausea. The majority of the subjects tolerated each session well. After the first session, nine of the eleven individuals declared their willingness to repeat the exercises that they had just performed. This number had increased to ten of the eleven by the last session. The realism of the scenario and the added value of using 3D glasses were evaluated near the midpoint of the scale.

Various aspects were considered to evaluate the current status of the framework and to identify potential future interventions in the system. In particular, a further usability analysis was conducted to complement the questionnaire results by means of manual annotations provided by the therapist to assist in identifying possible areas of improvement. Four categories of possible improvements were identified: sensory-motor interaction, stimuli presentation, task difficulty, and resolution of technical issues or bugs.

4.4.2 Preliminary Neurocognitive Investigation

This section reports the preliminary results regarding the participants' cognitive improvement, which were collected with the intent of fine-tuning the current configuration of the overall system. The pilot study included neuropsychological observations recorded immediately before the beginning and at the end of the rehabilitation protocol. These results were gathered to guide the design of a rigorous clinical trial, which is planned as the next step of the development of the devised platform.

Table 4.2 shows an overview of the evaluation for each patient (identified by a number in the first column). The values represent the raw scores obtained by individual patients in different evaluation tests. Columns two to seven report the results for reaction time (RT), which were collected via a test for the automated measurement of attentional disorders (Misurazione Informatica Disturbi Attentivi, MIDA) [12]. Each test, i.e., for simple reaction time (SRT) and complex reaction time (CRT), was repeated twice. The labels indicating the tests performed at the end of the rehabilitation protocol are preceded by the prefix r. Columns eight and nine report the results of attentional matrix (AM) tests [47]. Trail making test (TMT) [27] results measuring alternate attention are reported in columns ten and eleven. The next four columns report the results of short-term memory (STM) assessments, for both verbal content (V) and digits (D). Finally, results for working memory (WM), Corsi span (SC), and Stroop color-word interference (SCWI) [49] tests are given in the last six columns.

With the aim of simplifying the analysis of the results, Table 4.3 and Fig. 4.5 introduce a number of performance indicators obtained as combinations of the raw results and visual plots of the improvements achieved by the population sample. In Table 4.3, an indicator I for each test (column) reported in Table 4.2 has been devised, following the same order of presentation. Each indicator is constructed such that performance improvement is achieved when $I > 1$. The second row of the header of Table 4.3 explains the formulas used. All participants received high scores in several tests, with the exception of some participants who were not able to complete the pro-

Table 4.2 Neurocognitive investigation. Raw scores of neurocognitive tests for each patient before and after the rehabilitation protocol. The prefix *r* denotes the repeat of the test administered after treatment. RT: Reaction Time; AM: Attentional Matrices; TMT: Trail Making Test; STM: Short-Term Memory; WM: Working Memory; SC: Corsi Span; SCWI: Stroop Color-Word Interference

Pat.	RT						AM		TMT		STM				WM		SC		SCWI	
	SRT1	rSRT1	SRT2	rSRT2	RTC	rRTC			B-A	rB-A	V	rV	D	rD	WM	rWM	SC	rSC	3	r3
1	0.495	0.563	0.537	0.53	0.694	0.674	44	42	99	64	4	4	5	5	4	4	4	5	117	146
2	0.519	0.54	0.495	0.494	0.771	0.925	37	38	249	106	5	5	6	5	5	5	4	4	317	264
3	0.562	0.731	0.629	0.61	0.676	0.767	46	53	336	147	4	4	4	5	3	3	4	5	201	154
4	n.a.	n.a.	n.a.	n.a.	n.a.	n.a.	43	43	176	173	1	3	4	4	3	5	5	5	n.a.	n.a.
5	0.535	n.a.	0.457	n.a.	0.604	n.a.	54	n.a.	69	n.a.	4	n.a.	5	n.a.	3	n.a.	5	n.a.	239	n.a.
6	0.485	0.474	0.479	0.472	0.631	0.631	58	54	69	41	4	5	7	7	4	6	7	7	126	104
7	0.484	0.515	0.488	0.501	0.605	0.679	21	31	285	182	3	4	5	6	3	3	3	4	642	334
8	0.484	0.61	0.503	0.555	0.762	0.41	37	33	137	56	3	4	4	4	2	3	4	5	n.a.	n.a.
9	0.474	0.535	0.535	0.549	0.721	0.701	52	52	66	74	4	4	3	4	3	3	4	5	234	180
10	0.48	0.456	0.477	0.477	0.654	0.622	56	59	84	44	3	3	3	3	3	3	4	5	176	172
11	0.461	0.444	0.422	0.444	0.555	0.547	60	59	59	46	4	5	4	6	5	6	6	7	147	128

Table 4.3 Neurocognitive investigation. Performance indicators

Pat.	I_{SRT1} $\left[\frac{SRT1}{rSRT1}\right]$	I_{SRT2} $\left[\frac{SRT2}{rSRT2}\right]$	I_{RTC} $\left[\frac{RTC}{rRTC}\right]$	I_{AM} $\left[\frac{rAM}{AM}\right]$	I_{B-A} $\left[\frac{B-A}{rB-A}\right]$	I_V $\left[\frac{rV}{V}\right]$	I_D $\left[\frac{rWM}{WM}\right]$	I_{WM}	$\left[\frac{rD}{D}\right]$ I_{SC} $\left[\frac{rSC}{SC}\right]$	I_{SCWI} $\left[\frac{SCWI}{rSCWI}\right]$
1	0.88	1.01	1.03	0.95	1.55	1.00	1.00	1.00	1.25	0.80
2	0.96	1.00	0.83	1.03	2.35	1.00	0.83	1.00	1.00	1.20
3	0.77	1.03	0.88	1.15	2.29	1.00	1.25	1.00	1.25	1.31
4	n.a.	n.a.	n.a.	1.00	1.02	3.00	1.00	1.67	1.00	n.a.
5	n.a.	n.a.	n.a.	n.a.	n.a.	n.a.	n.a.	n.a.	n.a.	n.a.
6	1.02	1.01	1.00	0.93	1.68	1.25	1.00	1.50	1.00	1.21
7	0.94	0.97	0.89	1.48	1.57	1.33	1.20	1.00	1.33	1.92
8	0.79	0.91	1.86	0.89	2.45	1.33	1.00	1.50	1.25	n.a.
9	0.89	0.97	1.03	1.00	0.89	1.00	1.33	1.00	1.25	1.30
10	1.05	1.00	1.05	1.05	1.91	1.00	1.00	1.00	1.25	1.02
11	1.04	0.95	1.01	0.98	1.28	1.25	1.50	1.20	1.17	1.15
Av.	0.93	0.99	1.07	1.05	1.70	1.32	1.11	1.19	1.18	1.24

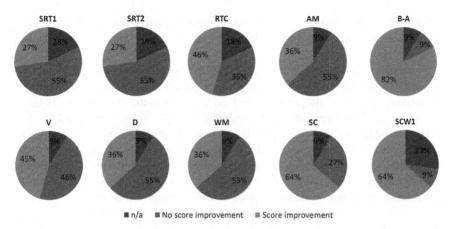

Fig. 4.5 Neurocognitive investigation. Percentages of participants who exhibited score improvements between the pre and post neurocognitive tests. RT: Reaction Time; AM: Attentional Matrices; TMT: Trail Making Test; STM: Short Term Memory; WM: Working Memory; SC: Span Corsi; SCWI: Stroop Color Word Interference

tocol. For instance, the fourth and eighth participants were not able to perform some of the tests because of neurological problems. In particular, the fourth participant felt strong shocks to the head during the MIDA test; he also suffered from aphasia for colors, which prevented him from completing the Stroop test. The eighth participant confused brown and purple colors because of vision problems. Moreover, during the administration of the protocol, the fifth participant underwent a surgical intervention and was unable to complete the tests. From the perspective of selective attention, four individuals reported higher scores for attentional matrices, whereas the others' scores remained constant or slightly decreased. This could be attributable to the fact that only some of the variables involved in the selection process were managed with the administration of the first developed scenario, but this hypothesis will be validated during clinical tests. Most participants' simple reaction times increased, whereas some showed decreases in their complex reaction times (50 % of the sample population). This probably occurred because after the training, the participants tended to process information in greater detail. Notably, almost every patient reported higher scores in alternate attention. In fact, nine of ten participants achieved significant increases in their train making test results. Similar observations can be made for the Stroop Test with regard to inhibitory control. Many participants achieved higher scores for both verbal and digit short-term memory, working memory (four of ten) and Corsi span (7 of 10). Figure 4.5 graphically shows the percentages of participants who exhibited score improvements between the tests administered before and after the rehabilitation protocol.

However, although the entire considered population reported simultaneous gains in scores related to more than one cognitive area after the VR training, no real conclusions can be drawn from the results. Indeed, at the time of writing, the conjectures formulated by observing the scores from the neurocognitive investigation will simply serve as guidelines for the creation of hypotheses to be confirmed via planned clinical trials.

In summary, a neuropsychological evaluation was performed using various tests before administering training. Ten of the eleven participants successfully completed the training, and the neuropsychological evaluation was then repeated to estimate the cognitive effectiveness of the VR treatment. After the treatment, all participants had improved in various cognitive skills addressed by the VR program. In particular, the participants showed significant improvements in attention (as assessed using the trail making test) and in verbal and spatial memory (as assessed using the bisyllabic word repetition test and the Corsi test, respectively). The improvements were higher in patients with chronic conditions (more than 30 months of illness).

4.5   Conclusions

This work presents a framework with the ultimate goal of improving cognitive functioning with the aid of VR technology. The framework is based on a theoretical model encompassing the development of several scenarios (comprising several tasks) for training a patient's cognitive abilities in different stages. A key feature of the devised VR system is the reconfigurability of the individual tasks, which provides the rehabilitation specialist with the ability to tune the system based on the patient's general conditions. In fact, the parameters affecting cognitive load can be tuned before each rehabilitation session. The usability of the rehabilitation tool was tested to evaluate whether the devised system can be effectively used by individuals with neurological disorders. The results indicate that the proposed tool can be used in the envisioned scenario. Because the complete theoretical model as presented in this paper has yet to be fully developed, future studies are planned to complete the VR framework. When completed, the system will be tested in a real clinical scenario by selecting patients and checking the results against a control group. With the objective of guiding these future activities and, in particular, guiding the planning of the clinical trials, a pilot test involving individuals with neurological disorders has already been conducted. The preliminary results collected through pre-tests and post-tests indicate a positive trend in the neurocognitive investigations. In particular, after the administration of a protocol involving a VR scenario enabling the manipulation of only a subset of the variables devised in the theoretical model, promising scores were achieved both for alternating attention and for verbal and spatial short-term memory. It should be noted that at the current stage, the performances in post-rehabilitation testing were affected by the fact that only a subset of the variables affecting the selection process were addressed during the experiments, considering that additional variables will be used in the future development of VR scenarios. Despite the encouraging positive results, in this phase of the implementation of the system, it is not possible to draw firm conclusions regarding the neuropsychological effectiveness of the system, which will be evaluated in planned future clinical trials.

Acknowledgements We thank Prof. Giuliano Geminiani, who supervised the work and provided insights and expertise that greatly assisted in this research. We are also immensely grateful to Dr. Marina Zettin, who supported the entire work and provided access to the host institution (Centro Puzzle), where we recruited the patients. This research was supported by the University of Turin and Centro Puzzle.

References

1. Aboalsamh H, Al Hashim H, Alrashed F, Alkhamis N (2011) Virtual reality system specifications for pain management therapy. In: 2011 IEEE 11th international conference on bioinformatics and bioengineering (BIBE), pp 143–147. doi:10.1109/BIBE.2011.69

2. Baldi M, Corr M, Fontana G, Marchetto G, Ofek Y, Severina D, Zadedyurina O (2011) Scalable fractional lambda switching: a testbed. J Opt Commun Netw 3(5):447–457

3. Baldi M, Giacomelli R, Marchetto G (2009) Time-driven access and forwarding for industrial wireless multihop networks. IEEE Trans Ind Inform 5(2):99–112

4. Baldi M, Marchetto G, Ofek Y (2007) A scalable solution for engineering streaming traffic in the future internet. Comput Netw 51(14):4092–4111

5. Bart O, Katz N, Weiss P, Josman N (2006) Street crossing by typically developed children in real and virtual environments. In: 2006 international workshop on virtual rehabilitation, pp 42–46. doi:10.1109/IWVR.2006.1707525

6. Borghetti M, Sardini E, Serpelloni M (2013) Sensorized glove for measuring hand finger flexion for rehabilitation purposes. IEEE Trans Instr Meas 62(12):3308–3314. doi:10.1109/TIM.2013.2272848

7. Burdea G, Cioi D, Martin J, Fensterheim D, Holenski M (2010) The rutgers arm ii rehabilitation system: a feasibility study. IEEE Trans Neural Syst Rehabil Eng 18(5):505–514. doi:10.1109/TNSRE.2010.2052128

8. Camporesi C, Kallmann M, Han J (2013) VR solutions for improving physical therapy. In Virtual reality (VR). IEEE, pp 77–78. doi:10.1109/VR.2013.6549371

9. Carassa A, Morganti F, Tirassa M (2014) Movement, action, and situation: presence in virtual environments. Proc Presence 2004:7–12

10. Connelly L, Jia Y, Toro M, Stoykov M, Kenyon R, Kamper D (2010) A pneumatic glove and immersive virtual reality environment for hand rehabilitative training after stroke. IEEE Trans Neural Syst Rehabil Eng 18(5):551–559. doi:10.1109/TNSRE.2010.2047588

11. Davies R, Lfgren E, Wallergrd M, Lindn A, Boschian K, Minr U, Sonesson B, Johansson G (2002) Three applications of virtual reality for brain injury rehabilitation of daily tasks. In: Proceedings of the 4th international conference on disability, Virtual reality and associated technologies, Veszprm, Hungary, pp 93–100

12. De Tanti A, Inzaghi M, Bonelli G, Mancuso M, Magnani M, Santucci N (1998) Normative data of the mida battery for the evaluation of reaction times. Eur Med Phys 34:221–220

13. Deutsch J, Deutsch D (1963) Attention: some theoretical consideration. Psychol Rev 70:80–90

14. Deutsch J, Lewis J, Burdea G (2007) Technical and patient performance using a virtual reality-integrated telerehabilitation system: preliminary finding. IEEE Tran Neural Syst Rehabil Eng 15(1):30–35. doi:10.1109/TNSRE.2007.891384

15. Gourlay D, Lun K, Lee Y, Tay J (2000) Virtual reality for relearning daily living skills. Int J Med Inform 60(3):255–261. doi:10.1016/S1386-5056(00)00100-3

16. Graefe A, Schultheis M (2013) Examining neurocognitive correlates of risky driving behavior in young adults using a simulated driving environment. In: 2013 International conference on Virtual rehabilitation (ICVR), pp 235–241. doi:10.1109/ICVR.2013.6662089

17. Hazrati M, Hofmann U (2013) Avatar navigation in second life using brain signals. In: 2013 IEEE 8th International symposium on intelligent signal processing (WISP), pp 1–7. doi:10.1109/WISP.2013.6657473

18. Hodges L, Anderson P, Burdea G, Hoffmann H, Rothbaum BO (2001) Treating psychological and phsyical disorders with vr. IEEE Comput Graph Appl 21(6):25–33. doi:10.1109/38.963458
19. i Badia SB, Morgade AG, Samaha H, Verschure P (2013) Using a hybrid brain computer interface and virtual reality system to monitor and promote cortical reorganization through motor activity and motor imagery training. IEEE Trans Neural Syst Rehabil Eng 21(2):174–181. doi:10.1109/TNSRE.2012.2229295
20. Jack D, Boian R, Merians A, Tremaine M, Burdea G, Adamovich S, Recce M, Poizner H (2001) Virtual reality-enhanced stroke rehabilitationm. IEEE Trans Neural Syst Rehabil Eng 9(3):308–318
21. Jacoby M, Averbuch S, Sacher Y, Katz N, Weiss P, Kizony R (2013) Effectiveness of executive functions training within a virtual supermarket for adults with traumatic brain injury: a pilot study. IEEE Trans Neural Syst Rehabil Eng 21(2):182–190. doi:10.1109/TNSRE.2012.2235184
22. Josman N, Hof E, Klinger E, Marie R, Goldenberg K, Weiss P, Kizony R (2006) Performance within a virtual supermarket and its relationship to executive functions in post-stroke patients. In: 2006 International workshop on virtual rehabilitation, pp 106–109. doi:10.1109/IWVR.2006.1707536
23. Klinger E, Kadri A, Le Guiet J, Coignard P, Lac N, Joseph P, Sorita E, Fuchs P, Leroy L, Servant F (2013) Agathe: A tool for personalized rehabilitation of cognitive functions. In: 2013 International conference on virtual rehabilitation (ICVR), pp 214–215. doi:10.1109/ICVR.2013.6662123
24. Krch D, Nikelshpur O, Lavrador S, Chiaravalloti N, Koenig S, Rizzo A (2013) Pilot results from a virtual reality executive function task. In: 2013 International conference on virtual rehabilitation (ICVR), pp 15–21. doi:10.1109/ICVR.2013.6662092
25. Lahiri U, Warren Z, Sarkar N (2011) Design of a gaze-sensitive virtual social interactive system for children with autism. IEEE Trans Neural Syst Rehabil Eng 19(4):443–452. doi:10.1109/TNSRE.2011.2153874
26. Lavie N, Hirst A, De Fockert J, Viding E (2004) Load theory of selective attention and cognitive control. J Exp Psychol Gen 133:339–354
27. Lezac M (ed) (1983) Neuropsychological assessment. Oxford University Press, Oxford
28. Lozano-Quilis J, Gil-Gomez H, Gil-Gomez J, Albiol-Perez S, Palacios G, Fardoum H, Mashat A (2013) Virtual reality system for multiple sclerosis rehabilitation using kinect. In: 2013 7th International conference on pervasive computing technologies for healthcare (Pervasive Health), pp 366–369
29. Mishra A, Foulds R (2013) Detection of user intent in neurorehablitation using a commercial eeg headset. In: 2013 39th Annual Northeast bioengineering conference (NEBEC), pp 21–22. doi:10.1109/NEBEC.2013.14
30. Myall DJ, MacAskill MR, Davidson PR, Anderson T, Jones R (2008) Design of a modular and low-latency virtual-environment platform for applications in motor adaptation research, neurological disorders, and neurorehabilitation. IEEE Trans Neural Syst Rehabil Eng 16(3):298–309. doi:10.1109/TNSRE.2008.922676
31. Neisser U (1978) Memory: what are the important questions? In: Practical aspects of memory. Academic Press, London, UK, pp 3–24
32. Nogueira K, Souza E, Lamounier A, Cardoso A, Soares A (2013) Architecture for controlling upper limb prosthesis using brainwaves. In: 2013 International conference on virtual rehabilitation (ICVR), pp 178–179. doi:10.1109/ICVR.2013.6662114
33. Paravati G, Gatteschi V, Carlevaris G (2013) Improving bandwidth and time consumption in remote visualization scenarios through approximated diff-map calculation. Comput Vis Sci 15(3):135–146. doi:10.1007/s00791-013-0201-8
34. Paravati G, Sanna A, Lamberti F, Ciminiera L (2011) An adaptive control system to deliver interactive virtual environment content to handheld devices. J Spec Top Mob Netw Appl 16(3):385–393. doi:10.1007/s11036-010-0255-5

35. Paravati G, Sanna A, Lamberti F, Ciminiera L (2011) An open and scalable architecture for delivering 3d shared visualization services to heterogeneous devices. Concurrency Comput Pract Experience 23(11):1179–1195

36. Popescu V, Burdea G, Bouzit M, Hentz V (2000) A virtual-reality-based telerehabilitation system with force feedback. IEEE Trans Inf Technol Biomed 4(1):45–51

37. Riva G (1998) Virtual environments in neuro science. IEEE Trans Inf Technol Biomed 2(4):275–281

38. Riva G, Gaggioli A, Villani D, Preziosa A, Morganti F, Corsi R, Faletti G, Vezzadini L (2007) A free, open-source virtual reality platform for the rehabilitation of cognitive and psychological disorders. Virtual Rehabil 2007:159–163. doi:10.1109/ICVR.2007.4362158

39. Rizzo A, Parsons TD, Buckwalter JG (2012) Using virtual reality for clinical assessment and intervention. In: Handbook of technology in psychology, psychiatry, and neurology: theory, research, and practice (2012)

40. Roosendaal T (2014) Blender 3d. http://www.blender.org/

41. Rose F, Brooks B, Attree E, Parslow D, Leadbetter A, McNeil J, Jayawardena S, Greenwood R, Potter J (1999) A preliminary investigation into the use of virtual environments in memory retraining after vascular brain injury: Indications for future strategy? Disabil Rehabil 21(12):548–554

42. Roy A, Soni Y, Dubey S (2013) Enhancing effectiveness of motor rehabilitation using kinect motion sensing technology. In: Global Humanitarian Technology Conference: South Asia Satellite (GHTC-SAS), 2013 IEEE, pp 298–304. doi:10.1109/GHTC-SAS.2013.6629934

43. Schultheis MT, Mourant RR (2001) Virtual reality and driving: The road to better assessment for cognitively impaired populations. Presence: Teleoper Virtual Environ 10(4):431–439. doi:10.1162/1054746011470271

44. Sherman W, Craig A (2002) Understanding virtual reality: interface, application and design. Morgan Kauffman, California

45. Silva L, Dantas R, Pantoja A, Pereira A (2013) Development of a low cost dataglove based on arduino for virtual reality applications. In: 2013 IEEE International conference on computational intelligence and virtual environments for measurement systems and applications (CIVEMSA), pp 55–59. doi:10.1109/CIVEMSA.2013.6617395

46. Snider J, Plank M, Lee D, Poizner H (2013) Simultaneous neural and movement recording in large-scale immersive virtual environments. IEEE Trans Biomed Circ Syst 7(5):713–721. doi:10.1109/TBCAS.2012.2236089

47. Spinnler H, Tognoni G, Bandera R, Della Sala S, Capitani EEA (1987) Standardizzazione e taratura italiana di test neuropsicologici. Ital J Neurol Sci 7(6)

48. Spyridonis F, Gronli TM, Hansen J, Ghinea G (2012) Evaluating the usability of a virtual reality-based android application in managing the pain experience of wheelchair users. In: 2012 Annual international conference of the IEEE engineering in medicine and biology society (EMBC), pp 2460–2463. doi:10.1109/EMBC.2012.6346462

49. Stroop JR (1935) Studies of interference in serial verbal reactions. J Exp Psychol 18(6):643–662

50. Sucar L, Orihuela-Espina F, Velazquez R, Reinkensmeyer D, Leder R, Hernandez-Franco J (2014) Gesture therapy: an upper limb virtual reality-based motor rehabilitation platform. IEEE Trans Neural Syst Rehabil Eng 22(3):634–643. doi:10.1109/TNSRE.2013.2293673

51. Takacs B, Simon L (2007) A clinical virtual reality rehabilitation system for phobia treatment. In: Information visualization, 2007. IV '07. 11th International conference, pp 798–806. doi:10.1109/IV.2007.7

52. Tao G, Archambault P, Levin M (2013) Evaluation of kinect skeletal tracking in a virtual reality rehabilitation system for upper limb hemiparesis. In: 2013 International conference on virtual rehabilitation (ICVR), pp 164–165. doi:10.1109/ICVR.2013.6662084

53. Treisman A (1998) Feature binding, attention and object perception. Philos Trans R Soc 353:1295–1306

54. Tunik E, Saleh S, Adamovich S (2013) Visuomotor discordance during visually-guided hand movement in virtual reality modulates sensorimotor cortical activity in healthy and hemiparetic subjects. IEEE Trans Neural Syst Rehabil Eng 21(2):198–207. doi:10.1109/TNSRE. 2013.2238250

55. Ueki S, Kawasaki H, Ito S, Nishimoto Y, Abe M, Aoki T, Ishigure Y, Ojika T, Mouri T (2012) Development of a hand-assist robot with multi-degrees-of-freedom for rehabilitation therapy. IEEE/ASME Trans Mechatron 17(1):136–146. doi:10.1109/TMECH.2010.2090353

56. Villiger M, Hepp-Reymond MC, Pyk P, Kiper D, Eng K, Spillman J, Meilick B, Estevez N, Kollias SS, Curt A, Hotz-Boendermaker S (2011) Virtual reality rehabilitation system for neuropathic pain and motor dysfunction in spinal cord injury patients. In: 2011 International conference on virtual rehabilitation (ICVR), pp 1–4. doi:10.1109/ICVR.2011.5971865

57. Weiss P, Naveh Y, Katz N (2003) Design and testing of a virtual environment to train stroke patients with unilateral spatial neglect to cross a street safely. Occup Ther Int 10(1):39–55

58. Zhou J, Malric F, Shirmohammadi S (2010) A new hand-measurement method to simplify calibration in cyberglove-based virtual rehabilitation. IEEE Trans Instr Measur 59(10):2496–2504. doi:10.1109/TIM.2010.2057712

Gianluca Paravati received an M.Sc. degree in Electronic Engineering and a Ph.D. degree in Computer Engineering from Politecnico di Torino, Italy, in 2007 and 2011, respectively. Currently, he holds an Assistant Professor position with the Control and Computer Engineering Department at Politecnico di Torino. His research interests include image processing, 3D visualization, distributed architectures, and human-computer interaction. In these fields, he has authored or co-authored several papers published in international journal and books. He is an editorial board member and also serves as a reviewer for several international journals and conferences. He is a member of the IEEE and the IEEE Computer Society.

Valeria Maria Spataro received an M.Sc. degree in Psychology from Università degli Studi di Torino, Italy, in 2011, receiving the award for the best thesis concerning the design of a virtual reality program for the rehabilitation of selective attention. She is collaborating with Università degli Studi di Torino and Politecnico di Torino on the design of a virtual-reality-based cognitive rehabilitation system.

Fabrizio Lamberti received M.Sc. and Ph.D. degrees in Computer Engineering from Politecnico di Torino, Italy, in 2000 and 2005, respectively. Since 2014, he has held an Associate Professor position with the Dipartimento di Automatica e Informatica at Politecnico di Torino. He has published a number of technical papers in international books, journals and conferences in the areas of computer graphics, visualization, human-machine interaction and mobile computing. He has served as a reviewer or a program or organization committee member for several conferences. He is a member of the editorial advisory boards of several international journals. He is a senior member of the IEEE and the IEEE Computer Society. He is a member of the IEEE Computer Society Technical Committee on Visualization and Graphics.

Andrea Sanna received an M.Sc. degree in Electronic Engineering and a Ph.D. degree in Computer Engineering from Politecnico di Torino, Italy, in 1993 and 1997, respectively. Currently, he is an Associate Professor with the Dipartimento di Automatica e Informatica at Politecnico di Torino. He has authored and co-authored several papers in the areas of human-machine interaction, computer graphics, scientific visualization, augmented and virtual reality, distributed computing, and computational geometry. He is a member of several scientific committees and a reviewer for journals and conferences. He is a senior member of the ACM.

Claudio Demartini is a professor at Politecnico di Torino, Italy, where he teaches Information Systems and Innovation and Product Development. His research interests lie in the areas of software engineering, architecture and web semantics. Currently, he is a member of the Academic Senate of Politecnico di Torino and a consultant on vocational education and training for the Ministry of Education, Universities and Research. He is a Senior Member of the IEEE and the IEEE Computer Society

Chapter 5
Technology for Standing up and Balance Training in Rehabilitation of People with Neuromuscular Disorders

Imre Cikajlo, Andrej Olenšek, Matjaž Zadravec and Zlatko Matjačić

Abstract People with neuromuscular disorders often suffer for gait and balance problems. Most of them are unable to stand up. Recent technologies do not specifically address these problems but offer an excellent tool for expanding possibilities and accelerate rehabilitation process. In this chapter we present virtual reality supported balance training, a haptic floor as an add-on for enhanced balance training and a postural response assessment and progressing from sit-to-stand with controlled active support. Experimental studies have been carried out with our developed novelties in healthy and neurologically impaired persons in order to demonstrate the feasibility of the approach. The haptic floor causes postural strategy changes when adding visual feedback, while the virtual reality balance training appeared to be also clinically effective in the small scale study. On the other hand the sit-to-stand trainer enables adjustable training conditions and facilitates the subject's voluntary activity. The device can also mimic the natural like standing up. All technological novelties have been tested in the preliminary case and proof-of-concept studies in the clinical environment.

5.1 Introduction

Recently repetitiveand intensive rehabilitation tasks, that require two or more physiotherapists to keep the patient in the standing position, have been replaced by modern technology. Passive or active standing frames enable upright posture so that the physiotherapist can focus on the targeted therapeutical work. The therapeutic exercise differs for each motor and control task, but may be similar for various neurological disorders or diseases. Various tasks for balance are designed for weight transfer only, while some of the tasks also require standing on a single leg.

Static balance task are often low dynamics and until the projection of the centre of gravity is within the support frame, the person's posture is statically stable.

I. Cikajlo (✉) · A. Olenšek · M. Zadravec · Z. Matjačić
University Rehabilitation Institute, Ljubljana, Slovenia
e-mail: imre.cikajlo@ir-rs.si

© Springer International Publishing AG 2017   87
A.L. Brooks et al. (eds.), *Recent Advances in Technologies*
for Inclusive Well-Being, Intelligent Systems Reference Library 119,
DOI 10.1007/978-3-319-49879-9_5

However, dynamic balance requires a quick and coordinated response from muscles to stabilize the body and return it to the static balance position. Most of the therapeutic tasks are carried out in the clinical environment and require additional objects, therapeutic aids or accessories. When some of these target based tasks are designed with novel technologies of virtual environments, the rehabilitation devices can enable a single physiotherapist to supervise the therapeutic procedure and intervene only when is needed. The tasks created within the virtual reality environment offer a capability to provide real time feedback to the patient during practice and is proven to be effective in patients suffered for stroke [1]. Patients' action is immediately seen after the intervention and the physiotherapists may also change the difficulty level or a strategy to achieve the goal [2]. Successful applications with virtual reality (VR) technologies can be found in static and dynamic balance training in post-stroke [3–5]. Most of the studies [4] reported on significant improvement of dynamic balance following VR-based therapy in comparison with other intervention.

The importance of repetition and immediate feedback were so far emphasized, however also adaptability and other aspects of training need to be explored and could play an important role in the future, e.g. postural responses in dynamic VR environment combined with haptic interfaces [6]. Postural responses appear as a reaction to the postural perturbation—the translating floor [7] provokes postural perturbations and imbalances the patient. The patient's response requires coordinated muscle activity [8, 9] and also coordinated visual activity when the floor's action was synchronized with virtual environment [10]. Such combination enabled study of various postural strategies depending on the visual or haptic stimulus [11]. Besides that, VR based balance training intervention has been enhanced by dynamic balance training and postural response assessment device.

But before one may take advantage of VR balance training he or she must stand up. And exactly the sit-to-stand (STS) transfer is probably the most demanding movement task in humans when considering the required muscle forces and powers that need to be executed in a short period of time to propel centre-of-mass (COM) upwards while simultaneously also controlling the movement of COM in the transverse plane. Rehabilitation of STS movement in neurologically impaired individuals in clinical practice is particularly challenging because with the current methods ranging from extensive manual support provided by several therapists to use of various lifting mechanisms that provide assistive forces at buttocks while trainees use extensively their arms to aid in controlling the movement of COM in all three planes of movement do not establish training conditions that would facilitate restitution of motor skills required for performing functional STS transfers [12, 13]. We have therefore conceived a mechanical apparatus where kinematic structure and control approach aimed at providing training conditions where kinematics of STS transfer would closely resemble the one seen in neurologically intact individuals.

5.2 Virtual Reality Enhanced Balance Training

Modern neurotherapeutical approaches emphasize the importance of intensive repetition of certain functional activities [14]. An apparatus for dynamically maintaining balance (DMB) is a mechanical device that enables hands-free and safe upright posture and balancing in clinical or home settings. The person is fixed at the level of pelvis and as balances the standing frame moves together with the person. The device can hold the person in the upright position, thus the physiotherapist can focus on various tasks of training of static and dynamic balance under controlled conditions. The applied tasks should be repeatable and target-based. In conventional balance training these tasks comprise of handling a ball, leaning to the specific direction or following the object in the physiotherapist's hand. However, such exercises have been recently implemented in computer graphics using VR technology [1–3] and designed in the way that enable subject's interaction with the therapist, setting up the difficulty level and adapting to the user needs.

The DMB (Fig. 5.1) is a mechanical device for maintaining dynamic balance that enables limited movement in all directions of the transversal plane. The DMB standing frame consists of two parallel load-bearing tubes fixed with two mechanical springs of cylindrical shape mounted in a steel cylinder on the base plate. On top of aluminium frame is a table for arm rest. The stiffness of both mechanical springs and thus the standing frame can be adjusted by raising or lowering the level (based on BalanceTrainer, medica Medizintechnik GmbH, Germany). The person inside the standing frame must use the back safety support to prevent fall and can use the additional knee support if needed. The DMB limits the inclination of the standing frame in the sagittal and frontal plane to $\pm15°$. The inclination of the standing frame was measured by double-axis inertial unit (Toradex AG, Switzerland) and the information was converted into suitable formation for OpenGL based VR environment (blaxxun contact, USA) and graphical user interface (designed in Matlab, MathWorks Inc., USA).

Fig. 5.1 Walk through the virtual city by changing the posture/balancing. A virtual/real therapist can guide the patient and provide advice, if needed

5.2.1   Clinical Pilot Study

The person was standing in vertical position in the balance trainer with his hands placed on the wooden table in front of him and controlled the VR game by leaning the body forward, backward or side by side. Leaning forward increased the velocity of the movement and backward decreased it, while lateral movement was equal to turning left or right. The VR game started at the tree avenue and continued avoiding benches and trash cans to the summer outdoor table on the left and returned back to the building. When the person entered the door, the task restarted. The time of such task was recorded at 3–5 iterations in six persons with stroke (58.5 SD 12.1 years) before and after the balance training. The participating persons practiced balance for 4 weeks, 3 days a week for 20 min. The persons with stroke fulfilled the following criteria: (1) first stroke, (2) can maintain upright posture, (3) can stand in the standing frame, (4) completed a short-term memory test, (5) no other neurological or musculoskeletal impairments or heart failure, (6) willing to participate and had no prior experience. Patients with neglect were excluded from the experimental group [3].

The participants managed to master the VR game and significantly reduced the time required to finish the task (Fig. 5.2). We may confirm that balance capabilities of the participants have improved as also the clinical outcomes of the pilot study demonstrated significant improvements of functional capabilities; Berg Balance Scale (BBS) [15] for 5 points, standing on the affected side for 2.4 s (300 %), Timed-Up-Go (TUG) test [16] for 10 s (30 %) and 10 m (10MWT) test for 5.6 s. A statistical comparison with other groups of patients who were using balance training without VR did not demonstrate significant differences. However, the physiotherapist's work was considerably easier with VR support in terms of physical effort [3].

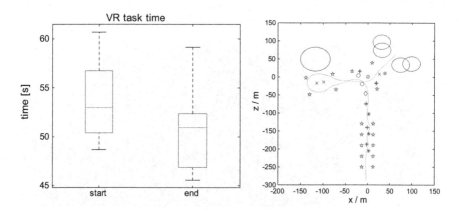

Fig. 5.2 *Left* The average VR game task time has significantly decreased with training. *Right* The accomplished path in the VR environment

5.3 Haptic Floor for Enhanced VR Experience

Horizontally translating standing plates named as "haptic" floor were applied during balance training. It has been reported that such information may have an important role in enhancing computer-enabled activities that involve action/reaction on lower extremities [6]. The participating persons received in addition to the visual feedback from virtual reality environment also a haptic interaction with the VR augmented environment. This interaction required from the person to take immediate action. For the particular case the haptic floor was activated at the interaction with the VR objects, such as collision with the obstacle or slippery floor. The stimulus caused by such sudden action had a similar effect as generation of postural perturbations at the surface level [17]. Postural perturbations at the foot level caused imbalance of the participating person. As the study requested from the person not to lift or move the foot, the person would activate the postural mechanisms to return to the equilibrium [18]. A combination of "real-feel" haptic information and VR based target based task may besides improvement of balance capabilities also have positive effect on postural responses [19].

5.3.1 Equipment Design

The horizontally translating standing plates were designed as an add-on for DMB. The designed plates interact with computer augmented environment and enable the person to "feel" the virtual impact. Thus the name "haptic floor" (HF) was used. HF was designed to fit the commercially available balance training device (BT, Medica Medizintechnik GmbH, Germany) and enable simultaneous movement of both standing plates in all directions of the transversal plane (Fig. 5.3). Two DC motors (Maxon motor AG, Switzerland) with gearboxes (Maxon, Planetary Gearhead GP

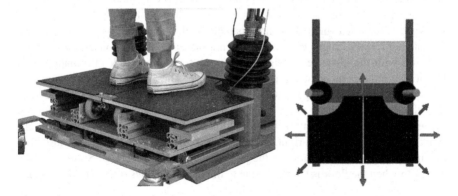

Fig. 5.3 Haptic floor translates in all directions in the horizontal plane

52, Switzerland) were connected with steel-wire to move the lower plate in medial-lateral (left-right) direction and the upper standing plate in anterio-posterior (forward-backward) direction. The HF and the chassis were made of aluminum and the upper standing plate was covered with hard surface rubber to prevent slipping of the feet [10]. The signal from encoders (A, B, I) were connected to the high-speed digital I/O by quadrature encoder (National Instruments (NI) 9403, USA) and decoded with software quadrature decoder (Labview 8.5 FPGA, NI). The control of the HF was achieved by embedded real-time controller (NI cRIO-9014, USA) sending data through analogue output module (NI 9263 AO, USA) to the Maxon 4-Q-DC servoamplifier (ADS 50/10, pulsed (PWM) 4-Q-DC Servoamplifier 50 V/10 A).

HF was position controlled using a two way PID controller; a steep step response at the time of perturbation and slow dynamics without overshoot (<5 %) when returning to the initial position [20]. A reference positions for both directions (sagittal, frontal) were set as inputs to the controller. The reference positions were determined from the velocity of the movement in the VR, position of the avatar and the angle of the impact at collision with the VR object. Such information consisted of the direction of HF movement (forward/backward/left/right) and the amplitude for both planes; sagittal and frontal. The information was sent to the HF controller over the UDP or TCP/IP. However, the displacement of the HF was limited by hardware and a soft limit was set due to the specific users, persons with stroke. The vertical component of the ground reaction force and centre of gravity (CoG) were assessed with Wii Balance Board (Nintendo Inc. USA).

5.3.2 Haptic and Visual Feedback

The person standing in the DMB had almost total control over his/her balance and could see immediately the consequences of the action in the VR game. When a collision with any VR object is unavoidable, the person can prepare for reaction and sometimes can predict and avoid the collision. However, if the person is aware that no action is taken at the instant of collision, often no response to that visual feedback is recorded. In such manner patients with balance problems can practice and also efficiently improve balance, but not also postural responses. When each collision with the obstacle or slippery floor causes biofeedback information with HF, it also induces an adequate postural response. The system calculates the angle and the velocity of the impact and determines the dynamics of the DC motor and thus the direction and the speed of the horizontal translation of the HF [10]. The movement of the HF in one of the 8 specific directions (forward, backward, left, right, forward-left, forward-right, back-left, back-right) would cause a perturbation and induce a postural response of the person standing in the DMB. Similar actions are taken when the floor is slippery. Due to the high dynamics of the HF the persons responded to the perturbations with automatically generated postural response by central neuromuscular system.

Obviously, the person must have changed the balance and postural strategy with the presence of different feedbacks. The main goal of the preliminary research in persons with stroke was to observe how the postural strategies change with absence or presence of the visual (VR) feedback or/and haptic feedback. Besides, we are already aware that changes in muscle coordination took place in neurologically intact participants [17].

5.4 Postural Strategy Changes with Visual Feedback

5.4.1 Pilot Study

A pilot study (Fig. 5.4) with two individuals, both stroke survivals (hemiparesis, ischemic, 53 and 61 years old, in average 172 cm, 82.5 kg, right side affected, Table 5.1) were carried out in order to identify the changes in the postural strategy and examine the short term clinical effects of haptic floor supported balance training. The inclusion and exclusion criteria were the same as for the VR supported balance training.

Fig. 5.4 A pilot study of postural responses w/o and with VR feedback information in person with stroke

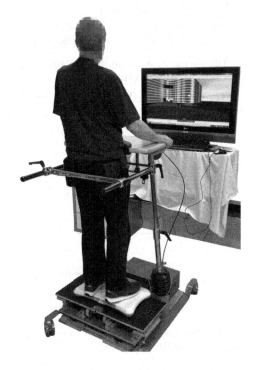

Table 5.1 Participants and the outcomes of clinical tests

Patient	Age	Height	Weight	Clinical outcomes					
				Before therapy			After therapy		
				BBS	10MWT	TUG	BBS	10MWT	TUG
ID1	53	165	75	52	18.1	25.9	52	14.3	17.9
ID2	61	179	90	51	13.7	14.3	51	13.9	12.1

5.4.2 Experimental Protocol

Both subjects participated in three stage postural responses assessments and 14 days of HF supported balance training. The HF supported balance training was similar to the virtual reality enhanced balance training (VRBT) using DMB, but using HF to provide horizontal translations at collisions with virtual objects or virtual slippery ground. Then the task restarted and the subject repeated the task as many times as possible in 5 min. In 14 days regularly 30 sessions were carried out, considering that each day the VRBT consisted of 3 sessions each lasting 5 min.

Before and after the training a three stage postural response assessment was carried out (Fig. 5.5):

(a) Assessment of the postural responses to the horizontal perturbations where the direction and timing of the perturbation was randomly selected. Without VR.
(b) Assessment of the postural responses during the VRBT game lasting only 80 s, without HF feedback.
(c) Assessment of the postural responses during the VRBT game lasting only 80 s, with HF feedback at collisions or slippery floor.

During the assessment the movement of the HF, centre of gravity (CoG) and the movement of the DMB were synchronized and measured.

The clinical tests (BBS, TUG and 10MWT) were carried out before and after the balance training for both participating subjects.

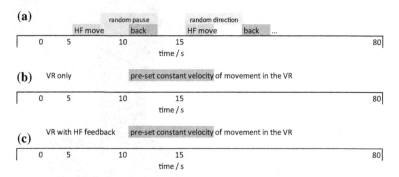

Fig. 5.5 Protocol of the assessment: **a** random generated postural perturbations, **b** VR feedback only, **c** VR and HF feedback

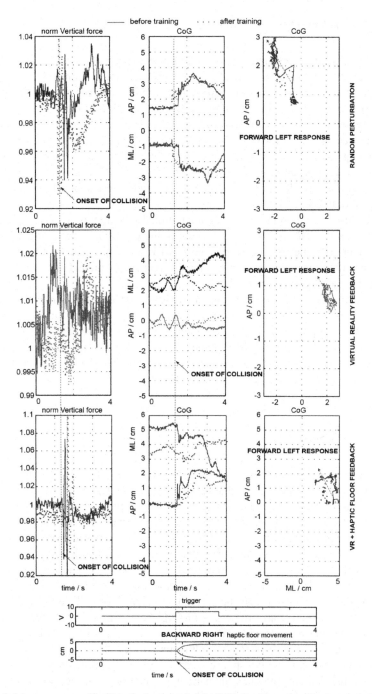

Fig. 5.6 Mean responses (CoG) to the backward right (BR) direction of postural perturbation for various conditions: Perturbation without feedback, virtual reality (VR) feedback and virtual reality feedback with haptic support. Sudden perturbation required immediate unprepared reaction and became faster with training. VR feedback only did not expect any perturbation thus the reaction was weak. While VR enabled preparation to the perturbation in advance, the response was slightly delayed and expected

5.4.3 Experimental Results

Unexpected random perturbations caused by horizontal movements of the HF in 8 different directions required immediate and rapid postural response (Fig. 5.6 upper diagrams present a response to the BR perturbation). However, the response in the FL direction was much smother after the balance training (dotted line).

At the second scenario the collision took place in the VR environment, but the HF remained still. Thus the response was rather weak and predicted as it was related to avoiding the collision with the VR object only. After the balance training the person managed to avoid the particular VR object without having to correct the path. This was recorded in the CoG time course and presented at Fig. 5.6 (central diagrams, dotted line).

At the third scenario the persons got additional HF support at the collision with the object in the VR environment. As seen in the CoG time course of the Fig. 5.6 (lower diagrams) the subjects responded rapidly before the balance training (solid line), but after the balance training they were able to predict the collision and prepared in advance (dotted line).

The analysis of the CoG demonstrated significant changes, while the clinical parameters could not confirm significant changes in the balance capabilities of the participating subjects. The BBS did not changed at all; however both participating subjects had high entry score. More changes were noticed at mobility tests 10MWT and TUG; both participants significantly improved their TUG score and the participant ID1 also improved his score at 10MWT.

5.5 Discussion

The outcomes are similar to the findings in healthy, neuromuscularly intact persons. Cikajlo et al. [17] reported on changes in postural strategies due to the different sensory information available. The authors successfully demonstrated how the visual feedback (VR task) enabled the person to "prepare" for the impact and thus choose different postural response strategy than at random and unpredictable perturbations. The outcomes of the experimental study in two patients with stroke are in line with our finding in neuromuscularly intact subjects. The mean CoG time course demonstrated that the person who took a balance training program will most likely predict the collision and unconsciously prepare for collision. Therefore we haven't noticed significant differences in CoG between the two participants with stroke and healthy subjects when HF provided perturbations at the collision with the object in the VR environment. In healthy individuals the major muscle groups' latencies during the postural responses were assessed by surface electromyography. The outcomes demonstrated that the participants choose different postural response strategies by synchronously activating bi-articular and distal muscle groups [17].

5.6 Sit-to-Stand Trainer

Figure 5.7 shows the conceptual presentation of the conceived mechanism (upper panel) and photographs of the actual apparatus with neurologically intact subject in four discrete postures (from sitting position through initial trunk forward inclination followed by well-coordinated extension movement of lower extremities to standing posture) while performing STS movement [21]. The kinematics of the proposed mechanism was designed such that closely resembles kinematics of STS of neurologically intact individual where important features are initial trunk inclination in the beginning of the STS movement combined with adequate support of anterior shank movement thus resulting in "natural-like" standing-up movement. Even though the mechanism provides "exoskeleton" type of support to shanks, buttocks and trunk this is accomplished with only one actuated degree-of-freedom where AC servo motor actuates/controls the movement of a wheeled segment that slides along the base of the apparatus as depicted in the upper panel of Fig. 5.7.

The control of the movement of the STS apparatus is illustrated in Fig. 5.8. The movement of the AC motor (TECO TSB08-90) follows a pre-determined velocity profile that spans from the motor orientation corresponding to sitting (folded) position of the mechanism to the motor orientation corresponding to standing (extended) position of the mechanism. Four levels of maximal speed of movement can be selected (Fig. 5.8a). Level of support provided by the mechanism to a standing-up subject is achieved by using the torque/current limit profile that can be pre-determined within the AC motor controller (TECO TSTA-30C). Ten levels of

Fig. 5.7 The *upper panel* shows illustrations of the conceived mechanism while the *lower panel* shows photographs of the device and neurologically intact subject in four discrete postures during standing-up

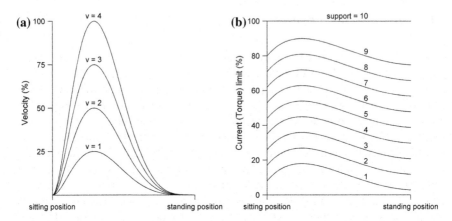

Fig. 5.8 Conceptual presentation of the control approach. **a** Four speed profiles that AC servo motor follows between sitting and standing posture of the mechanism. **b** Ten torque/current limit profiles that can be selected to determine adequate level of support during standing-up

support can be selected (Fig. 5.8b). This control approach requires from a person attempting STS movement within the apparatus, depending on the level of support selected, to provide additional propelling forces/torques since the movement of the motor is stopped when the upper limit of torque/current is reached. In this way a therapist can experimentally determine the most appropriate level of support which helps a trainee to successfully perform STS movement.

Kinematics, kinetics and electromyography of a neurologically intact subject when standing-up within STS apparatus.

5.6.1 Experimental Methods

In order to explore biomechanics of standing-up when using STS apparatus we first investigated kinematics, kinetics, energetics and electromyography in a single neurologically intact individual performing STS movement in three experimental conditions: (1) during natural standing-up from the chair of STS while the subject was not fastened to the STS device—FREE; (2) during STS apparatus assisted standing-up—SUPPORT-5; (3) during STS apparatus (support level 10—maximal level of support) standing-up—SUPPORT-10. In total six reflective markers were positioned to metatarsal joint, ankle joint, knee joint, hip joint, ASIS point on the pelvis and on the sternum. Movement of these six markers in space was captured with VICON Motion System (Oxford Metrics Ltd.) and subsequently used to monitor kinematics of movement. The person was standing on a force platform through which we measured the ground reaction forces in vertical direction (z axis), sagittal direction (y axis) and lateral direction (x axis). Additionally we were able also to compute center-of-pressure (COP) underneath the feet of the subject.

In order to estimate the power delivered to STS movement separately by the subject and by the STS device we used a simple biomechanical model. This model assumed that there are two forces acting on the center-of-mass (COM) of the subject: the first force is the ground reaction force (GRF) while the second force is the reaction force resulting from mechanical interaction of the subject with the STS device. As the estimate of the COM position, velocity and acceleration we used the data from the marker placed at the ASIS of the pelvis.

Muscular activity during STS movement was monitored through surface electromyography. Seven muscle groups were selected for monitoring: tibialis anterior (TA), soleus (SOL), gastrocnemius (GAS), quadriceps (QUAD), hamstrings (HAM), gluteus medius (GMED) and erector spinae (ESPIN). The EMG electrodes were positioned over the palpated muscle bellies, the area underneath the electrodes was properly cleaned and the electrical impedance was checked to assure optimal EMG recordings. The sampling frequency of EMG signals was set to 1 kHz. All EMG signals were appropriately processed (de-mean the raw signal-subtract out the mean; band-pass filter, 20–300 Hz; notch filter, 59–61 Hz; full-wave rectification—take absolute value; moving average window—150 ms). All EMG signals were normalized to the maximal value for each particular muscle group when standing-up in FREE experimental condition. This enabled meaningful comparison of muscle activity in both SUPPORT-5 and SUPPORT-10 experimental conditions.

The subject was asked to have the arms crossed on the chest and to perform standing-up in a most comfortable way with a self-selected speed in the experimental condition FREE. In the experimental condition SUPPORT-5 (the speed of movement of STS was set to 4—maximal speed) the subject was asked to "co-operate" with the STS device in a way that felt the most comfortable. In the experimental condition SUPPORT-10 the subject was asked not to volitionally participate during the standing-up process but to let the STS device to lift him. In every experimental condition first ten introductory trials were made to allow the subject to accommodate, followed by additional ten trials where the data was collected.

The data representing kinematics (angles), kinetics (ground reaction forces of the subject and STS reaction forces), energetics (power done by the ground reaction forces of the subject and STS reaction forces) and electromyography were averaged across the ten trials and the standard deviation was calculated to compare the variability of the standing up-process when standing-up in a natural way and when supported by the STS device. Also, for a better visualization the stick diagrams were plotted to indicate posture throughout the whole standing-up process.

5.6.2 Experimental Results

Figure 5.9 shows EMG profiles of standing-up for all three experimental conditions. When standing-up naturally the EMG profiles are consistent with in the literature well-documented patterns where the "prime movers" are muscles TA and

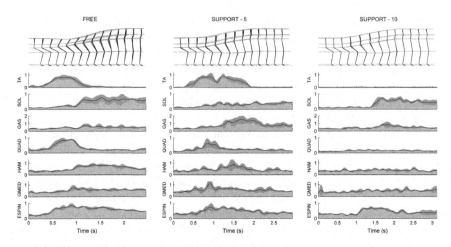

Fig. 5.9 Electromyography of selected lower limb and trunk muscles during standing-up in neurologically intact individual. *Left column* shows kinematics—stick diagram and muscle activation profiles during completely unhindered standing-up. *Middle column* shows kinematics—stick diagram and muscle activation profiles during standing-up while assisted with STS (SUPPORT-5 experimental condition). *Right column* shows kinematics—stick diagram and muscle activation profiles during standing-up while assisted with STS (SUPPORT-10 experimental condition)

QUAD [13]. TA moves the knee forward and into flexion, while strong QUAD activity provides the necessary power in the first phase of standing-up. In the second phase SOL and GAS take over and provide power for finishing the extension in the knees while HAM and GMED provide activity to extend the hip. ESPIN shows activity only of the small portion of the whole spine so it cannot be used for analysing the functional activity in the trunk, its activation shows certain degree of co-contraction which is always present. Similar EMG profiles with slight variations can be observed also in the partially supported standing-up (SUPPORT-5). This result indicates that STS device does not alter the "natural" EMG synergies required for efficient STS movement. When STS movement was fully supported (SUPPORT-10) TA and QUAD displayed almost no activity and also no modulation can be noticed in HAM and GMED muscles. In the second phase of STS movement we observe activity in SOL and GAS, which are necessary to complete the movement into full extension in the knees and hips. This is because the STS device is programmed such that it provides the propelling power in the first phase of standing-up movement while in the second phase its action is limited only to maintain the achieved posture. In this way the standing subject is fully supported only in the first, most demanding period of standing up process.

Figure 5.10 shows the vertical acceleration of COM and the corresponding vertical forces as well as power contribution from both vertical forces: The vertical ground reaction force Fz under the feet of the subject and STS force. Here, the maximal vertical accelerations are similar across the experimental conditions. When

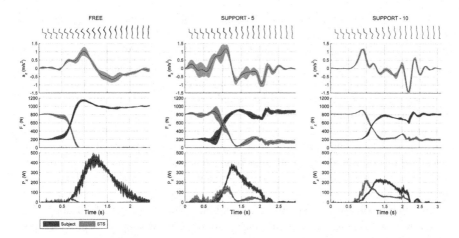

Fig. 5.10 Vertical acceleration, vertical ground reaction and STS forces and associated mechanical power during standing-up of neurologically intact individual. *Left column* shows kinematics—stick diagram and vertical acceleration, ground reaction force and power profiles during completely unhindered standing-up. *Middle column* shows kinematics—stick diagram and vertical acceleration, ground reaction force and power profiles during standing-up while assisted with STS (SUPPORT-5 experimental condition). *Right column* shows kinematics—stick diagram and vertical acceleration, ground reaction force and power profiles while assisted with STS (SUPPORT-10 experimental condition)

observing the forces in the experimental condition FREE it can be observed the propelling source accelerating COM upwards comes from the observed peak in the vertical F_z (weight of the subject was 1000 N) and correlates well with the peak in the vertical acceleration. This is further confirmed in the graph showing power contributions where all the power comes from the vertical GRF and thus by a subject. In both supported experimental conditions also the contribution from the vertical STS force can be observed, however also the subject must actively participate in propelling the COM in the upward direction. These results are in a good agreement with the observed EMG patterns.

5.7 Standing-up of Neurologically Impaired Individual When Supported with STS Apparatus

We have explored feasibility of the developed STS apparatus for assistance during training of STS movement in a selected subject that has suffered stroke ten years ago resulting in left-sided hemiparesis. Two experimental conditions with two levels of STS apparatus support were tested: (1) support level was set to 10—SUPPORT-MAX; (2) support level was set the critical value that provided just enough support to enable the subject to accomplish standing-up—SUPPORT-CRITICAL. The speed of STS movement was set to 2.

The subject in each of the experimental conditions first attempted five trials to familiarize with the device and standing-up in the device, followed by five trials during which data acquisition took place. In SUPPORT-MAX experimental condition the subject was instructed to be relaxed and to let the apparatus to lift him up while in SUPPORT-CRITICAL experimental condition he was instructed to initiate standing-up as soon as he heard the sound signalling the beginning of the supporting action from the STS device.

During all standing-up trials all three components of ground reaction force and movement of COP under the feet were recorded and averaged across trials. These results are presented for both experimental conditions and contrasted with standing-up of a neurologically intact individual in FREE experimental conditions in Fig. 5.11.

In comparison to the experimental condition where maximal STS support is given, the experimental condition where minimal-critical support is provided displays noticeable rise in vertical component of ground reaction force meaning that the patient had to provide volitional effort and power to accomplish standing-up. The shape of the vertical GRF is very similar to similar graph recorded in neurologically intact individual. From both horizontal components of GRF it can be noticed tendency of leaning toward the sound side, which is in the case of the selected subject to the right. This tendency is even more visible when comparing the COP trajectories which were in both cases shifted toward the right side. However, when critical support was provided the direction of COP movement was similar to the one observed in neurologically intact individual.

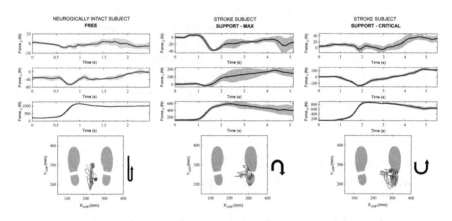

Fig. 5.11 Ground reaction forces and centre-of-pressure trajectories during standing-up in neurologically intact individual (*left column*) and in stroke subject when fully supported (*middle column*) and when minimally supported (*right column*)

5.8 Discussion

The results show that the kinematic structures as well as the developed speed-torque control of the STS device enables adjustable training conditions where the active participation of an individual is facilitated. Furthermore, the electromyography and energetics studied in different experimental conditions show that natural-like standing-up movement can be mimicked where the emphasis on the support delivered on the part of the STS apparatus is during the initial phase of standing-up process. This has proven to be appropriate since the initialization of the STS movement is the most demanding task associated with standing-up. Extensive results on a group of six neurologically intact and five chronic stroke subjects can be found in Matjačić et al. [22].

Acknowledgements The authors acknowledge the financial support of the Slovenian Research Agency (Grant No. P2-0228) and the Slovenian Technology Agency (Grant No. P-MR-10/07). The operation was partly financed by the European Union, European Social Fund.

References

1. Yang YR, Tsai MP, Chuang TY, Sung WH, Wang RY (2008) Virtual reality-based training improves community ambulation in individuals with stroke: a randomized controlled trial. Gait Posture. 28(2):201–206
2. Holden MK, Dyar T (2002) Virtual environment training: a new tool for neurorehabilitation. Neurol Report 26(2):62–71
3. Cikajlo I, Rudolf M, Goljar N, Burger H, Matjačić Z (2012) Telerehabilitation using virtual reality task can improve balance in patients with stroke. Disabil Rehabil 34(1):13–18
4. Darekar A, McFadyen BJ, Lamontagne A, Fung J (2015) Efficacy of virtual reality-based intervention on balance and mobility disorders post-stroke: a scoping review. J Neuroeng Rehabil 12(1):46
5. Deutsch JE, Merians AS, Adamovich S, Poizner H, Burdea GC (2004) Development and application of virtual reality technology to improve hand use and gait of individuals post-stroke. Restor Neurol Neurosci 22:371–386
6. Visell Y, Law A, Cooperstock JR (2009) Touch is everywhere: floor surfaces as ambient haptic interfaces. IEEE Trans Haptics 2(3):148–159
7. Moore SP, Rushmer DS, Windus SL, Nashner LM (1988) Human automatic postural responses: responses to horizontal perturbations of stance in multiple directions. Exp Brain Res 73(3):648–658
8. Garland SJ, Gray VL, Knorr S (2009) Muscle activation patterns and postural control following stroke. Mot Control 13(4):387–411
9. Henry SM, Fung J, Horak FB (1998) EMG responses to maintain stance during multidirectional surface translations. J Neurophysiol 80(4):1939–1950
10. Cikajlo I, Oblak J, Matjacic Z (2011) Haptic floor for virtual balance training. In: 2011 IEEE world haptics conference, pp 179–184
11. Keshner EA, Streepey J, Dhaher Y, Hain T (2007) Pairing virtual reality with dynamic posturography serves to differentiate between patients experiencing visual vertigo. J Neuroeng Rehabil 4:24

12. Burnfield JM, Shu Y, Buster TW, Taylor AP, McBride MM, Krause ME (2012) Kinematic and electromyographic analyses of normal and device-assisted sit-to-stand transfers. Gait Posture 36:516–522

13. Burnfield JM, McCrory B, Shu Y, Buster TW, Taylor AP, Goldman AJ (2013) Comparative kinematic and electromyographic assessment of clinician- and device-assisted sit-to-stand transfers in patients with stroke. Phys Therapy 93(10):1331–1341

14. Langhammer B, Stanghelle JK (2000) Bobath or motor relearning programme? A comparison of two different approaches of physiotherapy in stroke rehabilitation: a randomized controlled study. Clin Rehabil 14(4):361–369

15. Berg K, Wood-Dauphinee S, Williams JI (1995) The balance scale: reliability assessment with elderly residents and patients with an acute stroke. Scand J Rehabil Med 27(1):27–36

16. Ng SS, Hui-Chan CW (2005) The timed up & go test: its reliability and association with lower-limb impairments and locomotor capacities in people with chronic stroke. Arch Phys Med Rehabil 86(8):1641–1647

17. Cikajlo I, Krpič A (2014) Postural responses of young adults to collision in virtual world combined with horizontal translation of haptic floor. IEEE Trans Neural Syst Rehabil Eng 22 (4):899–907

18. Jacobs JV, Horak FB (2007) External postural perturbations induce multiple anticipatory postural adjustments when subjects cannot pre-select their stepping foot. Exp Brain Res 179 (1):29–42

19. Marigold DS, Eng JJ, Dawson AS, Inglis JT, Harris JE, Gylfadottir S (2005) Exercise leads to faster postural reflexes, improved balance and mobility, and fewer falls in older persons with chronic stroke. J Am Geriatr Soc 53(3):416–423

20. Krpič A, Cikajlo I, Savanović A, Matjačić Z (2014) A haptic floor for interaction and diagnostics with goal based tasks during virtual reality supported balance training. Slov Med J 83(2)

21. Oblak J, Matjačić Z (2013) Aufstehtrainer: Patentschrift DE 10 2012 102 699 B4 2013.10.17. [München]: Deutsches Patent- und Markenamt

22. Matjačić Z, Zadravec M, Oblak J (2016) Sit-to-stand trainer: an apparatus for training "Normal-Like" sit to stand movement. IEEE Trans Neural Syst Rehabil Eng [Epub ahead of print] 24(6): 639–649, doi: 10.1109/TNSRE.2015.2442621

Chapter 6
Exergaming for Shoulder-Based Exercise and Rehabilitation

Alvaro Uribe-Quevedo and Bill Kapralos

Abstract Exercise is often encouraged, preferably under physician supervision, to help overcome the various musculoskeletal disorders that can often hinder the execution of daily personal and work-related tasks. However, ones motivation to exercise typically decreases after a short period of time, particularly when considering repetitive exercise routines. Furthermore, assessing ones performance within an exercise program is important, particularly when considering rehabilitation-based exercise routines, yet assessment can be problematic as it consists of qualitative measures only (observation, questionnaires, and self-reporting). Obtaining quantitative information has traditionally required cost prohibitive specialized measuring equipment, scenario that is changing with current immersive technologies (virtual reality and gaming). Exergaming couples video games and exercise whereby playing a video game becomes a form of physical activity. Exergaming takes advantage of the engaging, interactive, and fun inherent in video games to promote physical activity and engagement applicable to physical training or rehabilitation. Furthermore, recent technological advances have led to a variety of consumer-level motion tracking devices that provide opportunities for novel interaction techniques for virtual environments and games and the ability to generate quantitative data (feedback) dynamically. In this chapter, we outline our experience in dynamic design, development, and testing of two independently developed exergames that have been specifically developed for shoulder rehabilitation. Shoulder injuries are very common, particularly with the college-aged population.

A. Uribe-Quevedo (✉)
Mil. Nueva Granada University, Cra 11#101-80, Bogota, Colombia
e-mail: alvaro.j.uribe@ieee.org

B. Kapralos
University of Ontario Institute of Technology, 2000 Simcoe St N, Oshawa, ON L1H 7K4, Canada
e-mail: bill.kapralos@uoit.ca

© Springer International Publishing AG 2017   105
A.L. Brooks et al. (eds.), *Recent Advances in Technologies for Inclusive Well-Being*, Intelligent Systems Reference Library 119,
DOI 10.1007/978-3-319-49879-9_6

6.1 Introduction

Approximately 15 % of the global population suffers from some form of musculoskeletal disorders that often times hinders the execution of daily personal and work-related tasks [49]. Aside from the loss of productivity that this leads to, it can also lead to a reduction in the quality of life. Treatment for musculoskeletal disorders often includes preventive, corrective, and maintenance exercises, each performed during a specific stage of the disorder treatment, often times under the care of a medical expert. Preventive exercises aim to reduce risk, improve and adjust a movement or posture, while maintenance-based exercises focus on physical activity required to sustain functionality in order to carry out daily activities [41]. When prevention has failed and an injury has occurred, patients are typically required to engage in some form of physical therapy/exercise (cryotherapy, electrotherapy, and kinesiotherapy). Often times, this is in conjunction with pain killers, and anti-inflammatory medication, although in some cases, surgical intervention may be required [32]. When pain persists and medication loses its effect, distraction and endorphin stimuli resulting from engaging activities may reduce pain perception; the endomorphin stimuli can be obtained from the production of dopamine that a user/player experiences while winning and overcoming challenges within a virtual environment that incorporate gaming elements [24].

Physiotherapy-based exercise requires the execution of repetitive movements (often times within a time limit), whose ranges of movement are set by a medical expert. The practice of physiotherapy exercise is primarily constrained to medical centres/clinics with specialized equipment and capable personnel (e.g., physiotherapists). Attendance at such facilities often requires transportation (assisted or not assisted, depending on the musculoskeletal disorder or disability). Sometimes the exercises are prescribed to be executed at home without any supervision or feedback regarding how the exercises (movements) are actually being performed. An additional problem relates to assessing a patient's performance within an exercise program. Traditionally, assessment has been mostly qualitative including observation by the medical expert/physiotherapist (when present), and the results of questionnaires that the patient completes. Given these considerations, patients are often not motivated (or lose motivation quickly), to engage in physical activities even though these physical activities can help prevent, correct, or maintain their quality of life. However, the degree that a patient adheres to physiotherapy exercise treatment requirements (home- or clinic-based) can be a large factor in the success of the physiotherapy program [5]. Other factors that may hinder a patient from maintaining a rehabilitation-based exercise program includes lack of time, confusing guides, lack of knowledge, lack of interest, fear of incorrectly performing the exercises, and pain associated with performing the exercises [16].

Developing an understanding of human locomotion has been a longstanding topic of interest, driving research towards creating solutions to quantify movement for different purposes (e.g., movement pattern identification, joint force and energy expenditures, and treatment of war veterans, among many others) [10]. In recent years, technological advances have introduced various consumer level motion sensing technologies to many homes. For example, the Microsoft Kinect motion sensing vision-based sensor is capable of detecting and tracking user movements and allowing interactions with the games using natural gestures thus eliminating the game controller. Motion capture technology provides a suitable solution to the challenges associated with observation-based assessment (whether the exercises are performed at home or in the clinic). Capturing exercise-based motions can provide quantitative information regarding a patient's movements while exercising. This information can then be provided to the patient thus providing them with (dynamic) feedback regarding their movements that can ultimately be used to correct any problematic movements.

Tracking technologies can be grouped into three categories [12]: (i) active-target systems which incorporate powered signal emitters (e.g., magnetic [25], mechanical [14], optical, radio [21], and acoustic signals [46]), and sensors placed in a calibrated environment, (ii) passive-target systems employ ambient or naturally occurring signals (e.g., vision-based systems [48]), and (iii) inertial systems which rely on sensing linear acceleration and angular motion (e.g., accelerometers, and gyroscopes). Each of these three approaches has its own share of issues and challenges, particularly when considering motion capture for physiotherapy-based exercise. More specifically, active-target systems are prone to signal degradation and interference [63]. For example, the accuracy of a magnetic tracker degrades as the distance between the emitter and receiver increases while mechanical trackers typically require a direct physical connection with the object (e.g., patient's limb) being tracked. Aside from the physical attachment to the tracking device, a major problem with mechanical trackers is the limited range of motion [63]. Passive-target systems generally require the user to wear markers (often times these are incorporated into a special suit worn by the user), and this can potentially lead to possible alterations of the movements that are relevant in the diagnosis and assessment of musculoskeletal disorders [1]. Furthermore, many optical-based tracking systems are prone to changes in lighting conditions. Finally, inertial sensors will typically measure acceleration or motion rates, and these signals must then be integrated to produce position or orientation which requires constant calibration and are also susceptible of error accumulation during measurements [18]. Despite the associated issues, the most frequently used motion capture systems are passive and employ optical sensors that require the user to wear markers on the body segments that are to be detected and tracked by infrared cameras [7, 15]. That being said, the majority of such systems are not consumer-ready but are rather targeted to research and development (mainly in the entertainment industry). As previously described, there are a variety of motion capture systems available. Most common systems employ passive markers attached to the body of the person (or object) being tracked. The camera is capable of detecting the markers and through the use of image analysis, a skeleton and its movements depicting the person (or object) can be reconstructed. Given the setup required for such marker-based solutions, they are often used in movies, gait research labs, sports labs or medical facilities with physiotherapy programs.

Virtual reality- (VR-) based rehabilitation and training can improve the user experience by providing immersive and interactive scenarios that are compelling due to user interfaces (UIs), and narrative. In addition, recent advances in 3D stereoscopic vision with affordable consumer-level devices, can help to increase a patient's motivation when participating in any exercise program [11]. Early applications of VR-focused on stroke rehabilitation by assisting patients in recovering mobility and functionality in order to perform tasks and ultimately improve their quality of life [26]. These early applications were generally cost prohibitive, and limited technologically, and although these systems did offer clear goals and tasks, their success relied on the user's commitment and motivation to follow through the process. With the current advances in technology, particularly with respect to computing hardware and graphical-based rendering, VR has evolved tremendously and has become accessible to the general public through affordable consumer-level (gaming-based) devices such as the Microsoft Kinect, and the Oculus Rift [9], that are bringing VR into the average living room. Such devices are placing a greater emphasis on interaction. For example, Johnson and Winters modified a commercial joystick to extend the arm movement to accommodate a therapeutic range of motion and thus provide and accessible tool for rehabilitation performance evaluation [30]. This increased emphasis on interaction can be traced back to the introduction of the Nintendo Wii gaming console that included the Wii remote (also known as the "Wiimote"), as its primary controller. Through the use of accelerometers and optical sensing, the Wiimote provides the user/player the opportunity to interact with and manipulate items on the screen using gesture recognition and pointing. The Nintendo Wii (and the Wiimote in particular), demonstrated the popularity of motion-based casual gaming and was responsible for bringing together gamers from all ages (from the very young to the very old) into scenarios where the "interactions where everything" [61]. VR is being increasingly used in a variety of applications as it provides users the opportunity to interact with scenarios otherwise impossible in the real world due to safety, ethical, or cost concerns. An example can be seen in tourism-based applications where a user can visit a destination not easily available to them [23], or post-traumatic stress therapy where patients are exposed to simulated, safe, and controlled environments to help them overcome their health state [34].

Despite the benefits of VR-based exercise approaches, the novelty of VR can wear off as the user becomes acquainted with the equipment and the virtual environment [24]. Exergaming can help overcome this challenge by increasing and maintaining motivation to exercise. Exergaming couples video games and exercise whereby playing a video game becomes a form of physical activity allowing healthy behaviours and learning to be promoted [56]. Exergames take advantage of the engagement, interaction, and fun inherent in video games to promote physical activity and engagement within a training or rehabilitation program [35, 47], which can be mundane and repetitive [31]. When compared to traditional exercise, exergaming has proven effective in terms of increasing user motivation regarding the practice of physical activity whether for fitness, rehabilitation, training, or entertainment. Exergaming has been linked to greater frequency and intensity of physical activity and enhanced health outcomes [2], and more recently with commercial exergames, where heart rate and oxygen consumption equals real physical exercising [42]. According to Skiba, there are three emerging trends in exergaming [56]: (i) the increased use of games for therapy and rehabilitation, (ii) the use of games in gyms and other settings to promote physical activity, and (iii) the greater involvement in gaming of corporations and health care providers. With respect to physical rehabilitation, the integration of video games can provide greater guidance for teaching patients how to properly move their limbs as UIs based on motion interactions can track and provide feedback [59]. As Janarthanan describes [28], video games can help a patient learn a pattern of limb movement and reapply it. In other words, "what we learn with these types of games is how to get well soon". Exergaming has also been beneficial to those suffering from traumatic brain injury and cerebral palsy [4], and exergames have been employed to assist patients with their stroke rehabilitation, and to assist young adults with physical and intellectual disabilities. Recent evidence also suggests that exergaming may improve cognitive function and thus augment traditional rehabilitation of motor symptoms in people with Parkinson's disease [4].

In a review of the literature related to the use of video games in health care, Kato highlights the positive effects regarding the use of exergames, and interactive media in general, for example, patients undergoing physical therapy were able to better cooperate with their required physical therapy program [31]. Another review of the literature that was conducted by Primack et al., who examined whether video games are useful in improving health outcomes [44]. They found that video games improved 69 % of psychological therapy outcomes, 59 % of physical therapy outcomes, 50 % of physical activity outcomes, 46 % of clinician skills outcomes, 42 % of health education outcomes, 42 % of pain distraction outcomes, and 37 % of disease self-management outcomes leading to the conclusion that although greater rigour is required, there is potential promise for video games to improve health outcomes, particularly in the areas of psychological therapy and physical therapy [44].

In this work, we describe two independent upper-limb exergames for the purpose of shoulder rehabilitation. We have chosen to focus on shoulder rehabilitation given that shoulder-related injuries and problems are widespread. More specifically, it is estimated that approximately 7.5 million people have some form of shoulder problems with 4.1 million of them related to the rotator cuff [39]. There are also approximately 1.4 million shoulder-based arthroscopies performed each year [40], and according to the World Health Organization (WHO), 40 % of the working population suffers from some form of shoulders and upper limb ailment caused by repetitive strain on ligaments, muscles, bone or tendons surrounding the joint [27]. Shoulder injuries can be acute or painful (from sudden movements), or chronic and everlasting (from overuse). Common afflictions can happen throughout the lifespan of a person as a result of repetitive tasks and bad postures or static shoulder loads resulting in tendinitis. However, shoulder pain and tendinitis can also be the product of aging and sports practices [45]. Therefore, motivation in therapy is a topic of interest among physicians as common exercises focus on stretching and strengthening the shoulder's muscles to provide flexibility and support. These exercises require commitment as observation and follow-up is only performed during the early stages of recovery [50]. To facilitate proper exercising, both of the exergames presented here stimulate exercising through kayak-based motions and map the resulting motions using a vision-based sensor (Microsoft Kinect), into a kayak-based game world providing the user (player) an interactive and immersive environment where they are in control of a kayak moving through a number of courses (rivers, etc.).

The chapter begins with an overview of shoulder-based exergame development starting with a characterization of the shoulder anatomy, movement, and exercising followed by a discussion of the issues involved in the shoulder-based exergame design and development. As with any exergame and serious game in general, the instructional component of exergames must not be overlooked and clear objectives must be devised early in the development process. Each of the exergames will then be described. This will include a discussion regarding our experiences in developing the games (e.g., what worked, what did not work, what we could have done differently, and lessons learned). We also discuss the results of several experiments that were conducted to test the usability of the exergames. There are a number of issues and difficulties in conducting user-based experiments to gauge the effectiveness of exergames and as a result, quantifiable data providing evidence that the game does indeed meet its intended objectives is sparse. The information presented in this chapter will be useful to those interested in the field of exergaming and to those interested in developing and using their own exergame.

6.2 Exergame Development

To develop the shoulder-based exergames, the upper limb was characterized to determine the body parts involved, their ranges of motion, and the exercises suitable to be incorporated within the exergames. This characterization provides information that serves as an initial step in determining the hardware and software development requirements. Greater details regarding the characterization of the upper limb are provided in the following sections.

6.2.1 Anatomy Characterization

The upper limb is comprised of bones, muscles and ligaments; it is segmented into two main parts: (i) the arm (humerus), and (ii) the forearm (radius and ulna) joint at the elbow. Human motion is performed in a three-dimensional space and movements are described in terms of the body planes [55]. As presented in Fig. 6.1, three planes segment the human body: (i) the sagittal plane, (ii) the frontal plane, and (iii) the horizontal plane. Within these planes there are several combinations of motion depending on the body part and their degrees of freedom (DOF), resulting in flexion/extension and adduction/abduction movements.

Fig. 6.1 Human body planes of movement

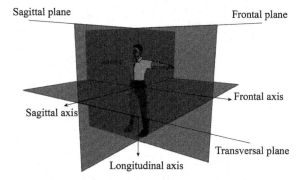

6.2.2   The Shoulder

The shoulder is comprised of seven joints, each of which is important to achieve fully functional movements [52]: (i) glenohumeral, (ii) superhumeral, (iii) acromio-clavicular, (iv) claviculae sternal, (v) scapulocostal, (vi) sternocostal and (vii) cos-tovertebral. Among the joints, the glenohumeral wears faster than the others as it allows rotations on the three coordinate axes, making it susceptible to wear and tear that may result in pain and limited motion. Muscles around the rotator cuff allow flexion/extension, adduction/abduction, circumduction, medial and lateral rotation movements and help keep the humerus head at the right place; this is the reason that while lifting heavy objects, abrupt movements can deteriorate the joint [17].

Shoulder movements involve flexion rotations with average angles varying from $0°$ to $180°$ moving the arm from the body along the sagittal plane as presented in Fig. 6.2. It also involves extension rotations ranging from $0°$ to $60°$ also shown in Fig. 6.2. It is worth noting that some people are capable of performing hyperflexion and hyperextension rotations that the average person cannot. Within the frontal plane, the shoulder allows abduction rotations from $0°$ to $80°$ and adduction from $0°$ to $45°$, similarly to flexion/extension some people may exceed these ranges. Since the shoulder can rotate about all three axes, there are additional combinations that can be achieved, such as the lateral and medial rotations commonly used when exercising the rotator cuff.

Fig. 6.2 Shoulder and elbow movements

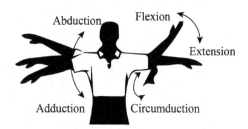

Fig. 6.3 Example shoulder exercises for medial and lateral rotations

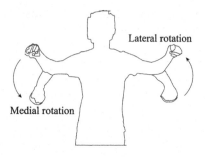

6.2.3 Common Afflictions and Exercises

Musculoskeletal disorders can result from overuse, bad posture, low muscular demanding activities, static work, recurrent vibrations, manipulation of heavy loads and repetitive actions all of which can result in pain, discomfort, and mobility limitations [27]. Exercises to prevent, correct, or maintain the shoulder are focused on moving the arm across the planes with or without weights depending on the condition of the patient. These movements can be related to several activities performed in multiple scenarios that can be within a VR environment. The goal of these movements (exercises) is to strengthen the muscles and tendons and thus help avoid strains and ruptures, and minimize the risk of muscle and tendon irritation, degeneration of bone and cartilages. Figure 6.3 illustrates two shoulder-based exercises recommended for the shoulder rehabilitation [32].

6.2.4 Design Overview

A general exergame system architecture is presented in Fig. 6.4. As shown, the game receives the camera-based motion capture information (here, using a Microsoft Kinect visual sensor), of the player performing the movements. This input informa-

Fig. 6.4 Example Exergame system architecture

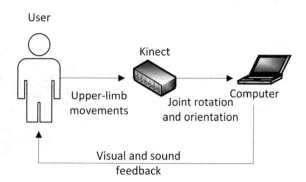

tion is then processed, typically mapped to game play (e.g., to corresponding motions of a virtual canoe), and compared to one or more reference motions. Based on this comparison, the player interacts with the virtual world through the game rules and goals, while being assessed on through motion capture through feedback (typically graphical and sound-based) to improve exercise execution. This cycle is continually repeated until the game ends.

6.2.5   Motion Capture

A cost effective, yet limited tool for measuring limb movements is the goniometer, which allows angle rotation measurements between two joints by being positioned alongside the body part to measure and its reference. However, currently, the most common form of assessment is through observation which leads to subjective analysis and becomes a challenge when the exercises are prescribed to be done at home without any supervision. These unsupervised exercises rely on the capacity of the patient to execute all movements as explained or as indicated by available media (e.g., pamphlets), without any feedback. Affordable solutions include the use of webcams and built-in device cameras. For example, BitGym is a tablet oriented application that uses the tablet's camera to capture body movements and allow for low latency tracking [8]. Prange et al. employed a webcam with a television to track horizontal movements in patients with hemiparesis [43]. For both of these examples, a quantitative assessment was performed outside the exergame due to the system's low accuracy resulting from the use of a single webcam. However, although inexpensive, these solutions present major challenges due to the current use of a single integrated webcam which requires tilting the computer screen (reducing visibility of the contents), and the lack of proper 3D motion tracked data associated with upper-limb movements (due to the abstraction of depth information from only one image). To overcome the difficulties of motion tracking with a single camera, stereoscopic cameras provide a potential solution as they can obtain depth information through a combination of red/green/blue and infrared cameras. Several depth cameras are available at the consumer level including the Microsoft Kinect [37], AsusXtion Pro-live [3], SoftKinectic [58], and the ZED 2K Stereo Camera [33], among others, with varying features regarding camera resolution, motion capture area, size, prices and ultimately availability on the market.

Both of our exergames employ a Microsoft Kinect sensor due to its low cost, decent specifications, and widespread use in addition to the freely available resources/information regarding it. Within three years of its launch in 2010, 24 million Microsoft Kinect (V1) units have been sold worldwide [57]. The Microsoft Kinect features have been widely studied and they have been incorporated into many applications across a variety of areas [64]. Taking advantage of the Microsoft Kinect's motion-capture capabilities, the Jintronix system for physiotherapy was patented and approved by the Federal Drug and Administration office in the USA [29]. This commercial solution provides configurable scenarios for physiotherapy and telerehabilitation with custom designed exercises and assessment based on metrics capture during the interaction. For the development of the motion capture subsystem, the Microsoft Kinect SDK [36] and Unity3D [60] were used to program user interactions and responses from the system within the game mechanics. Ranges of motion were considered and included within the code to provide feedback and quantifiable data from the performed movements.

6.2.6 Game Design

The design and development of an exergame follow a process similar to the design of an entertainment game; define the goals, rules, and feedback along with its formal, dramatic and dynamic elements [19]. However, designers and developers of exergames may not necessarily be afforded as much freedom and flexibility with respect to adjusting various aspects of the design throughout the development process if these adjustments interfere with the rehabilitation requirements and goals. For example, if an exergame requires the user to move their arm following an up and down motion, the designers and developers may not necessarily be able to change these motions to side-to-side. Rather than providing simple interactive environments, our approach to the development of the two exergames involved brainstorming for scenarios where a person requires shoulder movements to accomplish a goal. From this process, two kayaking shoulder exergames were developed by two different teams independently.

With the participation of medical experts, the design process resulted in identifying key elements to encourage a series of properly executed shoulder movements.

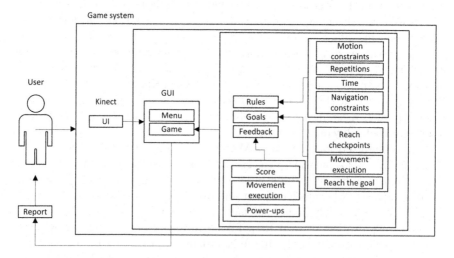

Fig. 6.5 Game system

The information allows diagraming the game system in terms of the rules, goals and feedback as presented in Fig. 6.5. A key element while developing any type of game is the fun component, which in our case was approached through challenges promoting self-competition and improvement once users actually play the game.

6.2.7 The Two Exergames

The first exergame focuses on lateral and medial rotations, which are important movements for strength and functionality while the second exergame focuses on more complex movements involving the shoulder's three DOF to kayak with a paddle-like tool that brings more reality to the game interaction. A description of each exergame along with a discussion of preliminary tests that were conducted to gauge user perceptions regarding the exergames is provided in the following subsections.

6.3 Exergame 1: Kayaking Through Lateral and Medial Rotations

The lateral and medial rotation exergame was developed as a tool for encouraging shoulder care. The goal of the game is to paddle from an origin (starting) position in a virtual lake to the finish line. Correct paddling results in swift navigation that allows collecting power bars to restore energy depletion caused by movement. To reach the finish line, appropriate paddling and power bars are required as otherwise the player stays adrift in the lake [38].

The exergame system receives several inputs from the user as follows: (i) menu navigation and selections are accomplished through hand tracking across the screen. To access the game, settings, and quit buttons, the user needs to hold the position over the menu for five seconds. (ii) once in the game, the user moves the virtual kayak through the upper limb's lateral and medial rotations mapped as suitable inputs representing kayaking movements to going forward, backward and steering left or right; and (iii) finally, after completing the exergame, the user can access the motion data and score stored in a plain text file separated by commas.

The engine receives the user interactions and executes the kayak animations from a third person point of view across the lake. The collision system provides feedback when collecting the energy capsules and also limits where the user can navigate within the lake. A collision may cause the kayak to sink and may prevent the user from capturing collectibles that are required to reach the goal. All interactions are programmed in response to xyz joint information obtained from the Microsoft Kinect's SDK while performing the medial and lateral rotations to recognize the number of repetitions, forearm trajectories, and the time spent while playing the game. Various audio and visual cues were introduced to the game to provide immer-

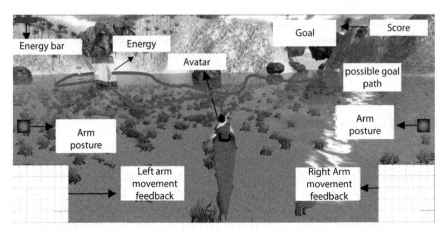

Fig. 6.6 Visual game elements (avatar, energy bar, rock obstacle, arm feedback and energy status)

sion within the environment. More specifically, visual cues were added to account for water reflections on the surface of the lake, vegetation in the bottom, an animated kayaking character, energy bars to boost the kayaker stamina, rocks around the lake, and energy status. Audio cues included paddling sounds, movement across the water and collision sounds with the energy capsules.

The player's paddling motion data (the upper-limb medial and lateral tracked rotation data), is captured and deemed to be valid if it was within 10 % of the actual (correct) motion. error from, This 10 % threshold was chosen in consultation with health care (rehabilitation) specialists after reviewing samples and measurements obtained with a virtual skeleton captured with the Microsoft Kinect of various users going through the required motions can account for light variations in user movements in addition to slight changes in lighting conditions. When a captured movement is classified as valid, they are mapped to the game's avatar (e.g., the game executes the corresponding animation to move the kayaker). If three consecutive invalid movements are detected a pop-up window appears to remind the user to: (i) check the movements and posture, (ii) check room lighting, and (iii) check the feedback motion graphics placed at the bottom of the screen; if working properly, these should render a rotation curve of the forearm. To encourage the player to pay attention to shoulder care and how proper movement execution is achieved, checkpoints were programmed displaying random information about anatomy, shoulder care, and the purpose of the exercise, so once the game is completed, a pop-up window prompts the user to answer a question related to the presented information. Figure 6.6 presents a screen-shot of the exergame with the required movements to paddle and the motion capture data.

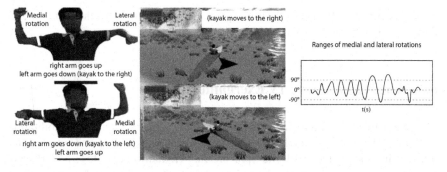

Fig. 6.7 Medial and lateral rotations for paddling with motion data

6.3.1 Exergame Presentation

Once the exergame was developed, to gauge user reactions regarding the exergame and its use, a preliminary user study was conducted with 22 participants (ages between 19 and 35). All of the participants were required to perform shoulder exercises to diminish the risk of developing rotator cuff afflictions due to continuous heavy lifting objects, or due to a sedentary lifestyle. Participants played the game for a brief period of time (minimum 1 min, maximum 2 min for every 45 min for one full work day (8 h)) as indicated by occupational health care guidelines to prevent musculoskeletal disorders [22].

Before playing the game, participants completed a survey that gauged their shoulder care awareness. Results of the survey showed that 77 % were aware of the importance of shoulder care, while the rest were not; 23 % believed shoulder pain is a common problem, 36 % occasional, 32 % frequent and 10 % non-existent; when asked about how to prevent shoulder afflictions, all participants ranked the solutions acknowledging physiotherapy as the best, followed by occupational health care, surgery, sports and exergames. Finally, with respect to their frequency of exercise, one person exercised every day, nine frequently, nine rarely, and two never. Figure 6.7 presents a user interacting with the exergame where medial and lateral rotations result in kayak movement with the corresponding motion capture data.

The exergaming session was overseen by two health care specialists from an occupational health and a fitness facility. After using the exergame, participants completed a second survey to gauge their perceptions of the game. Seventy seven percent found the exergame easy to use while the rest found it difficult to use. Those that found it difficult to use mentioned that it was the result of their lack of familiarity with the Microsoft Kinect sensor, and that they were not gamers. When asked about any future goals that they identified after playing the exergame, their responses were: to exercise, learn about shoulder care, learn how to properly exercise, make the best score, and finally reach the goal. Once all of the participants completed their game play session, the two health care specialists supervising the activity provided feedback on the experience. They highlighted the importance of motion tracking to better assess patient progression and the use of game elements to motivate the user (patient) into engaging physical activity.

6.4   The Rapid Recovery Exergame

The Rapid Recovery exergame was developed to facilitate shoulder rehabilitation and physical fitness. It couples the strong levels of engagement and interactivity inherent in video games with the Spincore Inc. Helium 6 baton [51, 54]. Players take on the role of a kayaker who must paddle their canoe through a pre-defined course. The game is played from a third-person perspective so that the player can see their real-world actions affecting the player avatar. Prior to beginning, the player must choose one of the three modes of operation (see Fig. 6.8): (i) kayaking (sitting or standing), (ii) canoeing, and (iii) stand-up paddling. Finally, the player chooses the length of their workout session in minutes. The player is then placed at the beginning of the course, the timer begins, and then they must paddle their canoe through various checkpoints throughout the level until the timer expires. The player grips the Helium 6 baton (physically, in the real world), just as they would grip a kayak paddle, and then goes through the identical motions as they would if they were paddling a real kayak. The player's Helium 6 baton movements are continuously captured using a Microsoft Kinect video sensor and mapped to movements in the game world. A sample screenshot (canoeing mode of operation), is provided in Fig. 6.9. Included in Fig. 6.9 is an inset which illustrates the player holding the Helium 6 baton. In addition to the kayaking motions, the game is also responsive to the speed of motions. In other words, faster motions of the Helium 6 baton will result in faster movement of the virtual kayak.

The player is provided with feedback throughout the duration of the game (see Fig. 6.9). This feedback provides them with information regarding the elapsed time on the course (game timer), current score, and the direction indicator that lets the player know the direction that they should be steering towards to reach the next checkpoint in the game. The player's score is linked to the player's physical motions with the Helium 6 baton. More specifically, the score is increased depending on whether the player's motions match (within a predefined threshold value), the reference motions (performed by expert kayakers and canoers), and how many laps they have completed. The reference motions were collected from two experienced kayakers who completed the required motions for each of the three modes of operation (seated and standing kayaking, canoeing, and stand-up paddling) while being recorded via a Natural Point Optitrack motion capture system (with eight infrared cameras).

Rapid Recovery also includes a tutorial level which the player may complete at any time from the main menu. This helps ensure that the player has a basic understanding of the game requirements (Helium 6 baton motions, etc.). The tutorial level includes a video of a professional kayaker demonstrating the proper paddle rotation (using the Helium 6 baton). This video is followed by a test round where a series of paddle motions are presented to the player (individually) which the player must then replicate. The player is allowed to proceed to the next motion in the test round only if their motions match the reference motion to within a certain threshold. Throughout the test round, feedback is provided to the player letting them know how they are doing (whether their motions are correct or incorrect, and if they are incorrect, how they can improve on a per limb basis).

6.4.1 Alpha Testing of Rapid Recovery

A preliminary alpha test of Rapid Recovery was conducted to examine the initial functionality of the game and to gauge the game's clarity of content, ease of use, and the user interface [51]. Participants completed the tutorial level and then played the game for a 15-min exploratory period. Playing the game involved using the game and exploring its interface/options (no restrictions were placed on what they did within the game during this period). After completion of this exploratory period, participants completed a brief questionnaire comprised of a subset of questions from the Questionnaire for User Interaction Satisfaction (QUIS); whose purpose is to "assess users' subjective satisfaction with specific aspects of the human-computer interface and several open-ended questions" [53]. In addition to the QUIS-based questions, the survey contained 12 open-ended questions where participants were asked for feedback regarding the Rapid Recovery exergame. Results showed that generally, the

Fig. 6.8 Rapid recovery supports three modes of operation: kayaking (sitting or standing), canoeing, and stand-up paddling. The user must choose one mode of operation at the start of the game

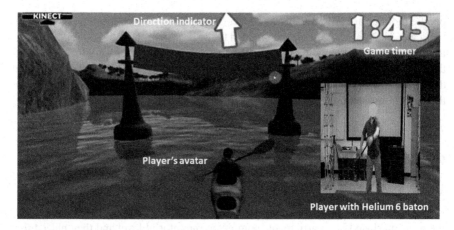

Fig. 6.9 Sample gameplay screenshot (canoeing mode of operation) along with the inset that illustrates the use holding the Spincore Fitness Helium 6 baton

game was enjoyable and well received. Participants also provided valuable feedback that will assist in future development of Rapid Recovery.

6.5 Conclusions

In this chapter, we presented an approach to develop upper-limb exergames focused on shoulder physical activity, and rehabilitation. This process yielded to two different and independently developed kayak-based exergames where a player's motions are captured via a Microsoft Kinect vision-based sensor and mapped to motions of a virtual kayaker within the game. Although the Microsoft Kinect sensor proved to be reliable for both of the games, with any vision-based sensor), the lighting within the environment it is being used in and (perhaps to a lesser degree depending on the colour of the players' clothing can have serious consequences on motion tracking, and thus negatively affect the exergaming experience. Although for the two exergames presented here lighting issues were not a big problem, both exergames were tested in settings with relatively controlled lighting. However, if these games are to be used in homes or health care centres, there is no guarantee regarding how consistent the lighting will be (Microsoft recommends that the Kinect is kept away from direct sunlight). Testing of an exergame should thus take into account the environment(s) it will be used in and testing should consider a wide variety of clothing colours. Aside from lighting issues, other problems can include space requirements and constraints. For both of our exergames, space was not a constraint and the distance between the Microsoft Kinect and the user was approximately 2 m although, when used in the home, this cannot always be guaranteed (greater details regarding the correct placement of the Microsoft Kinect sensor are available by Microsoft via

Xbox support website[1]). Furthermore, regarding the Microsoft Kinect, it has a limited accuracy (approximately 10 % of the movement [20]). If more accurate measures are required, additional sensors such as accelerometers and gyroscopes may have to be added and the data between the various sensors integrated. This may also lead to the use of more intrusive marker-based approaches were the user, for example, must wear some form of tracking device; this can, of course, affect the user's natural movements. Such a sensor integration approach may also lead to greater computational requirements.

Preliminary usability studies conducted with the exergames provided us with valuable information that will guide future iterations of the games. Aside from the feedback that was specific to each game (e.g., comments regarding the user interface, etc.), these preliminary tests have also shown that an important aspect of an exergame is the availability of multiple scenarios to limit the possibility of users becoming bored with the game and thus losing interest over time. A possible solution to this is the development of an exergame framework that allows the rehabilitation experts (and perhaps the end-user to some degree), to develop their own scenarios or modify existing scenarios. Of course, developing such a framework is not a trivial task as one must balance ease of use with respect to scenario development and the technical (programming) skills of the user developing them; there is no on-size-fits-all solution here.

Although designing and developing exergames is in many respects similar to designing and developing traditional "entertainment games", there are also important differences. More specifically, designers and developers of exergames must generally adhere to the corresponding content/knowledge base, while ensuring that their game is not only fun and engaging, but is also an effective rehabilitation tool. In other words, the motions/movements that a player/patient must perform within the game cannot necessarily be modified as this could lead to injury. Designers and developers must also consider the implications of game mechanics within the exercise context that allows for the addition of goals and rules such that in combination with computer generated environments, provide users (patients) with a far more engaging experience that can help overcome some of the limitations associated with traditional exercise and rehabilitation approaches (e.g., the inherent decrease in motivation to maintain an exercise and rehabilitation program over time). To this end, care must be taken to ensure that exergames are properly designed and developed to meet their intended goals. Part of this implies that a proper and thorough needs/task analysis be conducted. A needs/task analysis involves a significant research component typically involving interviews with subject matter experts, and the potential users of the exergame and is often done poorly or neglected altogether. Furthermore, once developed, their effectiveness must be tested to ensure that they do indeed meet their intended goals.

Assessing the effectiveness of an exergame is not a trivial task but it should be considered early on in the design process. We believe this involves a two-stage process, an initial "alpha" test to examine the usability and user interface of the

---

[1]Kinect setup http://support.xbox.com/en-US/xbox-360/accessories/kinect-sensor-setup.

exergame, followed by an assessment of its effectiveness. Assessing the effectiveness of an exergame can be aided by the availability of common off-the-shelf technology. For example, microphones and cameras can be used to capture any audio (e.g., voice of the participant if think-aloud techniques are employed), and the video of the participant as they are performing in the test. Furthermore, various sensors can be used to monitor various physiological responses (e.g., heart rate, galvanic skin response, amongst others), that can provide further insight into a player's state and engagement during gameplay. A complete discussion regarding the assessment of exergames is beyond the scope of this chapter but greater details are provided in a review article regarding the assessment of serious games by Bellotti et al. [6].

In addition to knowledge and expertise in game design and development, exergame designers/developers must also be knowledgeable in the specific content area covered by the exergame and should ideally possess some knowledge of teaching methods and instructional design. In other words, the development of effective exergames is not a trivial task and knowledge in game design solely is not sufficient to develop an effective exergame. As a result of potentially competing for interests between the rehabilitation (and potentially educational) and entertainment value of the product, exergame development is an inherently interdisciplinary process, bringing together experts from a variety of fields including game design and development, and experts from the content domain in question (e.g., medical professionals, rehabilitation and physical fitness experts). Although exergame designers are not expected to be experts in medicine and rehabilitation/physical fitness, possessing some knowledge in these areas will, at the very least, promote effective communication between members of the interdisciplinary team.

The future of exergames is rapidly evolving. Technological advancements such as open electronics, and 3D printing, are providing designers and developers of exergames with great opportunities to develop immerse and interactive experiences. However, care must be taken to ensure that the end product (the exergame) does indeed meet its intended objectives. To this end, although not a trivial task, future work should examine the development of a set of rules and guidelines (e.g., best practices) that can guide exergame designers and developers. Finally, rather than focusing on the visual (graphical) scene and perhaps to a smaller degree, the auditory scene, designers and developers should also consider, and take advantage of, multiple sensory channels [13]. Technological advancements have provided us with consumer level devices that allow for a variety of non-traditional cues (e.g., haptics) to be incorporated into exergames and video games in general and help create more compelling user experiences. An example of such rapid change is the launch of the Kinect V2 that exceeds the first version used in both exergames described here. Improvements can be seen in characteristics such as accuracy distribution, depth resolution, depth entropy, edge noise and structural noise, providing grounds for future works [62].

Acknowledgements The authors acknowledge the support of the Virtual Reality Center from Mil. Nueva Granada University and students Camilo Rincón, Wilson Nava and César Ramos. The authors also acknowledge the support of the Natural Sciences and Engineering Research Council of

Canada (NSERC), and the Social Sciences and Humanities Research Council of Canada (SSHRC), in support of the Interactive & Multi-Modal Experience Research Syndicate (IMMERSe) initiative.

References

1. Andriacchi TP, Alexander EJ (2000) Studies of human locomotion: past, present and future. J Biomech 33(10):1217–1224
2. Annesi JJ, Mazas J (1997) Effects of virtual reality-enhanced exercise equipment on adherence and exercise-induced feeling states. Percept Mot. Skills 85(3):835–844
3. ASUS (2016) Xtion-pro live. http://www.asus.com/3D-Sensor/
4. Barry G, Galna B, Rochester L (2014) The role of exergaming in parkinson's disease rehabilitation: a systematic review of the evidence. J Neuroeng Rehabil 11(1):1
5. Bassett SF (2003) The assessment of patient adherence to physiotherapy rehabilitation. NZ J Physiother 31(2):60–66
6. Bellotti F, Kapralos B, Lee K, Moreno-Ger P, Berta R (2013) Assessment in and of serious games: an overview. Adv Hum Comput Interact 2013(Article ID 136864):1–11
7. Benedetti M, Cappozzo A (1994) Anatomical landmark definition and identification in computer aided movement analysis in a rehabilitation context. internal report. Universita Degli Studi La Sapienza, pp 1–31
8. BitGym (2016) Unlock the power of the front camera on smart devices! http://www.bitgym.com/developer
9. Bolas M, Hoberman P, Phan T, Luckey P, Iliff J, Burba N, McDowall I, Krum DM (2013) Open virtual reality. Proceedings of the 2013 IEEE virtual reality conference. Lake Buena Vista, FL, USA, pp 183–184
10. Braune W, Fischer O (2013) Determination of the moments of inertia of the human body and its limbs. Springer Science & Business Media, Berlin, Germany
11. Burdea G (2002) Keynote address: Virtual rehabilitation-benefits and challenges. Proceedings of the 1st international workshop on virtual reality rehabilitation (mental health. Physical, Vocational). Neurological, VRMHR, Lausanne, Switzerland, pp 1–11
12. Burdea G, Coiffet P (2003) Virtual reality technology. Presence Teleoperators Virtual Environ 12(6):663–664
13. Butler DP, Willett K (2010) Wii-habilitation: is there a role in trauma? Injury 41(9):883–885
14. Carignan C, Liszka M, Roderick S (2005) Design of an arm exoskeleton with scapula motion for shoulder rehabilitation. Proceedings of the 12th IEEE international conference on advanced robotics. Seattle, WA, USA, pp 524–531
15. Charbonnier C, Chagué S, Kolo FC, Lädermann A (2015) Shoulder motion during tennis serve: dynamic and radiological evaluation based on motion capture and magnetic resonance imaging. Int J Comput Assist Radiol Surg 10(8):1289–1297
16. Cropley M, Ayers S, Nokes L (2003) People don't exercise because they can't think of reasons to exercise: an examination of causal reasoning within the transtheoretical model. Psychol Health Med 8(4):409–414
17. Ferreiro Marzoa I, Veiga Suarez M, Guerra Pena J, Rey Veiga S, Paz Esquete J, Tobio Iglesias A (2005) Tratamiento rehabilitador del hombro doloroso. Rehabilitación 39(3):113–120
18. Foxlin E (2002) Motion tracking requirements and technologies. In: Hale KS, Stanney KM (eds) Handbook of virtual environment technology: design, implementation, and applications. CRC Press, Mahwah, NJ, USA, pp 163–210
19. Fullerton T (2014) Game design workshop: a playcentric approach to creating innovative games. CRC press
20. Galna B, Barry G, Jackson D, Mhiripiri D, Olivier P, Rochester L (2014) Accuracy of the microsoft kinect sensor for measuring movement in people with parkinson's disease. Gait Posture 39(4):1062–1068

21. Garber L (2013) Gestural technology: moving interfaces in a new direction [technology news]. IEEE Comput 46(10):22–25
22. Granada UMN (2016) Ergo info. http://www.umng.edu.co/documents/10162/120721/Ergoinfo.exe
23. Guttentag DA (2010) Virtual reality: applications and implications for tourism. Tour Manage 31(5):637–651
24. Hoffman HG, Patterson DR, Carrougher GJ, Sharar SR (2001) Effectiveness of virtual reality-based pain control with multiple treatments. Clin J Pain 17(3):229–235
25. Huang J, Mori T, Takashima K, Hashi S, Kitamura Y (2015) Im6d: magnetic tracking system with 6-dof passive markers for dexterous 3d interaction and motion. ACM Trans Graph 34(6):217
26. Jack D, Boian R, Merians AS, Tremaine M, Burdea GC, Adamovich SV, Recce M, Poizner H (2001) Virtual reality-enhanced stroke rehabilitation. IEEE Trans Neural Syst Rehabil Eng 9(3):308–318
27. Jäger M, Steinberg U, für Arbeitsschutz B (2003) Preventing musculoskeletal disorders in the workplace
28. Janarthanan V (2012) Serious video games: games for education and health. Proceedings of the 9th international IEEE conference on information technology: new generations. Las Vegas NV, USA, pp 875–878
29. Jintronix (2016) Jintronix. http://www.jintronix.com
30. Johnson LM, Winters JM (2004) Enhanced therajoy technology for use in upper-extremity stroke rehabilitation. In: Proceedings of the 2004 IEEE engineering in medicine and biology society conference, San Francisco, CA, USA, vol 2, pp 4932–4935
31. Kato PM (2010) Video games in health care: closing the gap. Rev General Psychol 14(2):113
32. Kolar P, et al (2014) Clinical rehabilitation. Alena Kobesová
33. Labs S (2016) Zed 2k stereo cameras. http://www.stereolabs.com
34. Lange B, Koenig S, McConnell E, Chang CY, Juang R, Suma E, Bolas M, Rizzo A (2012) Interactive game-based rehabilitation using the microsoft kinect. Proceedings of the 2012 ieee virtual reality short papers and posters. Costa Mesa, CA, USA, pp 171–172
35. Lieberman DA (2009) Designing serious games for learning and health in informal and formal settings. In: Ritterfeld U, Cody M, vorderer P (eds) Serious Games: mechanisms and effects, Routledge, New York, NY, USA, chap 8, pp 117–130
36. Microsoft (2016a) Kinect for windows sdk. http://www.microsoft.com/en-ca/download/details.aspx?id=40278-
37. Microsoft (2016b) Meet kinect for windows. http://dev.windows.com/en-us/kinect
38. Nava W, Mejia CAR, Uribe-Quevedo A (2015) Prototype of a shoulder and elbow occupational health care exergame. In: HCI International 2015-Posters Extended Abstracts, Springer, pp 467–472
39. of Orthopaedic Surgeons AA (2009a) Common shoulder injuries. http://orthoinfo.aaos.org/topic.cfm?topic=a00327
40. of Orthopaedic Surgeons AA (2009b) Shoulder surgery. http://orthoinfo.aaos.org/topic.cfm?topic=a00066
41. O'Sullivan SB, Schmitz TJ, Fulk G (2013) Physical rehabilitation. FA Davis, Philadelphia, PA, USA
42. Peng W, Lin JH, Crouse J (2011) Is playing exergames really exercising? a meta-analysis of energy expenditure in active video games. Cyberpsychol Behav Soc Netw 14(11):681–688
43. Prange G, Krabben T, Molier B, van der Kooij H, Jannink M (2008) A low-tech virtual reality application for training of upper extremity motor function in neurorehabilitation. Proceedings of the 2008 IEEE virtual rehabilitation conference. Vancouver, British Columbia, Canada, pp 8–12
44. Primack BA, Carroll MV, McNamara M, Klem ML, King B, Rich M, Chan CW, Nayak S (2012) Role of video games in improving health-related outcomes: a systematic review. Am. J. Prev. Med 42(6):630–638

45. Putz-Anderson V, Bernard BP, Burt SE, Cole LL, Fairfield-Estill C, Fine LJ, Grant KA, Gjessing C, Jenkins L, Hurrell Jr JJao (1997) Musculoskeletal disorders and workplace factors. National Institute for Occupational Safety and Health (NIOSH)
46. Qi Y, Soh CB, Gunawan E, Low KS, Thomas R (2015) Lower extremity joint angle tracking with wireless ultrasonic sensors during a squat exercise. Sensors 15(5):9610–9627
47. Read JL, Shortell SM (2011) Interactive games to promote behavior change in prevention and treatment. J Am Med Assoc 305(16):1704–1705
48. Schubert T, Gkogkidis A, Ball T, Burgard W (2015) Automatic initialization for skeleton tracking in optical motion capture. Proceedings of the 2015 IEEE conference on robotics and automation. Seattle, WA, USA, pp 734–739
49. Shakespeare T (2011) World report on disability. Disabil Rehabil 33(17–18):1491
50. Sheps D (2013) The shoulder book. Workers Compensation Board of Alberta, Alberta, Canada
51. Shewaga R, Rojas D, Kapralos B, Brennan J (2015) Alpha testing of the rapid recovery kayaking-based exergame. Proceedings of the 2015 IEEE Games entertainment and media conference. Ontario, Canada, Toronto, pp 1–5
52. Shier D, Butler J, Lewis R (2001) Human anatomy and physiology, 9th edn. McGraw-Hill, Boston, MA, USA
53. Shneiderman B (1988) Designing the user interface: strategies for effective human-computer interaction. Addison Wesley Longman, Reading MA, USA
54. Shroeder B, Kunanec M, Kroese B, Pollard O, Sadoon O, Kapralos B, Brennan J, Leach E, Jenson J (2014) Rapid recovery: a kayaking-based exergame for shoulder rehabilitation and physical fitness. Proceedings of the 2014 IEEE games entertainment and media conference. Ontario, Canada, Toronto, pp 1–4
55. Sinclair J, Hingston P, Masek M (2007) Considerations for the design of exergames. Proceedings of the 5th International ACM conference on computer graphics and interactive techniques in Australia and Southeast Asia. Perth, Australia, pp 289–295
56. Skiba DJ (2008) Emerging technologies center: games for health. Nurs Educ Perspect 29(4):230–232
57. SMeisner J (2016) Xbox beyond the box. http://blogs.microsoft.com/blog/2012/05/29/xbox-beyond-the-box
58. Sony (2016) Depth sense cameras. http://www.softkinetic.com/Products/DepthSenseCameras
59. Staiano AE, Calvert SL (2011) The promise of exergames as tools to measure physical health. Entertain Comput 2(1):17–21
60. Unity (2016) Unity 3d. www.unity3d.com
61. Wingrave CA, Williamson B, Varcholik PD, Rose J, Miller A, Charbonneau E, Bott J, LaViola JJ Jr (2010) The wiimote and beyond: spatially convenient devices for 3d user interfaces. IEEE Comput Graphics Appl 30(2):71–85
62. Yang L, Zhang L, Dong H, Alelaiwi A, El Saddik A (2015) Evaluating and improving the depth accuracy of kinect for windows v2. Sensors J IEEE 15(8):4275–4285
63. You S, Neumann U, Azuma R (1999) Hybrid inertial and vision tracking for augmented reality registration. Proceedings of the 1999 IEEE virtual reality conference. Houston, TX, USA, pp 260–267
64. Zhang Z (2012) Microsoft kinect sensor and its effect. IEEE MultiMed Mag 19(2):4–10

Chapter 7
Development of an Occupational Health Care Exergaming Prototype Suite

Alvaro Uribe-Quevedo, Sergio Valdivia, Eliana Prada, Mauricio Navia, Camilo Rincon, Estefania Ramos, Saskia Ortiz and Byron Perez

Abstract Work-related health afflictions are a major concern as they may affect the ability for one to carry out daily activities and negatively affecting the quality of life. Possible causes have been associated with sedentary, repetitive movements, bad postures, and lack of occupational health exercises. Aside from the health consequences, there is also concern about the loss of productivity and its associated costs, in addition to the treatment costs (which in some cases may require attending to specialized facilities). The standard approach to minimize such risks involves preventive exercises at the workplace and annual medical examinations. Typically, employees are provided with printed or multimedia guides to educate them regarding work-related afflictions and preventive steps that can be taken to prevent them. However, such educational material generally does little to motivate employees to start and maintain a preventive exercise program. Exergaming (that is video games that incorporate exercise), are capable of motivating users into starting and maintaining an exercise program. In this book chapter, we present the development of an exergaming prototype suite to address common work-related preventive exercises associated with the lower-limb, upper-limb, hands, and eyes. By using a gaming scenario we are capable of providing the player with compelling interactions and game play experiences that allow increasing motivation to exercise, while tracking the player's movements and using that information as part of a medical assessment.

7.1 Introduction

Occupational health care is a worldwide concern given the increasing number of people suffering from work-related diseases, as these several serious social and economic consequences, such as work absenteeism [1]. Current trends in technology are changing how people and companies interact with the world, pushing for further

A. Uribe-Quevedo (✉) · S. Valdivia · E. Prada · M. Navia · C. Rincon ·
E. Ramos · S. Ortiz · B. Perez
Mil. Nueva Granada University, Cra 11#101-80, Bogota, Colombia
e-mail: alvaro.j.uribe@ieee.org

© Springer International Publishing AG 2017
A.L. Brooks et al. (eds.), *Recent Advances in Technologies for Inclusive Well-Being*, Intelligent Systems Reference Library 119,
DOI 10.1007/978-3-319-49879-9_7

use of information communication technologies [2]. However, the widespread use of computers and mobile-based technologies (e.g., laptop computers, tablets, and hand held devices), has also led to an increase in various health-related problems. Some of these problems include technology addiction [3], stress [4], and body injuries due to prolonged, forceful, low amplitude, repetitive use of hand held devices in conjunction with bad posture and sedentary behaviours [5–7].

To minimize the risks associated with prolonged use of computers and mobile devices (e.g., overweight [8], sleep disorders [9], and musculoskeletal disorders [10]), occupational health care policies outlining various exercises that focus on the following have been devised: (i) prevention, (ii) maintenance of mobility, and (iii) motion recovery. Depending on the country/state, these exercises can be performed with or without the supervision of a physiotherapist or other medical professional. Preventive exercises can help reduce health complications that may result in various mobility issues, resulting in the use of assistive devices, treatments for chronic afflictions, and possibly surgery. Currently, information regarding preventive exercises and exercise programs is relayed to employees by employers through the use of various print-based documents (e.g., pamphlets, brochures) and digital (e.g., web-based, apps). Cooper et al. [11], analyzed the impact of using web-based interventions over face-to-face, printed-based and telephone-based interventions for physical activity and concluded that although cost-effective, web-based solutions have small positive effects given the limitations associated with self-reporting and assessment. Rasotto et al. [12], examined the effect of providing tailored physical activity, rather than exercise guides whereby a one-size-fits-all approach is taken. After conducting a nine-month intervention on 34 participants, reduced pain symptoms on neck, shoulders, elbows and on wrists were documented, along with improvements in grip strength and upper-limb mobility.

The evolution of computer hardware, computer graphics, and virtual reality (VR), has provided the tools and technologies to create immersive and interactive controlled scenarios across a variety of areas including medicine and rehabilitation in particular. Burdea addresses the benefits and challenges of employing VR-based rehabilitation [13], acknowledging its potential to provide engagement through tailored experiences that can be monitored to obtain quantitative data for better medical assessment. Hoffman et al. [14], described the difficulties of adopting technologies from physicians and patients alike when facing newer and complex forms of interactions. Perry on the other hand, discusses the costs associated with VR-based rehabilitation as a barrier to provide massive solutions, but also describes how VR-based solutions are becoming mass-market ready thanks to technology advances [15].

The application of VR-based physiotherapy, or the use of immersive and interactive devices does not guarantee engagement or motivation to follow through the exercises. That being said, the incorporation of proper game design can provide a mean to create experiences to accomplish the exercising goal [16]. Exergames are games that provide fun and engaging interactive scenarios that allow acquiring exercise data with motion tracking with feedback to quantify and assess the user's progression [17]. Best [18], discusses how exergames can have a positive impact on youth, based on the engagement possibilities due to physical and cognitive challenges. Proper engagement can help overcome the lack of motivation, and boredom caused by repetitive movements and even discomfort due to either pain or poorly executed exercises that ultimately affect the exercising experience [19].

Exergames have been used to help stroke patients recover movement while providing physiotherapists reports to better assess the patient's progression, rather than relying on the patient's self-reports or observation [20]. Additional examples include an exergame for the elderly using a Wii Balance Board as input device to maintain balance skills and reduce the risk of falling [21], the general use of exergames in rehabilitation as a tool for gathering biofeedback [22], exercise promotion in youth to raise awareness [23], and fitness applications where data collection and engagement goes beyond basic calories burning measurements [24]. Commercial (mass-market) exergames are also available for the various video game consoles featuring sports and dancing games such as Nintendo's WiiUSports, Ubisoft's Your Shape Evolved, and Harmonix's Dance Central [25]. Video game user input interfaces have provided affordable hardware that can be employed as motion tracking devices in rehabilitation. Pogrzeba, Wacker and Jung analyzed the use of the Microsoft Kinect as a tool to monitor and engage users, highlighting its impact in sports and physiotherapy scenarios [26]. The emergence of such user interfaces also holds potential in the workplace where, for example, motion capture systems can provide guidance regarding the proper method of performing certain tasks [27]. The use of exergames in occupational health care contexts can provide a means to address sedentarism with motivational and engaging scenarios. Sra and Schamandt developed a set of three mini-exergames focusing on low-intensity physical exercises taking place during short breaks at work [28]. From this process, they highlight four key design considerations: (i) awareness, (ii) simple game mechanics, (iii) social interactions, and (iv) workplace constraints.

In this chapter, we present the development of an occupational health care exergaming suite comprised of modules to address upper and lower limb, and eye exercising in the workplace. The majority of exergaming work has focused on rehabilitation although some work has focused on prevention, targeting small children and the elderly. However, little work has examined exergames that focus specifically on work environments for prevention. More specifically, our approach differs from existing work by focusing on a solution that can be employed within a workspace setting with minimum hardware requirements using a smartphone, a wearable sensor, an eye tracker and a Microsoft Kinect. The chapter is organized in sections presenting, (i) the exergame suite development with a description of the characterization and general design considerations, (ii) game development for each body part (arms, hand, legs, and eyes), and finally, (iii) a brief discussion about the development, challenges, future works and recommendation for those interested in exergames for health.

Table 7.1 Common body parts movement in occupational health exercises

Body part	Movement	Motion range
Shoulder	Flexion/Extension	180°/60°
	Adduction/Abduction	45°/180°
Leg	Flexion/Extension	110°/30° (thigh)
		130°/30° (knee)
Hand	Radial/Ulnar deviation	20°/30°
Eyes	Horizontal/Vertical gaze	Depending on the screen size

7.2 Suite Development

The development of the exergaming suite encompasses various stages involving the characterization of the exercises, design, development, and testing.

7.2.1 Exercise Characterization

Exercise-at-work guides provide pictures or illustrations portraying a step-by-step procedure regarding how to perform an exercise. A typical exercise session is comprised of different exercises executed in sequence for a short period of time every hour.[1] No previous warming or stretching is required and the movements should be focused on proper execution rather than speed and should last at least 30 s. These exercises are commonly practised within an office environment in front of a desk, where space constraints may impose difficulties with respect to arm and leg movement and, therefore, space is an important design parameter to consider.

Work-related musculoskeletal disorders can cause pain, discomfort and weakness, thus, exercises play an important role in prevention and managing of the related conditions [1]. Exercises can be classified in (i) limbering up exercises, (ii) exercises for general health and fitness, (iii) exercises for rehabilitation, and finally, (iv) exercises to reduce fatigue [29]. Exercises taking place in the workplace are limbering exercises which, focus on short light to moderate warming up or stretching physical activity, performed various times in a day [30]. Limbering exercises at the workplace are comprised of eye, neck, shoulder, upper back, hands [29], and lower limbs [1]. These exercises are expected to be performed at work without any supervision but employees are encouraged to self-report any discomfort.

For our exergaming suite, we considered a working space of 1.5 m^2 focusing in exercises for: (i) the eyes, (ii) shoulders, (iii) legs, and (iv) hands. To guarantee that the exercises can be performed within the limited space, body movements where compared to the ranges of motion presented in Table 7.1 [31].

---

[1] Workplace exercises, http://www.ehs.utoronto.ca/services/Ergonomics/exercise.htm.

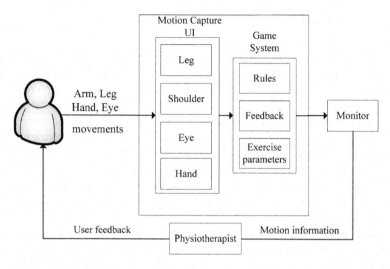

Fig. 7.1 System architecture

7.2.2 Exergame Design

The exergame design process follows the same design procedure as any other (entertainment-based) video game. More specifically, we first we define the goal and design the game mechanics. Taking Table 7.1 as a reference, the goal of each exergame is to execute repetitive movements within the ranges of motion in a compelling interactive manner that will lead to engagement. With this in mind, the game elements (obstacles, resources, user actions, rewards, and the corresponding scene components [32]) are designed based on the system the input and output. Relevant design aspects can be drawn from the target user, in our case, the player who is typically working in front of their computer without any movement limitations. Since age can vary widely within a company, our design approach allows for the customization of ranges of movement and the ability to add scenarios thereby allowing broader target audiences.

A general system architecture considering the basic elements to start the design process is presented in Fig. 7.1. As shown, the user interactions (movements) are captured by the various sensors, and sent to the game system so its components react to them and provide visual feedback to the player. The purpose of the feedback is to allow the players and medical experts who may be overlooking the exercise program to obtain quantifiable data on the gaming experience and more specifically, the movements to better assess user progression.

In the following sections, the design considerations for each of the exergames is presented, beginning with the lower-limb and followed by the upper-limb, eyes, and hand exergames. In each section, a description regarding common exercises, their associated ranges of motion, and how the motions were incorporated into to each mini game is provided.

7.3 Lower-Limb

Afflictions of the lower-limb can result in reduced mobility, which, depending on the severity may require tailored treatments and in some cases the support of motion assisting devices such as canes or walkers. Supporting devices can help provide the user stability and balance, but as noted by Bateni and Maki [33], such devices may require the user to possess a level of strength which increases the metabolic demand. Approaches to lower-limb exercising with exergames have employed various user interfaces. For example, Jordan, Donne and Fletcher modified a PlayStation 2 controller to map the lower-limb movements which, in conjunction with a treadmill, formed a novel input mechanism for playing racing games, where the player movements controlled the speed of the cars, highlighting the importance of having proper exercise stimuli [34].

Another gaming interface, the Wii-Balance board, has also been used in exergames as it tracks pressure over four points requiring the players to exhibit muscle control relevant to posture and walking while using it. Diest et al. [35], employed the Wii-Balance board to provide a fun and engaging exercising scenario to the elderly as a tool to train posture control to avoid falls. Diest et al., acknowledged the positive effects on motivation when using exergames and concluded that further research is required on sensors to obtain quantifiable data.

Microsoft's Kinect has also been studied as a tool to support decision making in rehabilitation. Levac et al. [36] acknowledged the importance of using off-the-shelf solutions such as the Microsoft Kinect to support clinicians in adopting, implementing, and promoting better rehabilitation practices.

7.3.1 Shank Exergame

The Shank game focuses on the lower-limb so that the user can execute flexion and extension movements while seated. The goal of the game is to reach fully extended and half flexion movements that result in audio feedback and score points to track good motion. The game takes advantage of a smartphone with its built-in accelerometer so that the user can exercise in the workplace. The game requires the user to strap the smartphone to the shank using any commercial phone holder. The game reacts to user movements within the rotational ranges and configured by default to account for a $\pm 10°$ variation in the movement to compensate for inertia. However, ranges of movement and tolerances can be adjusted depending on the mobility of the user. The user is awarded points for performing proper movements and the game ends when the user completes a pre-defined number of (repetitive) movements. The game was developed for Android-based smartphones and it features information about lower-limb anatomy and care to complement the exercise experience. The Graphical user interface (GUI) is configured to respond to touch commands and buttons associated with the start, stop, and view results from the exergaming session [37].

7.3.1.1 Motion Tracking

To track movement, Eq. 7.1 was used to calculate the amount of tilt (in degrees) using the smartphone's accelerometer information, where pg_x, pg_y and pg_z are the values obtained from the three coordinate axes (the smartphone is placed with its back facing the user). Once the user reaches the target position, the game counts it as a correct execution and provides a cheering sound from a crowd to provide positive feedback regarding the movement. In contrast, when the movement is performed incorrectly, no sound is presented. After completing the exercise, the user can review the tracked data that it is displayed on the screen (the tracked data is also saved as a text file that can be viewed using any spreadsheet software).

$$cos_{pitch} = \frac{pg_z}{\sqrt{pg_x^2 + pg_y^2 + pg_z^2}} \tag{7.1}$$

7.3.1.2 User Experience

To examine the user's perception of the game, we recruited 20 students who were enrolled in courses that collectively included a six hour minimum computer-based lab component at the Mil. Nueva Granada University, to play the exergame for two minutes and then complete a survey. 68 % of the users found the game interesting while the other 32 % focused on the correct execution of the movements that allowed them to advance to other levels. An unexpected outcome was that players engaged in indirect competition as they accumulated points during the game play session. The main drawback during each session was securing the smartphone correctly If the phone was placed loosely, the tracked data was not valid, but if placed too tight, it caused discomfort (See Fig. 7.2).

Due to the wide variations in smartphone hardware, and more specifically, the inertial sensor quality varying across manufacturers, we experienced a major challenge with low-end devices whose data was not usable due to noise and inaccuracy. To overcome this issue, we employed a Samsung Galaxy S4 with all participants. Some usability concerns were mentioned by some players, particularly with respect to incoming calls or other notifications as these would alter the game play by compelling the players to respond to such notices. Figure 7.3 presents a screen capture of the GUI menu and the motion capture data.

7.3.2 Thigh Exergame

The thigh game focuses on flexion/extension and adduction/abduction exercises. The goal of the game is for the user to kick a ball by swinging the leg onward and sideways depending on the exercise. The objective of kicking the ball is to make it land in a

Phone strapped to the leg   Upper view of the exercise

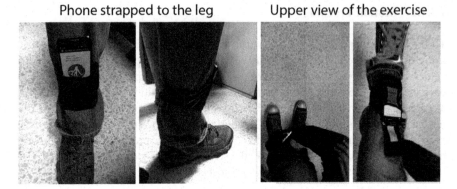

Fig. 7.2 Smartphone placement, and movements

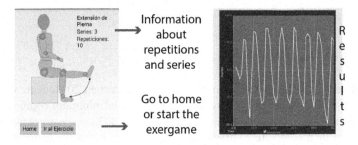

Fig. 7.3 Game GUI

bucket. The ball physics behaviour was programmed to act as a parabolic shot using Eq. 7.2 (where x is the ball's final position; y is the height; x_o and yo are the initial positions; v_o is the initial velocity; g is the gravity; and t is time of the movement). The distance that the ball is travels, is thus dependent on how the user moves their leg to kick the ball. The game is configured to end when the user runs out of balls to kick.

$$x = x_0 + v_0 \cdot t$$
$$y = y_0 - v_0 \cdot t - 0.5 \cdot g \cdot t^2 \qquad (7.2)$$
$$y = y_{max} \cdot t + v_0 \cdot t + 0.5 \cdot g \cdot t^2$$

7.3.2.1 Motion Tracking

To track user movements, we implemented a wearable box comprised of an Arduino Uno,[2] a nine degree of freedom (DOF) inertial sensor (accelerometer, gyroscope and

---

[2] Arduino UNO, http://www.arduino.cc/en/Main/ArduinoBoardUno.

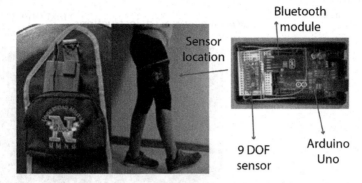

Fig. 7.4 Wearable placement, and game GUI

magnetometer),[3] Bluetooth module,[4] a four AA battery pack, and a sports holder for mp3 players.

The user wears the device around the thigh, with the connector facing outside the leg, as presented in Fig. 7.3 [38]. To develop the game, we chose Processing,[5] as it allowed us to develop a cross-platform game with good compatibility for Android platforms (data transfer between hardware and software). The GUI provides feedback regarding the executed movement and where the ball lands using 2D graphics as presented in Fig. 7.4. The best kicking movement involves proper extension rotation within a $\pm10\,\%$ threshold and an angular velocity estimated within the game depending on each target's distance.

7.3.2.2 User Experience

To analyze user perception, we recruited eleven participants from Mil. Nueva Granada University who spend more than six daily hours working with computers. The group was comprised of five students who practised sports, four students who did not practice sports, and two health care specialists. Participants used the game and were able to complete it under 1 min. After finishing the game, each participant completed a survey to gauge their interest regarding the game. From the survey, ten participants expressed interest in exercising by choice, and the remaining one expressed necessity as the main factor to exercise. All participants stated having fun while playing the game; while playing, four participants focused on good exercise execution, three focused on achieving the highest score, and the remaining five focused on the GUI. Although, none of the participants provided any negative comments, they all suggested that a 3D scenario would have been more compelling. However, only those who participated in sports manifested interest in the data as it

---

[3]IMU 9150 www.sparkfun.com/products/11486.

[4]Bluetooth module, www.sparkfun.com/datasheets/RF/BlueSMiRF_v1.pdf.

[5]Processing, http://processing.org/.

Fig. 7.5 Game GUI

Menu: play, report, instructions

Flextion/extension

Feedback

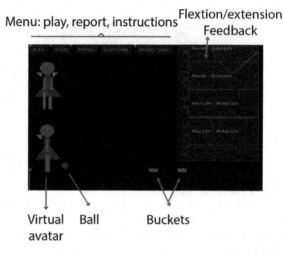

Virtual   Ball   Buckets
avatar

could help them improve their performance. The participants who did not actively participate in any sports concluded that gaming elements could motivate them into engaging with and continuously practice exercises at their workplace. A major concern expressed by the users was the size of the wearable device which took approximately 1 min to properly place it on their body. Figure 7.5 presents a screen capture of the GUI menu and the motion capture data.

7.4 Upper-Limb

The upper-limb is comprised of the shoulder (3 DOF—flexion/extension, adduction/abduction, and circumduction), elbow (1 DOF—flexion and extension) and wrist (2 DOF—ulnar and radial deviations). The upper-limb plays an important role in daily activities such as choosing and grasping objects. Upper-limb disorders involving the rotator cuff require abduction and internal/external rotations exercises while standing straight with the elbows horizontally aligned [39]. Solutions to address upper-limb reduced mobility using games have focused on exergames for rehabilitation of patients with stroke [40], home solutions to recover motor control for children with cerebral palsy [41], and physical rehabilitation for overcoming motor disabilities [42].

Similarly to the lower-limb scenario, various tracking devices have been incorporated into various upper-limb exergaming solutions. More specifically, exergames have employed exoskeletons [43], the Nintendo Wiimote [44], the Microsoft Kinect [45], and various inertial sensors [46]. We have designed two exergames for the upper-limb; one for the shoulder and other for the hand as described in the following subsections.

7.4.1 Shoulder Exergame

The shoulder exergame focuses on upper-limb flexion, extension and adduction, abduction movements. The player uses the arm rotations at the shoulder to control the direction where a snake-like character heads to. The user is required to stand 1.5 m from the sensor while extending the arm towards the screen with the hand facing the sensor[6]. The goal is to control a snake looking for food while avoiding touching the edges of the screen while it grows with time. The game ends once the user touches the screen borders, collides with the snake's own body, or reaches a defined score. The amount of playtime was limited to 1min to avoid discomfort resulting from having the arm extended too long. The snake growing rate increases over time adding difficulty to the experience. To further engage users, a second level was added in the form of a Breakout-like game, where users control a movable paddle through adduction/abduction upper limb movements to keep the ball bouncing without touching the bottom of the screen.

7.4.1.1 Motion Tracking

To guarantee proper arm movement and tracking, the Kinect sensor was programmed to detect valid movements within a configurable threshold so it can be adjusted to various players anthropometrics. Once the game starts, the snake-like character starts moving (direction is set randomly each time the game begins). The character can only move in four directions (up, down, left, or right), depending on where the player is moving their hand while executing flexion (down), extension (up), adduction (right) and abduction (left).

In this exergame, we employed the Microsoft Kinect due its affordability, availability at the time, and tracking accuracy (approximately 95.16 % [47]). Currently the Microsoft Kinect has been replaced with the Microsoft Kinect V2, which was launched in 2014. However, given the widespread popularity of the (original) Microsoft Kinect sensor, it is still widely being used in a variety of applications. For example, it is being used to measure physical activity [48], and usability assessments in exergames [49], among many other applications. Figure 7.6 shows the user playing the exergame with the Microsoft Kinect Sensor.

The development of the exergame was achieved using the Microsoft Kinect SDK and UNITY3D.[7] The GUI is comprised of a menu that guides the user through the mini-game and upper-limb care. To access each option, the user can move the cursor over the area of interest and wait three seconds to be activated.

---

[6]Kinect setup for Xbox360, https://support.xbox.com/en-US/xbox-360/accessories/kinect-sensor-setup#066732c46e6349e48f910e18a9bf106d.

[7]Unity3D, http://unity3d.com/.

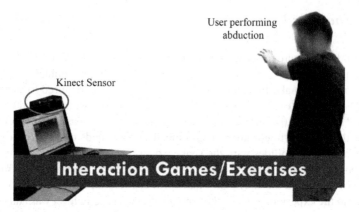

Fig. 7.6 Upper limb tracking movements

7.4.1.2 User Experience

To examine user perception, 33 participants who spent more than eight or more hours daily working with a computer within the Mil. Nueva Granada University were recruited to play the game. Each participant was asked to start the game and play each level for 1min. After playing, we asked the participants to answer a survey regarding their exercising habits, and game-related thoughts in terms of usability, engagement and interest. In terms of their exercising habits, none of the participants reported taking breaks to exercise while working, despite being aware of its importance. All of the participants expressed interest in such an application for exercising as they believed that the need, obligation, boredom, and even laziness can be overcome by game mechanics providing an entertaining activity with feedback to check their progression. In terms of usability, tracking presented some challenges primarily due to colours of the participants clothing, which, as the survey results suggest, affected engagement. More specifically, 92 % of the participants described the game as being engaging while 8 % disagreed. The gathered data was shown to an occupational health care specialist who found it interesting due to the possibilities of studying the user's progression, rather than relying on self-reporting or annual examinations. Figure 7.7 presents a screen capture of the GUI menu and the motion capture data.

7.4.2 Hand Exergame

This exergame focuses on flexion and extension hand movements. The user controls the legs of a virtual duck. Hand motion is mapped to the duck's legs and through each of the player's hand cycle, the duck is propelled through the water. This principle allows the user con control the swimming direction and speed of the duck.

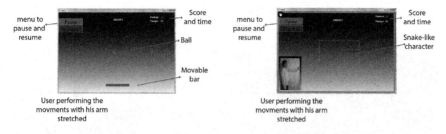

Fig. 7.7 Game levels

The objective of the game is for the user to reach the goal by swimming across the lake propelling the duck through scattered checkpoints. The game can be completed in between 30 s and 1 min, thus providing the player with time to explore the lake.

7.4.3   Motion Tracking

To track each hand, we employed the LeapMotion[8] 3DUI sensor, a device that allows non-invasive motion capture through a set infrared lights and cameras. The accuracy and reliability of the LeapMotion sensor has been studied in static and dynamic scenarios comparing the tracked data. Guna et al., highlighted the importance of tracking movment within 250 mm of the surface of the sensor, as otherwise the drop in accuracy makes de data unusable. Another remark from Guna et al. [50], is that the sampling frequency of 40 Hz is low and can provide challenges in dynamic motion tracking scenarios. However, slow movements and static scenarios provided the best accuracy results, reporting 0.01 mm or less in some tests. Another consideration is the range of movements and visibility from the sensor, in our case, given the focus on tracking the hand during flexion/extension rotations, which involves an ample movement the tracked data, the LeapMotion sensor proves to be sufficient and sufficiently accurate in our scenario [51].

To play the game, the user needs to position both hands parallel to the leapMotion sensor and at a distance of 10 cm. Optimum tracking is limited to a maximum flexion of 20° and an extension of 170°, as otherwise the tips of the fingers and the palm are not visible.

The development was achieved using LeapMotion's assets for UNITY3D. The GUI navigation is achieved through mouse-based interactions and the Leap-Motion is only required once the game starts. The player obtains visual feedback from the duck as it swims across the lake. The duck can swim straight (alternating both hand's movements), left (only right hand's movements) and right (only left hand's movements). Time is always visible and the ducklings are easily spotted given the yellow colour of their feathers providing visual references to follow as presented in Fig. 7.8.

---

[8]LeapMotion www.leapmotion.com/.

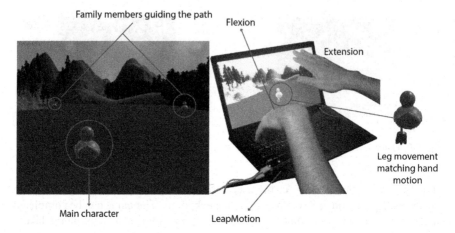

Fig. 7.8 Exergame movements and GUI

7.4.3.1 User Experience

We recruited a group of seven students who worked with computers in a virtual reality laboratory at the Mil. Nueva Granada University, to play the game in order to gauge their perception of using the exergame. The participants were asked to start the exergame, explore the lake, and reach the goal within a minute. After completing the game, we asked the participants to complete a survey to gauge motivation, engagement, and usefulness of the exergame. From the survey we observed that all participants were motivated to reach the goal and thus complete the game; without realizing it, participants acknowledged focusing more on the game than the exercises, being unaware that proper exercising was a key element to reach the goal in time. Finally, regarding its use, participants manifested that having affordable hardware as the LeapMotion that could be used for other tasks was of interest to them. A major challenge was encountered when the sensor lost track of the player's hand when overexceeding the flexion or extension rotation. To avoid having to restart the game, we programmed pop up messages along the lake to provide tips and inform when the hand is not being tracked. To avoid tracking issues due to fast moving hands, we paced the interactions with visual feedback for the user to synchronize.

7.5 Eye Exergame

According to the World Health Organization (WHO), there are approximately 285 million people globally suffering from low vision, affecting their quality of life.[9] The "computer syndrome" has been rising among computer users, thus, increasing the

---

[9]Visual impairment and blindness http://www.who.int/media-centre/factsheets/fs282/en/.

possibilities of suffering vision diseases such as the amblyopia or lazy eye [52]. However, vision therapy within the workplace concentrates on exercises to focus/defocus and relax the eye muscles.

Some interactive approaches can be found in the form of apps. For example, an interactive Android guide[10] presents digital information about eye exercises and how anyone can do them. An interactive approach can be seen in an eye exercising app that encourages the user to follow dots on the screen while keeping track of how much time did it take to perform the activity without eye tracking.[11] Other approaches involve modern 3D games that require the eye muscle's to adapt to the continuously changing images, such as 3D4amb.[12] A more interactive example of an eye training game is the FlashFocus[13] for the Nintendo DS console, which allows the user to complete several mini-games involving quick time events that require concentration, following objects and reactions to perform the action within the game's time frames. Other approaches have used devices and libraries such as OpenCV with webcams to detect eye movement and blinking [53], and infrared sensors that can detect corneal reflection, thus allowing to estimate gaze and map eye movements [54].

The eye exergame consists of three levels. The first level requires the user to save the seven Wonders of the World (e.g., The Great Pyramid of Giza, The Great Wall of China, Petra, The Colosseum, Chichén Itzá, Machu Pichu, The Taj Majal and Christ the Redeemer), from being abducted by space aliens using a tractor beam. The player needs to focus the gaze on each object to impede its abduction. The second level requires the user to place all of the rescued wonders at their corresponding places of origin. Finally, the last level presents a dog running from its master on a crossroad and the player is required to catch it by following it without being hit by traffic.

7.5.1 Motion Tracking

The objective of these games is to exercise the rectus muscles by following the objects from top to bottom and backward accordingly, and by default, each exercise running time is set to three minutes so users do not overexert. To accomplish proper eye tracking, we employed the Gazepoint[14] eye tracker (this tracker requires five to nine calibration points and includes an accuracy of $0.5°$). We developed the game using Gazepoint's SDK and UNITY3D. An important step that differs from the previous games that we described, is that in this case a calibration process is required per user to guarantee proper interactions with the eye tracker. The calibra-

---

[10]Eye exercises http://www.play.google.com/store/apps/details?id=com.remind4u2.eye.training. program\Źhl=es.

[11]Eye exercising http://www.play.google.com/store/apps/details?id=kr.Neosarchizo. EyeTraining\Źhl=es.

[12]3D4amb http://3d4amb.unibg.it/principles.html.

[13]FlashFocus http://www.nintendo.com/games/detail/jzDLc7ECGsrBaBRV9Ujs06hEWV-wJsj7.

[14]Gazepoint PT tracker www.gazept.com/product/gazepoint-gp3-eye-tracker/.

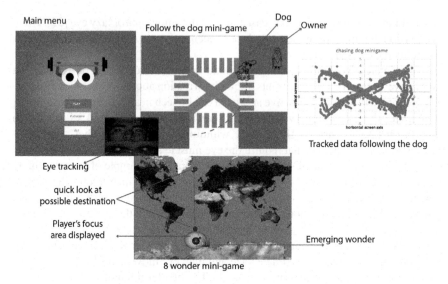

Main menu

Follow the dog mini-game

Dog

Owner

Tracked data following the dog

Eye tracking

quick look at
possible destination

Player's focus
area displayed

Emerging wonder

8 wonder mini-game

Fig. 7.9 Eye game levels and tracked data

tion requires the user to look at various dots on the screen. Navigation of the GUI is achieved through eye tracking so an improper calibration would affect all user interactions. At the end of the game, a text file containing the player's sight path within the game can be displayed, Fig. 7.9 presents example screenshots of the game and the tracked data.

7.5.2 User Experience

To measure interest on the developed exergame, a random sample of 23 participants including students, professors, and occupational health care specialists from Mil. Nueva Granada University were recruited to play the game. Participants played the game through its three levels of difficulty and answered a survey focused on engagement, usability, and motivation. After playing the game (all three levels), all of the participants agreed that exergames were engaging. 4 % of the participants found non-intuitive eye interactions, and 9 % of the participants were not able to play the exergame due to calibration issues associated with lightning and its skin reflection. Only 4 % found the games demotivating (these were the same participants that had issues with the interactions), and finally, 86 % found the experience fun and would consider playing the exergame further.

7.6 Conclusions

In this book chapter we presented a suite of exergames comprised of upper-lower limb, and eye games to perform in the workplace as possible complementary tools to traditional occupational health guides. We have conducted preliminary usability tests with surveys to gauge user perceptions regarding the games.

Exergames provide engaging exercising scenarios that may result in successful experiences if properly designed. From the experiences presented in this book chapter, exergames can be of interest in occupational health scenarios where employees can use as part of their work routine to minimize occupational health risks associated with sedentarism. During the tests, participants expressed interest in such solutions as a form of exercising that they would engage with, rather than the use of current printed and multimedia guides where no feedback and engagement is provided.

The use of affordable user interfaces allows implementing enriched interactions that are feasible to implement in constrained working places. This proved to be of interest to the participants as they experienced devices that are within reach and they could also used at home, or when considering the smartphone, everywhere. From our approach, the sensors presented in each of the mini-games in this chapter provided information that can be used to asses wide ranges of motion. Further research is required to examine long-term use and effects. As with newer forms of interaction, the novelty can wear off if no additional content is produced. Future work will be focused on the development of a cross-platform framework for the physician to define the game parameters and customize the game environment to respond to the exercising needs according the user's working scenario.

Finally, we would like to emphasize that although greater work and research is required, our preliminary results are promising and motivate us to continue and expand upon our work related to exergames with a focus on occupational health scenarios. In this chapter we presented a prototype of an occupational exergaming suite that caused attention to users and physiotherapists withholding potential to promote good health care habits in the workplace.

Acknowledgements The authors acknowledge the support of Mil. Nueva Granada University, its Faculty of Medicine, its Virtual Reality Centre and its Research division for funding thigh, shank and shoulder exergames under grant ING1545.

References

1. WH Organization (2003) Protecting workers health series (5)
2. Greene L, Mamic I (2015) The future of work: increasing reach through mobile technology. ILO
3. Global, international, and cross-cultural issues in IS (SIGCCRIS)
4. Wang J, Wang H, Gaskin J, Wang L (2015) Comput Hum Behav 53:181
5. Sharan D (2015) Proceedings 19th triennial congress of the IEA vol 9, p 14

6. Kietrys DM, Gerg MJ, Dropkin J, Gold JE (2015) Appl Ergonomics 50:98
7. İnal EE, Çetİntürk A, Akgönül M, Savaş S (2015) Muscle Nerve 52(2):183
8. Thomée S, Lissner L, Hagberg M, Grimby-Ekman A (2015) BMC Public Health 15(1):1
9. Hale L, Guan S (2015) Sleep Med Rev 21:50
10. Biswas A, Oh PI, Faulkner GE, Bajaj RR, Silver MA, Mitchell MS, Alter DA (2015) Ann Intern Med 162(2):123
11. Cooper AJ, Dearnley K, Williams KM, Sharp SJ, van Sluijs EM, Brage S, Sutton S, Griffin S (2015) BMC Public Health 15(1):1
12. Rasotto C, Bergamin M, Simonetti A, Maso S, Bartolucci GB, Ermolao A, Zaccaria M (2015) Man Ther 20(1):56
13. Burdea G (2002) 1st International workshop on virtual reality rehabilitation. (Mental health, neurological, physical, vocational) (VRMHR sn, 2002), pp 1–11
14. Hoffman HG, Doctor JN, Patterson DR, Carrougher GJ, Furness TA III (2000) Pain 85(1):305
15. Perry TS (2016) IEEE Spectr 53(1):56
16. Sinclair J, Hingston P, Masek M (2007) Proceedings of the 5th international conference on computer graphics and interactive techniques in australia and Southeast Asia. ACM, pp 289–295
17. Osorio G, Moffat DC, Sykes J (2012) Games for health: research, development, and clinical applications 1(3):205
18. Best JR (2015) Zeitschrift für Psychologie
19. Leary KCO, Pontifex MB, Scudder MR, Brown ML, Hillman CH (2011) Clin Neurophysiol 122(8):1518
20. Wüest S, van de Langenberg R, de Bruin ED (2014) Eur Rev Aging Phys Act 11(2):119
21. Gerling KM, Schild J, Masuch M (2010) Proceedings of the 7th international conference on advances in computer entertainment technology. ACM, pp 66–69
22. Giggins OM, Persson UM, Caulfield B (2013) J Neuroeng Rehabil 10(1):1
23. Loughlin EKO, Dugas EN, Sabiston CM, Loughlin JLO (2012) Pediatrics 130(5):806
24. Boulos MNK, Yang SP (2013) Int J Health Geographics 12(1):1
25. Thin AG, Brown C, Meenan P (2013) Int J Comput Game Tech 2013
26. Pogrzeba L, Wacker M, Jung B (2012) E-learning and games for training, education, health and sports. Springer, pp 125–133
27. Theng LB (2014) Assistive technologies for physical and cognitive disabilities. IGI Global
28. Sra M, Schmandt C (2015) arXiv:1512.02921
29. Leah C (2011) Exercises to reduce musculoskeletal discomfort for people doing a range of static and repetitive work. Printed. http://www.hse.gov.uk/research/rrpdf/rr743.pdf
30. da Costa BR, Vieira ER (2008) J Rehabil Med 40(5):321
31. Chaffin DB, Andersson G, Martin BJ (1999) Occupational biomechanics. Wiley, New York
32. Salen K, Zimmerman E (2004) Rules of play: Game design fundamentals. MIT press
33. Bateni H, Maki BE (2005) Arch Phys Med Rehabil 86(1):134
34. Jordan M, Donne B, Fletcher D (2011) Eur J Appl Physiol 111(7):1465
35. Van Diest M, Lamoth CJC, Stegenga J, Verkerke GJ, Postema K (2013) J Neuroeng Rehabil 10(1):1
36. Levac D, Espy D, Fox E, Pradhan S, Deutsch JE (2014) Physical Therapy
37. Valdivia-Trujillo S, Prada-Dominguez E, Uribe-Quevedo A, Perez-Gutierrez B(2014) Games Media Entertainment (GEM), 2014 IEEE. IEEE, pp 1–2
38. Ramos-Montilla E, Uribe-Quevedo A (2015) HCI international 2015-posters extended abstracts. Springer, pp 478–483
39. Green A (2003) J Am Acad Orthop Surg 2:365
40. Joo LY, Yin TS, Xu D, Thia E, Chia PF, Kuah CWK, He KK (2010) J Rehabil Med 42(5):437
41. Weightman A, Preston N, Levesley M, Holt R, Mon-Williams M, Clarke M, Cozens AJ, Bhakta B (2011) J Rehabil Med 43(4):359
42. Chang Y, Chen S, Huang J (2011) Res Dev Disabil 32(6):2566
43. Lo HS, Xie SQ (2012) Med Eng Phys 34(3):261

44. Saposnik G, Teasell R, Mamdani M, Hall J, McIlroy W, Cheung D, Thorpe KE, Cohen LG, Bayley M (2010) Stroke 41(7):1477
45. Lange B, Chang C, Suma E, Newman B, Rizzo AS, Bolas M (2011) Engineering in medicine and biology society, EMBC, 2011 Annual International Conference of the IEEE. IEEE, pp 1831–1834
46. Smith ST, Schoene D (2012) Aging Health 8(3):243
47. Antón D, Goñi A, Illarramendi A (2015) Methods Inf Med 54(2):145
48. Nathan D, Huynh DQ, Rubenson J, Rosenberg M (2015) PloS one 10(5):e0127113
49. Harrington CN, Hartley JQ, Mitzner TL, Rogers WA (2015) Human aspects of IT for the aged population. Design for everyday life. Springer, pp 488–499
50. Guna J, Jakus G, Pogačnik M, Tomažič S, Sodnik J (2014) Sensors 14(2):3702
51. Weichert F, Bachmann D, Rudak B, Fisseler D (2013) Sensors 13(5):6380
52. Wimalasundera S (2009) Galle Med J 11(1)
53. Grauman K, Betke M, Lombardi J, Gips J, Bradski GR (2003) Univ Access Inf Soc 2(4):359
54. Poole S, Ball LJ (2006) Encycl Hum Comput Interact 1:211

Chapter 8
Game-Based Stroke Rehabilitation

Mehran Kamkarhaghighi, Pejman Mirza-Babaei
and Khalil El-Khatib

Abstract Strokes are the most widely recognized reason of long-term disability of adults in developed countries. Continuous participation in rehabilitation can alleviate some of its consequences and support recovery of stroke patients. However, physical rehabilitation requires commitment to tedious exercise routines over lengthy periods of time, which often causes patients to drop out of therapy routines. In this context, game-based stroke rehabilitation has the opportunity to address two important barriers: accessibility of rehabilitation, and patient motivation. This chapter provides a comprehensive analysis on the advances in human-computer interaction (HCI) and development of games to support stroke rehabilitation. There are existing cases in the field of game-based stroke rehabilitation studied, for example the development of motion-based video games particularly addressing rehabilitation among stroke patients. This chapter discusses the effectiveness of design efforts in HCI and games research to support stroke rehabilitation by integrating findings from medical research that consider health outcomes of game based stroke rehabilitation. Based on these findings, challenges and opportunities in game based stroke rehabilitation are discussed. Critical components that influence the effectiveness of Tele-rehabilitation are identified and further design opportunities in the field of game-based stroke rehabilitation are explored as an important step toward the creation of games that are accessible, motivating and enjoyable for stroke patients. Finally, the chapter presents the challenges and opportunities in game-based stroke rehabilitation.

8.1 Introduction

Stroke is characterized as a sudden death of brain cells in a localized area due to insufficient blood flow; it is the most common cause of long-term adult disability in developed countries. Consequences of strokes include visual, cognitive and

M. Kamkarhaghighi · P. Mirza-Babaei (✉) · K. El-Khatib
University of Ontario Institute of Technology, Oshawa, ON, Canada
e-mail: pejman@uoit.ca

© Springer International Publishing AG 2017
A.L. Brooks et al. (eds.), *Recent Advances in Technologies
for Inclusive Well-Being*, Intelligent Systems Reference Library 119,
DOI 10.1007/978-3-319-49879-9_8

motor-skill losses, some patients may lose both memory and speech: up to 85 % of stroke patients suffer from hemiparesis—weakness on one side of the body—and between 55 % and 75 % of survivors experience motor-skill deficits; conditions such as hemiplegia—paralysis or weakness on one side of the patient's body including loss of control over one leg—can result in a physical disability that makes walking difficult or impossible. Often, these issues substantially limit an individual's abilities of interacting with the world, and reduce their ability of leading an independent life [2, 28]. In this context, increased dependency on the care of others negatively influences the physical and emotional well-being of stroke survivors, and has a negative impact of their quality of life [22].

Fortunately, rehabilitation and continuous participation in occupational therapy can alleviate some of these consequences and support the recovery and independence of stroke patients. However, physical rehabilitation particularly often requires commitment to tedious exercises routines over prolonged periods of time [3]. Many stroke patients could recover some physical functionality by performing daily repetitions of motions with affected limbs [1]. Successful rehabilitation programs implemented immediately after a stroke, may enhance the recovery process by minimizing functional disability [14], but rehabilitation programs are expensive, labour intensive and unavailable to patients in remote areas. In addition to these accessibility barriers, patients that do enrol in rehabilitation programs often do not follow through with therapy routines, leading to high dropout rates and a loss of potential benefits.

There are new technologies that have the potential to address these two main issues: accessibility of rehabilitation, and patient motivation [2, 43]. Computer-based stroke rehabilitation can be carried out in the patient's home independently while reducing treatment costs and increasing its accessibility for patients with mobility difficulties or those living in remote areas. Additionally, game-based solutions offer the potential of integrating playfully challenging elements and external motivations into therapy routines, thereby increasing patient inspiration to follow through with previously tedious tasks [1].

This chapter provides a comprehensive analysis of design efforts in human-computer interaction (HCI) and games research to support stroke rehabilitation. This chapter also reviews existing cases in the field of game-based stroke rehabilitation (e.g., the development of motion-based video games particularly addressing rehabilitation among stroke patients [1, 8]. In this chapter, the effectiveness of design efforts in HCI and games research to support stroke rehabilitation by integrating findings from medical research that consider health outcomes of game-based stroke rehabilitation are presented. Based on these results, challenges and opportunities in game-based stroke rehabilitation are discussed, and critical components that influence the effectiveness of Tele-rehabilitation to broad groups of stroke patients are identified. According to patients' motivation issues and medical records' needs, a game-based rehabilitation system for post-stroke patients should be non-invasive, safe and attractive, applicable at any time or place especially at home, able to record patient health data, and provide feedback for physicians as needed.

Further exploration of design opportunities in the field of game-based stroke rehabilitation is an important step toward the creation of games that are accessible, motivating and enjoyable for stroke patients; only if games can provide empowering experiences for patients while offering options to be effectively managed by medical staff, then they can be adopted into medical practice and contribute to patients' recovery from strokes.

8.2 Benefits of Game-Based Stroke Rehabilitation

This section discusses benefits of game-based stroke rehabilitation by relating efforts in the fields of human-computer interaction and games research to medical research.

8.2.1 Increased Patient Motivation and Engagement

Previous works in the field of game-based stroke rehabilitation aim to leverage the motivational power of games to increase patient commitment to follow through with therapy routines. Follow-up studies have demonstrated the effectiveness of games in this respect, showing that systems that have focused on factors such as entertainment are able to achieve increased engagement [1, 2, 31, 43]. For example, the study of Bach-y-Rita et al. [2] presents a system which uses the context of a digital game and proved to be efficient in entertaining users. Such entertaining systems may help users to be more focused on games' goals, rather than paying attention to the repetitive motions patterns, potentially achieving better progress in their recovery.

The required movements for the recovery of affected limbs in stroke rehabilitation patients involve a high amount of repetition, which can quickly become a boring and uninteresting activity for some [3]. However, the concern around increasing the patients' motivation on performing these exercises has not received enough attention [2]. One reason might be the diversity of stroke types and effected brain areas, which makes it difficult to design a single system that can effectively provide engagement for a large number of different patients.

8.2.2 Lower Access Barriers to Rehabilitation

As mentioned earlier, another concern related to stroke rehabilitation is the fact that it demands time and resources. Depending on the severity of the condition, the recovery process can take from weeks to years of constant daily practice of hundreds of exercises. As a consequence of this, the treatment goes beyond the rehabilitation centres and patients are given a large responsibility for continuing the treatment at home [3]. Various studies show that the clinical recovery period is often effective, however, it is slow or inefficient for most patients when the recovery continues at home, mostly due to the fact that just a small number of patients perform the exercises in the recommended way [1]. Keeping this in mind, it is important to find a solution that is not only reasonable for rehabilitation centres, but also affordable for patients. With the advancement of digital technology, there are more opportunities for cost-efficient digital systems to be used as supporting tools in rehabilitation, such as cameras, sensors, computers and video games. They are, however, required to be adapted in order to fulfil the needs of the rehabilitation plan [15].

8.3   Design Requirements in Game-Based Stroke Rehabilitation

To be an effective solution, there are a number of requirements that are needed for efficient game-based stroke rehabilitation, including:

8.3.1   Provide Rehabilitation Exercises for Different Parts of the Body

Game based stroke rehabilitation systems need to be designed in a way to perceive with precision when users are performing the necessary motions (for the comprehensive list of stroke rehabilitation please see: http://www.stroke-rehab.com). The help of a physiotherapist is necessary together with the development team to effectively design the right motions and mechanics [1]. The system must be precise on the way it tracks patient's motions, preventing patients from incorrect or inadequate performances [1]. It is also important to prevent users from exaggerating the amount of needed exercises, hence another opportunity when using a digital system is the potential of giving feedback about the necessary training intensity to achieve the best recovery progress through the activities [18].

8.3.2   Monitor Patient Progress and Provide Feedback and Information

One of the main advantages of using digital systems to assist with rehabilitation practices is the potential of the system's interaction with the user, such as the ability to monitor and track patients' progress to provide them with specific and targeted feedback. For example, following Nicholson [25], a simple use of a visual system showing body areas that are getting improvements due to the performed activity may increase motivation and the development of personal goals. Moreover, some patients may not feel motivated to perform certain activities or motions due to the lack of understanding of the benefits or perceiving little progress [22]. Early introduction of relevant information prior to playing shows the rehabilitation objectives of each exercise could help avoid discouragement (due to the lack of information) and provides greater ease on creating new personal goals. Also the generated data during the playing time, like earned points, duration of playing game, results, etc. can be monitored by a physician and show the patient's participation in the rehabilitation process.

8.3.3 Provide Diversity and Flexibility of Content to Accommodate Various Preferences

As mentioned before patients affected by strokes are a wide target audience, hence there is also a challenge on making design decisions when designing a single rehabilitation system that uses entertainment as one of the key pieces to provide user engagement [3]. One possible solution is to focus on the diversity of content. A game having different kinds of mini-games with different gameplay experiences can attract players that have different profiles with specific preferences. Furthermore, diversity in content also may increase the amount of different options that each user has to practice, preventing boredom with a single game. The topic of content diversity was also observed in a study in which a book reader system for a stroke patient was efficient at the beginning, but as it contained just a single book, it quickly became uninteresting, failing to engage the user over a longer period of time [3]. Also, different types of games or even different levels, can provide physicians with different tools, like a "Pharmacy of Games". A physician can prescribe one or more games with diverse levels for a specific patient and monitor the patient's activities and feedbacks and update their rehabilitation program accordingly. Flexibility of content, which is diversity of health games in the proposed repository, can open opportunity of using off-the-shelf input devices as hardware of the rehabilitation system, which is cheaper and more accessible than special purpose rehabilitation devices.

8.3.4 Allow for Patient Autonomy and Ability to Connect with Other Patients

The vast majority of stroke affected patients eventually become severely dependent on family members or friends. Due to motor-skill, cognitive and visual problems, there are few patients who can return to their full-time work, or perform common activities of daily life, such as eating or dressing with ease [1]. Therefore, it is important to consider a patient's autonomy when designing rehabilitation systems. This can be mainly reflected on making users feel in control when using the system. For example, games with multiplayer functionalities may offer a cooperative or competitive setup that can be used by families, friends or even other affected patients [1, 3]. However, it is important for the systems to provide a possibility for individual use, thus respecting their autonomy. It is also possible to balance between a cooperative scenario, where affected patients can play as a supportive role, limiting their interaction with the game goals, versus a competitive scenario, where they may feel an unfair competition.

Due to the diversity of affected limbs and the severity of strokes, it is not possible to design a game that aggregates all users. However, this issue can be softened through customization tools. The ideal situation could be to have the involvement of a physiotherapist to perform initial customization of the system, entering the necessary levels of precision, speed and other relevant factors, focusing on the needs of each separated patient [1]. Moreover, it is important for the game to offer some sort of adaptability functions that can be fully automatic or manual. As with the progress of the training, the patient's performance tends to improve, there should be an intelligent system to perceive the user's progress and automatically adapt its difficulty, or may have pre-sets of different levels that the patients can change when they feel the need. This adaptability is of vital importance to help the patients' recovery and maintaining their interest in the game [15, 1].

8.4 Literature Review

Technology-based rehabilitation has been highlighted as an opportunity for in-home rehabilitation that would allow patients to continuously engage in therapy even after being discharged from a hospital. It would also make therapy more accessible for persons living in remote areas.

In this context, research in the area of stroke rehabilitation systems suggests that maintaining patient motivation is an important step towards ensuring therapy success; however, poorly designed and implemented technology-based rehabilitation is often tedious, repetitive, and does not provide encouraging feedback, hence leading to a significant reduction in patient motivation.

Case studies have shown that games and similar technologies have the potential to provide significant external motivation for stroke patients. Improving patients motivation to participate in rehabilitation programs and longer therapeutic sessions may led to better functional outcomes over the treatment course [15]. Hence, games designed for home-based stroke rehabilitation could provide motivation for stroke patients to continue the rehabilitation exercises [1].

Games could be designed to decrease repeated motions monotony and provide performance feedback to help patients in recovery process from stroke. These games should ensure the patients are performing exercises correctly to maximize rehabilitation process impact. It is also suggested that games which provide a sense of social interaction with other similar patients can help decrease patients sense of isolation and increase their motivation to exercise [1].

8.4.1 Background

In the middle of 1970s, Bach-y-Rita et al. [2] adopted the newly introduced electronic pong games for the functional training for hemiparesis limb patients. Their customizable input controllers were developed to match variety of grips of the stroke patients. The second version of the pong-game device consisted of an actuator/sensor sliding lever, where the lever was attached to a "trolley car" that moves on ball-bearing wheels inside a box.

Wood et al. [43] examined the efficacy and feasibility of an approach called the Palanca ("the lever" in Spanish). They applied the WMFT (Wolf Motor Function Test), which they used to measure patient's ability to perform tasks with their affected arms and limb movements. They also designed a pre-test and post-test case series. Comparison of pre-treatment and post-treatment scores of WMFT showed motor-skill improvement and both adherence and satisfaction, were found to be high.

They argued that the Palanca is sufficiently motivating and interesting for patients to be prepared for rehabilitations in an enjoyable and low cost way. This study was based on results of playing electronic table. The study claimed that this approach appears to be motivating, efficacious, and enjoyable [43].

In 2007, the group of Goude et al. presented a model for a hepatic and stereo-vision immersive workbench in stroke patients' homes. A customized virtual reality framework linked to a haptic workbench was used to prototype and implement training games based upon different patterns. Twenty games have been produced including upper extremity reaching exercises, neglect assessment, and coordination training exercises. The variations of games provide a range of activities that can be used to personalize a rehabilitation session [16]. Flores et al. [15] studied different systems that implemented game-based rehabilitation, for example, Goude et al. [16], presented a game designing model based on association between the stroke rehabilitation taxonomy and gaming patterns. A group of stroke symptoms and treatments were explained in this model.

In more recent studies, Alankus et al. [1] reviewed games that were developed to assist stroke patients in the outset of recovery, as such came the development of haptic glove based games to improve the player's finger flexion and extension, where players scaring butterflies, playing the piano, and squeezing virtual pistons. Broeren et al. [5] by using a pen-like haptic device, created several 3D games. Burke et al., similar to whack-a mole, built two webcam based games. A series of 3D games involving manipulating objects developed by Yeh et al. [44] and Sanchez et al. [30] designed and implemented training exercises.

Following the review of some existing stroke recovery games, Alankus et al. [1] developed a series of games using various input devices (see Table 8.1). Their games aim to measure the effectiveness of different approaches by having different goals and differing from being single or multi-player, cooperative or competitive, and presenting varying degrees of cognitive challenges. For example, *Frog Simon*, a single-player game that requires precise input and provides a level of challenge on memory, and *Dirt Race*, a two-player cooperative game.

In other similar efforts, Todorov et al. [40] used virtual reality technology for two player table-tennis training. Hemiplegia patients were trained in a virtual environment. Shin et al. [34] developed a system called RehabMaster which is a safe and feasible virtual reality system for upper extremity function enhancement in patients who suffered from stokes. A Kinect-based system by Chang et al. [10] used

Table 8.1 Summary of game-based stroke rehabilitations

Rehabilitation body area	Input device	Game title
Balance and mobility	Wii	Wii Sports (Baseball, Bowling, Boxing, Golf, Tennis) and Wii Fit (Yoga, Strength, Aerobic, Balance) [12]
Cognitive and motor skills	2D input Device	Garden [1]
	Accurate 2D input Device	Frog Simon, Baseball Catch and Frogger [1]
	Accurate horizontal 1D input Device	Catch the Kitty [1]
	Basic 1D input	Dirt Race [1]
	Basic vertical 1D input	Under the Sea [1]
	Vertical 1D inputs	Pong [1]
Finger	3D Environment	Pinch [19] and Piano [24]
	Android-based tablet	FINDEX [9]
Gait ability	PlayStation 2 and EyeToy Play	Goal Attack, Table Tennis, Homerun, Knockout and Bowling [35]
Hand	Cyber-Glove and force feedback glove	VR Rehab System [18]
	Hand tracking device (Virtual Reality)	Handcopter [27]
Pronation and supination	3D Environment	Spatial Rotation [19]
Hand and Leg	Kinect (Virtual Reality)	Hill climbing and Running [29]

(continued)

Table 8.1 (continued)

Rehabilitation body area	Input device	Game title
Motor Skills	EMG Signals and Robot	EMG-driven [36]
	Kinect (Virtual Reality)	Animal Feeder, Fruit Catcher [4] and Kinerehab [10]
	Electromagnetic sensor	Table tennis [40]
	Pen-like haptic device	VR-task [6]
	Virtual Reality, Haptics and	A series of 3D games [44]
	Modern Sensing Technique	
Thumb and Finger (Range of movement)	CyberGlove	Window wiper, Traffic light [24]
	RMII force feedback glove	Graphical Pistons [24]
Upper Arm Motor	Kinect (Virtual Reality)	Kinematic game [23]
Upper Limb/Extremity	3D awareness sensor	RehabMaster [33]
	Actuator/Sensor sliding lever	Pong V2 [2]
	Elinor Game Console	Elinor (Taylor et al.)
	Joystick	Pong V1 [2]
	Passive antigravity arm orthosis	Java Therapy 2.0 Games [30]
	RehabMaster™ (Virtual Reality)	Goalkeeper, Underwater Fire, Bug Hunter and Rollercoaster [34]
	CyWee Z	Solitaire, Mah-Jong, FreeCell and Bejeweled [17]
	EXO-UL7 (robotic exoskeleton arms)	Flower Paint, Reach, Pong, Pinball, Circle and Handball [37]
	and Virtual Reality	
	Handheld roller-ball device and Webcam	Colour Ball and Mole Attack [11]
	Kinect (Virtual Reality)	5 Jintronix based games [26]
	Kinect (Virtual Reality) and Leap Motion	Paper Ball and Avatar Based Training [42]
	Palanca Device	Electronic table tennis [43]
	Webcam	Rabbit Chase and Arrow Attack [8]
	Wii Remote Device (Augmented Reality)	Little Treasure, Bowling and Growing Fruit [21]

(continued)

Table 8.1 (continued)

Rehabilitation body area	Input device	Game title
Wrist muscle activation	Surface Electromyographic (sEMG)	NeuroGame Therapy (NGT) [7]
Upper Limb/Extremity and Gait ability	Kinect, Mitt with timer,	Kayak and Fruit picking [23]
	and Glove (Virtual Reality)	
Arm	3D Environment	Octopus [41]
	Core:Tx Device	Core:Tx [32]
Coordinated arm and shoulder movement	3D Environment	Ball Shooting and Reaching [19]

as a rehabilitation tool for children with acquired muscle atrophy and cerebral palsy. Ustinova et al. [41] made a 3D videogame that could enhance arm postural coordination in traumatic brain injury patients.

In a study conducted by Taylor et al. in 2009 [39], a game based inexpensive rehabilitation tool (called Elinor) for stroke patients was used in patients' home. Their goal was to develop an accessible, cost effective, supervised, and robust system to motivate patients in their own home.

A book chapter published by Jung et al. [19] describes on-going interdisciplinary efforts, involving Communication, Electrical Engineering, Computer Science, Psychology, and Physical Therapy researchers, to develop three-dimensional game environments for post-stroke recovery. Their first step was to identify common everyday-life activities that people are unable to perform after suffering from a stroke. Then, the goals and patterns of movements in corresponding training tasks from traditional therapy were analysed in order to transform these training tasks into digital games in three-dimensional environments. In order to list important game features, this study pointed that, the interface is the crucial factor determining the forms and quality of interactions. Visual and audio feedback is one of the most essential game features that can make digital games interesting, interactive, and exciting [19].

The study presented by Hijmans et al. in 2010 aimed to determine the effectiveness of a movement based game controller in bilateral and upper-limb rehabilitation. They applied FMA-UL (Fugl-Meyer Assessment upper-limb section) for assessment. The study showed that this combination of controller and game could improve upper-limb function in patients with chronic stroke in short time [17]. In a study done in 2010, a computer game presented by Shusonget al. used EMG nodes as input devices and an avatar robot acting as a boy who asks the patients to do exercises. The robot uses movement and speech to communicate. Based on patient's motion and movement capabilities traced by EMG signals, the computer game suggests exercises. This method is effective in early rehabilitation, before muscles become too stiff [36]. In 2011, Schuck et al. presented the Core: Tx; which is a stroke rehabilitation game who provides a repetitive, non-invasive practice for rehabilitate affected arm impairment, and affected arm disability [32].

One important question in this area of research is whether researchers should focus on adapting available games that were created for entertainment or should focus to develop games specifically for rehabilitation use. Some studies suggest partnerships between serious game scientists and game console manufacturers (such as Microsoft and Sony) or game studios or publishers (such as EA and Ubisoft), should be formed to design and implement games especially for rehabilitation [12].

In 2014 Maung et al. discussed the improvement of upper extremity rehabilitation games for chronic hemiparesis patients. They used the Microsoft Kinect, a camera-based controller, to guarantee that the patients were being affable with the therapy. Using an off-the-shelf tracking system was a key point and freed up substantial developer time. They also designed a glove to measure extension and flexion. An outdoor river and an underground maze is the environment of the game. The therapy movements are: Rowing, Avoid Barriers, which requires the player to perform the move-right or left gesture, Rapids, Fishing, Bottles (the player retrieves the bottle from the water), Maze, Chests, Bats (who annoy the player), and Picking fruits [23]. In 2010 the group of Shin et al. studied the effect of rehabilitation exercise on chronic stroke patient's gait. They used PlayStation 2 and EyeToy Play, an off-the-shelf gaming system, to interact directly with television screen. The game scenarios were: Goal Attack, Table Tennis, Knockout Homerun, and Bowling which patients using their arms, hands, and body to play. The results showed the effective rehabilitation exercise by using video game at home [35]. In 2011 Di Loreto et al. proposed a serious game with MRS (Mixed Reality System) and an adaptation module for stroke rehabilitation. The main aim of this system was to increase frequency and intensity of training sessions. The developed MRS system for this study had three subsystems: the gaming for managing the game application, the motion capture for tracking the patient's movement, and the display for displaying the virtual game environment. Proposed technique focuses on a real-time adaptation module based on a user-cantered adaptation approach [13].

Saini et al. [29] presented game design principles for hand and leg rehabilitation. Improving the accuracy of stroke exercises was their main goal. They demonstrated several advantages of utilizing off-the-shelf games hardware in stroke rehabilitation like: increasing motivation, effectiveness, time efficiency, memory concentration and remote monitoring. They presented a system to store all data for analyse the patient's activities. They used games which especially designed for rehabilitation with hill climbing and running scenarios [29]. Simkins et al. in 2012 investigate games proposed for upper-limb rehabilitation. In this study, stroke patients used an upper-limb robotic exoskeleton unilaterally and bilaterally and then played seven different static games designed for them. The games were static and dynamic, static games were named: Flower, Paint, and Reach, and the Dynamic games were: Pong, Pinball, Circle, and Handball [37]. In 2012, the group of Lee et al. proposed an augmented reality-based game system for assessment of stroke rehabilitation. The proposed rehabilitation game system used the both Wii remote device and augmented reality technology. The main purpose was upper limb rehabilitation and physical motor function improvement. They used Motor Assessment Scale (MAS) for the assessment scale. This game was designed for rehabilitation of users with mild dyskinesia (people who have the ability to move independently). Several games developed: "Little Treasure" for the whole body activities. "Bowling Game", for upper extremity rehabilitation. "Growing Fruit", for upper extremity activities. These games enhanced the user's experience and increased the interaction and fun of the games content [21]. In 2012, the group of Pantoja et al. described the Handcopter Game, a health game that its input device was a low cost visual hand tracking system. The main goal of this study was to implement an inexpensive tool to supplement hand functions rehabilitation in stroke patients [27].

Borghese et al. [4] described a game engine to support home rehabilitation. According to this study, games should be customized for each patient based on functional status, and supervising the patient's activities is mandatory to prevent maladaptation. This study argued that monitoring, adaptation, and online evaluation of the movements should be in real time [4]. In 2013, in order to validating a low cost virtual reality system for upper limb rehabilitation, a protocol presented by Norouzi-Gheydari et al. Based on this protocol, during each session, patients be seated on a chair and facing the virtual reality system in a calibrated distance. The dynamic scope of movement of the arm within the context of the reaching space of the game was measured. Then the patient would interact with all the games at least once during a 20 min session. These activities were developed for unilateral movement, trunk control, and bilateral movement in a sitting position [26].

In 2014 Shin et al. developed RehabMaster™, an interactive game-based virtual reality system for post-stroke rehabilitation of the upper extremities. The motions were suggested by stroke rehabilitation specialists. The exercises designed to improve the range of motion, deviation, and strength [33, 34]. In a study by De Leon et al. [11] two augmented reality based games presented. The games simulated stroke patients rehabilitating with upper limb disabilities. During the game, a handheld roller-ball device was used. This device which focused on movements and muscles, changed the way the games are played. In another scenario, without this device, the games will be played with the webcam. The games attempt to simulate daily living activities. In "Colour Ball Game" the user should hit a coloured ball with the same colour by using the paddle [11] In 2014 Carabeo et al. from the Philippines introduced a tablet based game, FINDEX, which tracking the patient's progress for fine finger control, isolation and coordination, and. They used Rosenbusch Test assessment, which uses time, and accuracy measures in its assessment. FINDEX's designed based on daily functional activities, like household activities. During the game, the player must push piano keys by using particular fingers. In order to increase finger range of motion, the players water flowers by their thumb movements. Dragging, tapping, and stretching fingers address in this platform [9]. In 2014 Brown et al. introduced the electromyography-controlled video game system; called NeuroGame Therapy (NGT). It consists of a laptop computer, NGT console, sEMG leads, and disposable electrodes. The NGT system console uses a custom Neurochip circuit to amplify and digitize bipolar analog sEMG signals from two muscle groups and transmit these signals via universal serial bus to the computer [7].

Vogiatzaki et al. in 2014 reported on the results of their game-based training system project, FP7-StrokeBack. This approach used Personal Health Record (PHR) system for effective application of rehabilitation protocols. The StrokeBack concept system architecture contains a home deployed patient system and supporting physiological remote monitoring of patient wellbeing. This platform shared information between the patient and their supporting medical team [42]. In a 2002 study by Merians et al. three chronic stroke patients were trained on dexterity tasks on real objects and virtual reality exercises. The aim of the simulation was to improve range of motion, and movement speed. CyberGlove and Rutgers Master II-ND (RMII) were used as input devices. They used CyberGlove to exercise the speed, range of motion, and fractionation movement and the RMII force feedback glove used for finger strengthening exercises. The virtual reality exercises designed to rehabilitate range of movement, speed of finger fractionation, and strength of movement [24]. In a more recent review in 2014 Knüwer and Johannes did a systematic review to summarize the efforts of using Nintendo Wii for stroke rehabilitation. Their findings showed that Nintendo wii could be used as a feasible tool with traditional approaches of stroke rehabilitations.

A systematic review done by Souza et al. in 2011 to identify the functional results obtained in the rehabilitation of individuals with stroke by using the interface of the Nintendo Wii games Knüwer [20]. All studies showed benefits such as improved motor coordination and agility of upper limbs using the Wii associated with conventional therapies such as physiotherapy and occupational therapy [38].

Based on the review of existing design efforts in the field of game design for stroke rehabilitation (see Table 8.1 again), the most important benefits of game-based stroke rehabilitation summarized, and promising design opportunities that are associated with social and medical aspects of stroke rehabilitation identified.

8.5 Conclusion

This chapter provides a summary of game-based approaches to support stroke rehabilitation from the perspective of game design. From reviewing studies and implemented stroke rehabilitation games, a combination of mental supports, self and external motivation, and accessible interfaces are important considerations for designing game-based rehabilitation. It is important to emphasize that playing games at home can be done anytime and potentially empower patients to take ownership over their therapy by using cost-efficient digital systems for rehabilitation. Autonomy designed games can be reflected on making users to feel in control when using the rehabilitation process and customized patient center based games with personal goals can fit each patient's conditions, even with off-the-shelf hardware. Flexible and diversity of game contents and themes can support variety of patients' taste and keep them satisfied.

One potential development could be to leverage motivational pull of commercial and social available games to draw players in the long run, and not just in the context of tightly controlled research studies.

Acknowledgements We would like to thank NSERC and UOIT for their research support. We also would like to thank **Dr. Reza Talebi for his contribution to this study.**

References

1. Alankus G, Lazar A, May M, Kelleher C (2010) Towards customizable games for stroke rehabilitation. In: Proceedings of the SIGCHI conference on human factors in computing systems. ACM, pp 2113–2122
2. Bach-Y-Rita P, Wood S, Leder R, Paredes O, Bahr D, Bach-Y-Rita EW, Murillo N (2002) Computer-assisted motivating rehabilitation (CAMR) for institutional, home, and educational late stroke programs. Top Stroke Rehabil 8:1–10
3. Balaam M, Rennick Egglestone S, Fitzpatrick G, Rodden T, Hughes A-M, Wilkinson A, NIND T, Axelrod L, Harris E, Ricketts I (2011) Motivating mobility: designing for lived motivation in stroke rehabilitation. In: Proceedings of the SIGCHI conference on human factors in computing systems. ACM, pp. 3073–3082
4. Borghese NA, Pirovano M, Lanzi PL, Wüest S, de Bruin ED (2013) Computational intelligence and game design for effective at-home stroke rehabilitation. Games Health: Res Dev Clin Appl 2:81–88
5. Broeren J, Bjorkdahl A, Claesson L, Goude D, Lundgren-Nilsson A, Samuelsson H, Blomstrand C, Sunnerhagen KS, Rydmark M (2008) Virtual rehabilitation after stroke. Stud Health Technol Inform 136:77
6. Broeren J, Jalminger J, Johansson L-Å, Rydmark M (2014) A kinematic game for stroke upper arm motor rehabilitation-a person-centred approach. J Accessibility Des All 4:13
7. Brown EVD, McCoy SW, Fechko AS, Price R, Gilbertson T, Moritz CT (2014) Preliminary investigation of an electromyography-controlled video game as a home program for persons in the chronic phase of stroke recovery. Arch Phys Med Rehabil 95:1461–1469
8. Burke JW, McNeill M, Charles DK, Morrow PJ, Crosbie JH, McDonough SM (2009) Optimising engagement for stroke rehabilitation using serious games. Vis Comput 25:1085–1099
9. Carabeo CGG, Dalida CMM, Padilla EMZ, Rodrigo MMT (2014) Stroke patient rehabilitation a pilot study of an android-based game. Simul Gaming 45:151–166
10. Chang Y-J, Chen S-F, Huang J-D (2011) A Kinect-based system for physical rehabilitation: a pilot study for young adults with motor disabilities. Res Dev Disabil 32:2566–2570
11. De Leon N, Bhatt S, Al-Jumaily A (2014) Augmented reality game based multi-usage rehabilitation therapist for stroke patients. Int J Smart Sens Intell Syst 7
12. Deutsch JE, Brettler A, Smith C, Welsh J, John R, Guarrera-Bowlby P, Kafri M (2011) Nintendo wii sports and wii fit game analysis, validation, and application to stroke rehabilitation. Top stroke Rehabil 18:701–719
13. Di Loreto I, Abdelkader G, Hocine N (2011) Mixed reality serious games for post-stroke rehabilitation. In: 2011 5th International conference on, pervasive computing technologies for healthcare (PervasiveHealth). IEEE, pp 530–537

14. Duncan PW, Zorowitz R, Bates B, Choi JY, Glasberg JJ, Graham GD, Katz RC, Lamberty K, Reker D (2005) Management of adult stroke rehabilitation care a clinical practice guideline. Stroke 36:e100–e143

15. Flores E, Tobon G, Cavallaro E, Cavallaro FI, Perry JC, Keller T (2008) Improving patient motivation in game development for motor deficit rehabilitation. In: Proceedings of the 2008 international conference on advances in computer entertainment technology. ACM, pp 381–384

16. Goude D, Björk S, Rydmark M (2007) Game design in virtual reality systems for stroke rehabilitation. Studies in health technology and informatics 125:146–148

17. Hijmans JM, Hale LA, Satherley JA, McMillan NJ, King MJ (2010) Bilateral upper-limb rehabilitation after stroke using a movement-based game controller. J Rehabil Res Dev 48:1005–1013

18. Jack D, Boian R, Merians AS, Tremaine M, Burdea GC, Adamovich SV, Recce M, Poizner H (2001) Virtual reality-enhanced stroke rehabilitation. IEEE Trans Neural Syst Rehabil Eng 9:308–318

19. Jung Y, Yeh S-C, Mclaughlin M, Rizzo AA, Winstein C (2009) Three-dimensional game environments for recovery from stroke. In: Serious games: mechanisms and effects. New York: Routledge, Taylor and Francis, pp 413–28

20. Knüwer J (2014) Wii-habilitation: the use of motion-based game consoles in stroke rehabilitation; a systematic review

21. Lee, R-G, Tien S-C (2012) Augmented reality game system design for stroke rehabilitation application. In: 2012 4th international conference on computational intelligence, communication systems and networks (CICSyN). IEEE, pp 339–342

22. Maclean N, Pound P, Wolfe C, Rudd A (2002) The concept of patient motivation a qualitative analysis of stroke professionals' attitudes. Stroke 33:444–448

23. Maung D, Crawfis R, Gauthier LV, Worthen-Chaudhari L, Lowes LP, Borstad A, Mcpherson RJ, Grealy J, Adams J (2014) Development of recovery rapids-a game for cost effective stroke therapy. Found Digit Games (FDG)

24. Merians AS, Jack D, Boian R, Tremaine M, Burdea GC, Adamovich SV, Recce M, Poizner H (2002) Virtual reality–augmented rehabilitation for patients following stroke. Phys Ther 82:898–915

25. Nicholson S (2012) A user-centered theoretical framework for meaningful gamification. Games + Learning+. Society 8:1

26. Norouzi-Gheidari N, Levin MF, Fung J, Archambault P (2013) Interactive virtual reality game-based rehabilitation for stroke patients. In: 2013 International conference on virtual rehabilitation (ICVR). IEEE, pp 220–221

27. Pantoja AL, Pereira A (2012) A video-tracking based serious game for motor rehabilitation of post-stroke hand impairment. SBC 3:37

28. Rehana Z (2013) functional outcomes of lower limb of stroke patients after receiving physiotherapy. Department of Physiotherapy, Bangladesh Health Professions Institute, CRP

29. Saini S, Rambli DRA, Sulaiman S, Zakaria MN, Shukri SM (2012) A low-cost game framework for a home-based stroke rehabilitation system. In: 2012 international conference on computer & information science (ICCIS). IEEE, pp 55–60

30. Sanchez RJ, Liu J, Rao S, Shah P, Smith R, Rahman T, Cramer SC, Bobrow JE, Reinkensmeyer DJ (2006) Automating arm movement training following severe stroke: functional exercises with quantitative feedback in a gravity-reduced environment. IEEE Trans Neural Syst Rehabil Eng 14:378–389

31. Saposnik G, Teasell R, Mamdani M, Hall J, McIlroy W, Cheung D, Thorpe KE, Cohen LG, Bayley M (2010) Effectiveness of virtual reality using Wii gaming technology in stroke rehabilitation a pilot randomized clinical trial and proof of principle. Stroke 41:1477–1484

32. Schuck SO, Whetstone A, Hill V, Levine P, Page SJ (2011) Game-based, portable, upper extremity rehabilitation in chronic stroke. Top Stroke Rehabil 18:720–727

33. Shin J-H, Ryu H, Jang S, Kim S (2013) Task-specific interactive game-based virtual reality rehabilitation system for stroke patients: A usability test and two clinical experiments. J Neurol Sci 333:e569

34. Shin J-H, Ryu H, Jang SH (2014) A task-specific interactive game-based virtual reality rehabilitation system for patients with stroke: a usability test and two clinical experiments. J Neuroengineering Rehabil 11:32

35. Shin W-S, Lee D-Y, Lee S-W (2010) The effects of rehabilitation exercise using a home video game (PS2) on gait ability of chronic stroke patients. J Korea Acad-Ind Cooperation Soc 11:368–374

36. Shusong X, Xia Z (2010) EMG-driven computer game for post-stroke rehabilitation. In: 2010 IEEE conference on robotics automation and mechatronics (RAM). IEEE, pp 32–36

37. Simkins M, Fedulow I, Kim H, Abrams G, Byl N, Rosen J (2012) Robotic rehabilitation game design for chronic stroke. Games Health: Res Dev Clin Appl 1:422–430

38. Souza LB, Paim CDRP, Imamura M, Alfieri FM (2011) Uso de um ambiente de realidade virtual para reabilitação de acidente vascular encefálico. Use of interactive video game for stroke rehabilitation, Acta fisiátrica, p 18

39. Taylor A-S, Backlund P, Engstrom H, Johannesson M, Lebram M (2009) The birth of elinor: a collaborative development of a game based system for stroke rehabilitation. VIZ'09 2nd international conference in visualisation. IEEE, pp 52–60

40. Todorov E, Shadmehr R, Bizzi E (1997) Augmented feedback presented in a virtual environment accelerates learning of a difficult motor task. J Mot Behav 29:147–158

41. Ustinova KI, Leonard WA, Cassavaugh ND, Ingersoll CD (2011) Development of a 3D immersive videogame to improve arm-postural coordination in patients with TBI. J Neuroeng Rehabil 8:61

42. Vogiatzaki E, Gravezas Y, Dalezios N, Biswas D, Cranny A, ortmann S, Langendorfer P, Lamprinos I, Giannakopoulou G, Achner J (2014) Telemedicine system for game-based rehabilitation of stroke patients in the FP7-"StrokeBack" project. In: 2014 European conference on networks and communications (EuCNC). IEEE, pp. 1–5

43. Wood SR, Murillo N, Bach-Y-Rita P, Leder RS, Marks JT, Page SJ (2003) Motivating, game-based stroke rehabilitation: a brief report. Top Stroke Rehabil 10:134–140

44. Yeh S-C, Rizzo A, Zhu W, Stewart J, Mclaughlin M, Cohen I, Jung, Y. Peng W (20045) An integrated system: virtual reality, haptics and modern sensing technique (VHS) for post-stroke rehabilitation. In: Proceedings of the ACM symposium on virtual reality software and technology. ACM, p59–62

Part III
Touch and Wearables

Chapter 9
Multi-sensory Environmental Stimulation for Users with Multiple Disabilities

Cristina Manresa-Yee, Ann Morrison, Joan Jordi Muntaner and Maria Francesca Roig-Maimó

Abstract Multi-sensory environments can improve and maximize the well-being of individuals with multiple disabilities, that is, individuals who have more than one significant disability (one of which is a cognitive impairment). In this chapter we present a multi-sensory environmental stimulation system that combines different technologies such as computer vision and tactile cues in the form of vibrations. The system offers users control over the environmental stimulation by responding to their body movements. A vision-based interface detects the user's hand position and activates meaningful and motivational outcomes when the hand is positioned over specific regions. Further, we extended the system by including a wearable vibrotactile interface that encourages users to move their arms by using vibrations that exploit the saltation perceptual illusion known for inducing movement.

Keywords Multi-sensory environmental stimulation · Interactive environment · Multiple disabilities · Well-being · Vision-based interfaces · Vibrotactile interfaces · Saltation

C. Manresa-Yee (✉) · M.F. Roig-Maimó
Department of Mathematics and Computer Science, University of Balearic Islands, Crta. Valldemossa km 7.5, 07122 Palma, Spain
e-mail: cristina.manresa@uib.es; xisca.roig@uib.es

A. Morrison
Department of Architecture, Design and Media Technology, Aalborg University, Rendsburggade 14, 9000 Aalborg, Denmark
e-mail: morrison@create.aau.dk

J.J. Muntaner
Department of Applied Pedagogy and Education Psychology, University of Balearic Islands, Crta.Valldemossa km 7.5, 07122 Palma, Spain
e-mail: joanjordi.muntaner@uib.es

© Springer International Publishing AG 2017
A.L. Brooks et al. (eds.), *Recent Advances in Technologies for Inclusive Well-Being*, Intelligent Systems Reference Library 119, DOI 10.1007/978-3-319-49879-9_9

9.1 Introduction

People with multiple disabilities have severe or profound dysfunctions in different development areas (such as physical, sensorial…), and will always include some form of cognitive impairment. Consequently, these individuals have limited functionality, slow development processes and they need permanent support in all basic activities of daily living [1]. People with this profile often face a lack of control and opportunity in their everyday life [2] and they are likely to be severely impaired in their functioning in respect of a basic awareness and understanding of themselves, of the people around them and of the world they live in [3].

While it is recognised that technology can provide impaired users with opportunities to learn, share information, and gain independence [4, 5], users with multiple disabilities frequently find difficulties in using it due to their physical and/or cognitive impairments [6]. As a result, there is not a wide range of interactive applications for this group.

According to Vos et al. [7], people with multiple disabilities are "greatly at risk [of experiencing] low subjective well-being". Vos et al. [7] highlight the importance of finding ways to improve the subjective well-being of this group. Studies show that multi-sensory environments can improve and maximize the well-being of individuals with impairments [8, 9]. The goal of a mutli-sensory environment is to actively stimulate the individual's senses (vision, hearing, touch, smell and taste) with limited need of higher cognitive processing [10]. Further, early stimulation is also known to be a useful and necessary treatment aimed at developing as much as possible the social psychophysical potential of any person at high environmental and/or biological risk [11]. Multi-sensory stimulation for these individuals provides enjoyment, facilitates relaxation and may suppress self-stimulation [12]. As a result, in different therapies we find all kind of tasks related to multisensory stimulation (e.g., Snoelezen rooms, sensorial cards and bids, playing with different temperatures, tastes, smells, etc.).

With this in mind, we designed SINASense, a multi-sensory environmental stimulation system that combines different technologies such as computer vision and tactile cues-specifically vibrations. The system encourages users to execute particular body movements to trigger meaningful stimuli in their immediate environment and offers control of the situation to these users who experience a continual lack of a sense of ownership and/or control of their own interactions with environment in most of their daily activities.

In this chapter we describe the system's design and development, together with the evaluation carried out on children who presented with multiple disabilities. The chapter is structured as follows: after summarising the related work, we describe the system and its rationale. Then we present the results of an experiment that was conducted to evaluate the system, and discuss the findings and challenges of our experience.

9.2   Related Works

Work related to the SINASense project includes interactive systems for users with disabilities that offer control on the environmental stimulation and vibrotactile interfaces, specifically those that motivate users' actions.

9.2.1   Interactive Systems for Controlling Environmental Stimulation

In reviewing the literature, we found different design approaches of interactive systems aiming to offer interactive experiences and control over environmental multi-sensory stimulation to users with multiple disabilities. There are works that use single or multiple switches to trigger diverse responses using vocalization-detecting devices [13, 14], mechanical switches [15] or hand-tapping responses together with a vibration-detecting device [16] for users who lack specific motor coordination and spatial accuracy. Other standard commercial products have been used such as a mice [17], air-mice [18], the Nintendo Wii Balance Board [19] or the Nintendo Wii remote Controller [20]. Finally, we focus our attention on interactive systems based on computer vision techniques: Lancioni et al. [21] used an optic sensor to detect the deliberate blinking of a user with minimal motor behaviour to trigger motivational outcomes and Mauri et al. [13] detected the user's movement in regions of interest marked by the facilitator of the sessions to activate the feedback.

All these works, albeit presenting positive results, were evaluated with a small group of users. Aside from [13, 15] who tested their systems with seven and eight users, the others studies evaluated the system with one or two users. Further, there is a high variability in the users' characteristics, which complicates the result comparisons.

9.2.2   Vibrotactile Interfaces for Users with Disabilities

Vibrations are a form of cutaneous stimulation, as they produce a distortion on the cutaneous surface [22]. When vibration cues are included in interfaces, these can be used both as feedback–such as the vibration in a mobile phone– and as a supportive function to motivate users' actions [23–25].

Vibrotactile interfaces for disabled users have specially focused on solutions for users with vision impairments. In this case, the purpose is to improve users' interaction with the environment and with other people to enhance navigation abilities [26–28], interact with nearby objects [29], present graphical information non-visually [30], enrich interpersonal communication [31, 32], play with video-games [33, 34] or generate Braille on a mobile phone [35].

Users with hearing impairments can benefit from these interfaces too, as vibrotactile cues can translate sounds (music, speech or environmental noises) into physical vibrations [36, 37].

Although limited prior work exists that describes the use of vibrotactile cues as an interaction style for people with cognitive impairments, Grierson et al. [38] and Knudsen et al. [39] presented vibrotactile interfaces to help users with dementia or with mild cognitive impairments to navigate.

9.3   SINASense

SINASense is an interactive system addressing the needs of users with multiple disabilities that normally depend on others to interact with the environment, but when there is a meaningful stimulus in the environment, they can pay attention to this. We have experience designing accessible interfaces and collaborating with a centre for users with cerebral palsy [40–44], but this project was focused on those users with severe or profound impairments (physical, sensorial and cognitive impairments). The main aims of this system for therapists to work with the users were:

- Increase the intentional movements of their upper body limbs.
- Reduce their isolation.
- Control the interaction with the surrounding world.
- Achieve their active participation in the task.
- Suppress the self-stimulation by offering them external senses stimulation.

The system comprises different modules. First, a simple motion-based interface that uses computer vision to track a coloured band placed on the user's arm to detect its movement and position [45] by means of the Camshift probabilistic algorithm [46]. This module is combined with a set of action/reaction applications controlled by the user's body motion that will trigger meaningful and motivational outcomes–music, images, videos—from the system when the user maintains the arm in particular positions.

When evaluating this first module, we observed that the therapist conducting the session assisted users orally and sometimes physically (e.g., tapping their arm, helping them to carry out the movement) to help them to be aware of the reaction in the environment that their body motion was causing. Due to this observation and taking into account the importance of touch with these users, we extended the system with V-Sense to include vibrotactile cues aiming at avoiding the physical help and reduce the oral support.

V-Sense, is a wearable vibrotactile interface placed on the user's arm that encourages movement [47] by exploiting a perceptual illusion. Therefore, instead of

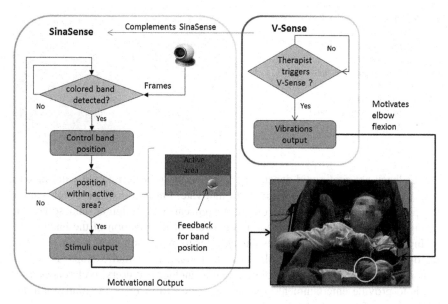

Fig. 9.1 Overview of the system. *Yellow Circle* is the colored band for SinaSense. *Red circle* is V-Sense

helping the user to physically raise their arm, the therapist would trigger the vibrations. An overview of the system is depicted in Fig. 9.1.

9.4 Motion-Based Module and Multi-Sensory Stimulation Applications

The motion-based interface is based on detecting and tracking a coloured band placed on the user's wrist with a standard webcam. This system is "invisible" to the users, as they are not aware of the presence of a computer or any input devices.

The user is located in front of the camera, and the system responds to his or her actions by triggering motivational stimuli into the environment. These actions are arm movements as for example raising the arm or moving the arm horizontally. A standard webcam is used, as tests with a Microsoft Kinect™ to obtain depth information to work other arm movements–towards the screen–were ineffective due to the proximity of the background (wheelchairs or prams) to the user's body.

The multi-sensory stimulation application detects whether the coloured band attached to the user's body is within a particular region, giving multi-sensory feedback when it is. The application enables the therapist to divide the screen in different regions whose colour backgrounds can be selected to adapt it to the user's vision skills, and to select the outcomes to be triggered when the user places the hand in that region: images, videos or music (see Fig. 9.2). The regions that offer

Fig. 9.2 Four regions with different feedback: *green* region shows an image, *pink* region plays a video, *blue* region plays a song and the *red* region has no feedback. The hand icon maps the position of the coloured band on the screen

the feedback are configured depending on the work to be carried out with the user and the kind of intentional movement to motivate. Further, the image representing the position of the user's band on the screen can be configured and serves the therapist (and the user) as feedback to be aware of the position of the hand.

In order to trigger the outcomes, the user has to move their hand within an active region. When the hand is not placed in that area, the stimulus stops. It is very important for the system to be configurable to include the users' preferences, as it needs to provide entertainment, joy and recreation to increase the motivation and engagement of the user.

9.5 Vibrotactile Module

The vibrotactile module is based on the saltation perceptual illusion, where a rapid vibrotactile pulse delivered first to one location and then to another on the skin produces the sensation of a virtual vibration between the two vibrators [48].

The saltation illusion has been used in vibrotactile interfaces for users with no disabilities to communicate motor instructions for snowboarding [23] or to incite users to perform fundamental movements, that is, flexion, extension, abduction, adduction and rotation [49]. In this last case, the vibrotactile patterns help the user to remember an already known set of movements. However, as far as we are aware, there are no studies that have used the saltation illusion to motivate movements in users with impairments [50].

In this stage of the project, due to the users' conditions, therapists wanted to specifically work two kinds of arm/forearm movement. On the one hand, elbow flexion for users with very little movement who found difficulties lifting totally their arms. On the other hand, shoulder flexion for people capable of lifting their arms completely (see Fig. 9.3).

In order to design both vibrotactile interfaces– for the elbow or shoulder flexion– we considered the next factors to achieve intuitive interfaces [51]:

- Pattern of vibration: the vibrotactile pattern should exploit the effects of saltation.

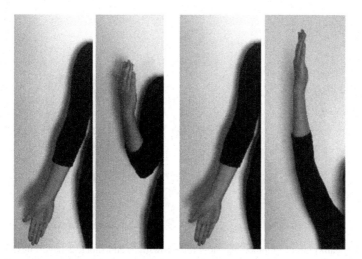

Fig. 9.3 *Left* Elbow flexion. *Right* Shoulder flexion

- Direction of vibration: for example the push/pull or "follow me" metaphors [52].
- Location of vibration: vibrators should be placed on the skin close to the area where the movement is going to occur.

To implement the interfaces, we used an Arduino Mega microcontroller and three vibrators (Precision Microdrives 310–103 Pico Vibe™ 10 mm). A switch was provided to trigger the vibrations whenever the therapist felt it was required.

9.5.1 Elbow Flexion

The vibrators should be placed on or near the muscle/joint/body part involved in the movement. In this case, to flex and extend the elbow, the vibration motors could be placed on the bicep muscle or across the elbow joint. We follow the location used for elbow flexion/extension described in [49], as it has been tested with successful results (with users with no disabilities). Therefore, three vibrators are set up in line on an armband and placed on the bicep muscle above the elbow joint, as shown in Fig. 9.4 (left).

The vibrotactile stimulations will travel up to simulate the pull metaphor of the forearm. The pattern will consequently pulse the three vibrators (V1-V1-V1-V2-V2-V2-V3-V3-V3), where each vibration cue will be repeated as recommended by McDaniel et al. [49] for improved user perception. The burst will be of a 100 ms, with a 50 ms inter-burst interval, which is considered optimal to elicit saltation and has been successfully evaluated in previous works.

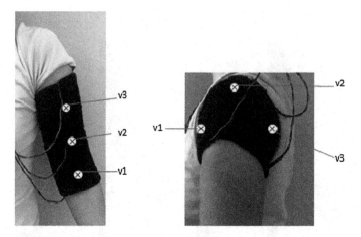

Fig. 9.4 Localization of vibration actuators for elbow (*left*) and shoulder (*right*) flexion

9.5.2 Shoulder Flexion

Three vibration actuators are set up on a shoulder pad embracing the shoulder as shown in Fig. 9.4 (right). The vibrators could be placed on an armband on the biceps but closer to the shoulder than the ones used for the elbow flexion. However, we chose the first approach to completely differentiate it from the elbow flexion.

The pattern of vibration will be: V1-V1-V1-V2-V2-V2-V3-V3-V3 to motivate the arm lifting. Once again the burst duration of 100 ms is used, with a 50 ms inter-burst interval.

9.6 Evaluation

The preliminary evaluation was conducted in a centre for users with cerebral palsy, and took place in two separate stages. In the first stage, the motion-based interface and the multi-sensory stimulation applications were tested. Then, a year later, the vibrotactile module was evaluated.

The system was evaluated with children with cerebral palsy. The children were selected by the therapists based on their conditions. Additionally, together with assistance from the users' parents, they collected motivational material to include in the system.

Two educational psychologists were hired specifically to conduct the sessions. At the beginning, the sessions were also supervised by the user's therapist to help the educational psychologists learn more efficiently about the user's preferences, skills, communication and behaviour.

9.6.1   Evaluation of SINASense

SINASense was evaluated during three months with seven users (two girls, five boys), whose ages ranged from 4 to 12 years. Therapists selected the users based on their characteristics, the impact the system could have in their daily activities and their schedule in the centre. All users have a unique profile regarding cognitive, physical and sensorial skills (none had hearing impairments). The users have severe to profound intellectual disabilities accompanied by impairment of sensation, communication, perception, behaviour, and/or by a seizure disorder. However, they show interest (to a greater or lesser extent) when stimuli are produced in the environment. They all use wheelchairs or prams. Therapists are still assessing the learning and communication skills of each child and their levels of attention and their levels of understanding are different.

9.6.1.1   Procedure

The evaluation took place over the course of three months. Each session had a duration of 15 min and was carried out in a private room set aside in the centre. The room had a large TV connected to a computer and the user was sitting in front of the TV (and a webcam). A minimum of nine sessions per user and a maximum of 23 were performed over the course of the three months. Due to physical decay, one user completed only nine of the sessions therefore, he did not participate in the evaluation.

Sessions were video recorded for posterior analysis and notes of important events were taken during the sessions, such as "the intentional movements are clear", "the user laughs" or "the user is very agitated".

Depending on the user, the lights were dimmed to allow him or her to concentrate and focus on the stimuli provided by the system. Therapists selected the most functional arm/hand for each user to train with the aim of being able to produce some impact with the use of this arm/hand in their daily lives (see Hand/App column in Table 9.1). A pink band bracelet was placed on the user's wrist in order for them to perform the movements.

In this case, the configuration of the screen was divided into two horizontal/vertical regions (see Fig. 9.5). The region that required more physical effort from the user was the one that gave the motivational feedback in form of audio (see Hand/App column in Table 9.1). This audio was configured to be pleasing and meaningful to the user.

9.6.1.2   Results and Discussion

At the beginning of the session, users were not really aware that their arm movement was the one causing the feedback. The educational psychologist had to work

Table 9.1 Results. In Hand/App column: LH: left hand, RH: right hand, HAM: horizontal arm motion, VAM: vertical arm motion. Frequency and Duration columns: darker colours mean better results

User	Hand / App	Frequency of intentional movement	Duration
1	RH, VAM	Very low increase	Very high increase
2	LH, VAM	High increase	Low increase
3	LH, VAM	Low increase	High increase
4	RH, HAM	High increase	High increase
5	RH, HAM	Low increase	Low increase
6	RH, VAM	Very low increase	Very high increase
7	RH, VAM	Very low increase	Very high increase

Fig. 9.5 System's configuration for the evaluation. The ball is the feedback of the user's hand. *Left* the zone further to the functional hand is the active region. *Right* The *blue* zone is the active region

with them by encouraging them orally and in some occasions touching or tapping their arm or even helping them to raise or move it.

However, the continued work with the system helped users to build the relationship between their actions and the reactions. Users seemed to enjoy the stimuli (audio) as several of them smiled when listening to the music, made noise and one even started laughing and tapping the table when he heard his mother's voice.

After three months of work, the educational psychologist reported an increase in the intentional movement of all users and she highlighted the increase in the duration of maintaining the arm in a specific region (see Table 9.1). The results in Table 9.1 are qualitative and based on the observation of the educational psychologist. It is important to highlight that usually, since users increase the duration of maintaining the arm in a desired position, they have a low increase in the number of times they move it (users 1, 3 and 6).

From this evaluation we also obtained important insights that should be considered in the development of an interface based on vision to detect the movement of a user with multiple disabilities and the multi-sensory applications.

First, the positive feedback has to be very clear, motivational and especially promptly. The feedback has to help the user to be aware of the relationship between the action (of the body) and the reaction (in the environment), but also provide him or her with entertainment and recreation to increase the engagement with the stimuli [41].

The tracking of the coloured band has to be very robust to not confuse the user with feedback not triggered by his or her movement. In this particular study, difficulties appeared when the user's dressing had similar colours to the band, which we solved by using other clothing. The colour of the band to track could be configurable in the system to offer more flexibility.

Finally, the system has to be highly-configurable [53], setting it up has to be fast and profiles have to be able to be saved. In this way, therapists can focus on the user and not on the system. Settings have to include background colour to adapt to the user's vision skills, number of regions and the selection of the image that will offer the feedback of the hand's position. This feedback will help the therapist to monitor the user's hand, but also to place correctly both the webcam and user depending on the arm movement range, the user's height and the proximity of the user to the TV, sometimes limited by the kind of wheelchair or pram the user has. Outcomes must also be configurable, to adapt the system to the user: e.g. users with hearing impairments will be more attracted to visual material and users with vision difficulties will pay attention to audio material.

In the future, other feedback could also be included by switching on/off any device (e.g. a vibrator mat, lights, a radio, an electric mirror ball) which works in binary mode and can be connected to a radiofrequency (RF) remote plug or to the USB connector.

9.6.2 Evaluation of V-Sense

The initial tests with V-Sense were to evaluate the interface to motivate the elbow flexion. In this case study, as we were working with children, their limbs are short and with two vibrators it was enough to cover the region from the elbow to the shoulder considering the distance needed to exploit the saltation illusion.

We worked with five children whose ages ranged from six to 14 (two girls, three boys). Four of the five users had participated in the SINASense evaluation. The selection of the users was due to their characteristics, the impact the system could have in their daily activities and their schedule in the centre. Once again, the users had unique and similar limitations as the ones mentioned in SINASense's evaluation. Regarding communication, two of them have gestures for YES and NO, one is learning them and for the other two there is still no way for them to answer simple questions. With respect to participant levels of attention and levels of understanding: three of them seem to understand the instructions on how the system works and for the other two it is difficult to assess.

9.6.2.1 Procedure

We conducted sessions during two periods: three weeks before summer holidays and four weeks after. The sessions had a duration of ten minutes and were carried out in the same private room as when testing SINASense. Users did between 6 and 7 sessions (one session per week).

Once again, therapists selected the most functional arm/hand for each user. The left arm was selected for all users, but with several children, therapists are still not sure which arm is more suitable. When testing SINASense, it may have been the case that they were using the right arm, and in this test they were testing the left one.

Before placing the vibrotactile interface or the coloured band, the educational psychologist instructed the user the movement to perform to trigger the outcomes and help him or her to perform it several times. Then, the coloured band and the V-Sense prototype were attached to the arm. Finally, V-Sense was activated several times to show the user the vibrations and what action was expected from them.

During the first three weeks there were no restrictions when triggering the vibrations or helping physically the user. Then, initially it was decided that during the first two minutes, the therapist could physically help the user and trigger the vibrations. But for the remaining eight minutes, the physical support would be replaced totally by the vibrations (and maximum two vibrations per minute for users not to get used to them).

Once again, sessions were video recorded and notes of important events were taken during the session, related to the interface and the system in general. Events were usually related to the behaviour or state of the user: e.g. "the user falls asleep", "the user does not respond with a physical arm movement, but smiles", "the user is tired/agitated" or "the user does not change the facial expression when V-Sense is triggered".

9.6.2.2 Results and Discussion

We achieved mixed results from the users. First of all, it is important to note that all sessions varied greatly depending on the user's behavioural state on that particular day. Until this point the educational psychologist has always had to motivate and encourage the user orally.

User 1 did not express with any sign whether he enjoyed the music. He usually raised his arm when the educational psychologist encouraged him, although, when vibrations were triggered he did not respond immediately raising his arm, but a few seconds later he raised it (not always). Therefore, it is difficult to confirm the connection between both events. His facial expression did not change much when vibrations were activated. And frequently he felt asleep in the evaluation sessions (this also happened in the classroom when working on any other activity).

We believed that user 2 did not have sensibility on the working arm, as his facial expression did not change at all when vibrations were triggered. However, the tutor suggested that maybe he was too used to vibrations as they already work with vibrations in the classroom (e.g. vibrating mats, vibrating objects). He raised his arm due to the therapist oral encouragement and enjoyed the audio-visual stimuli.

There were sessions with User 3 where he actively participated by raising his arm when the therapist motivated him orally. However, in other sessions he appeared to be in no mood to cooperate generally or to respond to the stimuli. However, he did not respond to the vibration in any way, not by raising his arm, nor by changing his facial expression.

User 4 understood the instructions (and she communicated with us when we asked her simple questions) but her response to the vibrations was variable. Some days she seemed not to acknowledge them, but most of the times she smiled when she felt them and few times she raised her arm directly after triggering the vibro-tactile interface. User 4 has more strength in the right arm, but the therapist wanted to strengthen the musculature of the left arm for it to be more functional. On the fifth session, she seemed to want to help herself to raise the left arm with the right arm, so we may assess again which arm to work with. She really enjoyed the stimuli offered by the system.

Finally, user 5 was very participative and engaged with the activity and the stimuli offered when she raised her arm. She laughed, made sounds and seemed to sing along to the songs. The motivational outcomes were enough to engage the user, so although she responded to the vibration, especially at the first activations, we did not trigger the vibrations frequently due to her active participation.

An excerpt of a video captured in one of the sessions using the vibrotactile interface can be seen in Fig. 9.6.

This preliminary study showed that haptic feedback to encourage limb movement can be suitable for some users with multiple disabilities. However, several issues have to be taken into account.

First, the selection of the arm can sometimes be unclear. In some cases, the user will prefer to work with an arm because he or she has less difficulties moving it, but therapists may decide to work with the other one due to health decisions (e.g. stereotypies) or user's skills assessment (especially with the young users).

It is very difficult to assess the use of vibrations with those users who do not express in any manner noticing the vibrations. The arm selected can lack of sensitivity or users may be too used to work with other vibrating devices.

Fig. 9.6 User working with the system: from *left* to *right*: Initial arm position. The user raises her arm due to the vibrations. Final arm position which triggers motivational outcomes

To use effectively the saltation illusion, users need to understand the instructions. With several users participating in this study, we think it would not make a difference to use the saltation illusion or to just activate all vibrators at the same time as their understanding is limited. But in the case where the user understands the vibrating signal, more sophisticated patterns can be developed to motivate different movements.

9.7 Conclusions

In this chapter we described a multi-sensory environmental stimulation system which provides control and enjoyment to users with multiple disabilities. We used a vision based interface to develop a system, which detects the user's body motion and uses it to activate meaningful stimuli in the environment when the body part is in particular positions. The material used in the system has to be motivational to engage users in the activity, increase their active participation and their intentional movements to reduce their isolation, promote their well-being (e.g. relaxation, pleasure) and suppress the stereotypies or self-injuries.

Due to the participants' conditions, we worked with arm movements. In the first evaluation sessions, we observed that the therapist conducting them had to be continuously supporting the user orally and sometimes even physically (e.g. tapping their arm, helping them to carry out the movement). Therefore, we decided to extend the system and include a vibrotactile interface that exploits the saltation perceptual illusion to motivate users to perform the arm movements on their own.

The evaluation with users with multiple disabilities varies in difficulty levels as each case presents unique intellectual, behavioural, physical and sensory conditions. The lack of communication skills hinders our interaction with them. Additionally, the user's state on a particular day results in unique sessions.

In these kinds of evaluations, we face diverse challenges including: therapists usually need to support the activity continuously, the users' learning and progress is very slow, we are not sure that all users understand the instructions, users may have difficulties when performing certain movements (inappropriate muscular strength, bad body posture) and they present involuntary movements which can disturb the main activity.

The preliminary results with SINASense showed that our proposal promotes the users' active participation and engagement with what is happening around them and increases their respond –increasing the number of times users perform the desired movement or by maintaining the arm position. Mixed results were achieved with the vibrotactile interface. Several users understood the actions expected when vibrations were triggered and they responded with an arm movement or acknowledging them by a change in their facial expression. With other users, it is complicated to understand what they feel when vibrations are triggered, as they do not express any sign of noticing them.

However, we are encouraged with the results to continue working in this research line. SINASense is intuitive and easy-to-use, but we realised that for using the vibrotactile interface users should understand the instructions. In this evaluation, we tested one set of connected vibrators, but in the future, other vibrators could be placed on the user to motivate different body movements such as flexions, extensions or even rotations. Testing different vibration intensities and frequencies could be also interesting to observe if these characteristics could influence the perception and the response of the users. Further, another future work line will be to try the interface with users who present a mild cognitive impairment to analyze their interaction.

Acknowledgements We thank T. Manresa, R. Mas, P. Palmer, J.V. Larsen, M.A. Menéndez and G. Moyà for the technical help. We thank also Marga, Magui and especially all the therapists, carers, management and users for their support, effort and time in ASPACE Baleares. The authors thank the anonymous reviewers for their constructive comments. This work was partially supported by TIN12-35427 granted by the Spanish Government, and by the EU funded project CultAR (FP7-ICT-2011-9 601139). C. Manresa-Yee acknowledges the support of the FUEIB agreement 3194 and the mobility grant CAS12/00199, Programa José Castillejo granted by the Ministerio de Educación, Cultura y Deporte, Spanish Government. M. F. Roig-Maimó also acknowledges the support of the FPI Grant BES-2013-064652.

References

1. Soro-Camats E, Rosell C, Basil C (2012) El alumnado con pluridiscapacidad: características, evaluación y necesidades educativas. In: Soro-Camats E, Basil C, Rosell C (eds) Pluridiscapacidad y context. Universitat de Barcelona, Interv, pp 5–32
2. Brown D, Standen P, Evett L et al (2010) Designing Serious Games for People with Dual Diagnosis: Learning Disabilities and Sensory Impairments. In: Zemliansky P, Wilcox DM (eds) Des. Implement Educ Games Theor Pract, Perspect, pp 1–16
3. Services SES (2010) Section 6: General Learning Disabilities. Signposts, A Resour. pack Teach
4. Seegers M (2001) Special Technological Possibilities for Students with Special Needs
5. Manresa-Yee C, Muntaner J, Sanz C (2012) Educational e-inclusion for students with severe motor difficulties. J Access Des All 2:165–177
6. Hersh M (2014) Evaluation framework for ICT-based learning technologies for disabled people. Comput Educ 78:30–47. doi:10.1016/j.compedu.2014.05.001
7. Vos P, De Cock P, Petry K et al (2010) What makes them feel like they do? Investigating the subjective well-being in people with severe and profound disabilities. Res Dev Disabil 31:1623–1632. doi:10.1016/j.ridd.2010.04.021
8. Cox H, Burns I, Savage S (2004) Multisensory environments for leisure: promoting well-being in nursing home residents with dementia. J Gerontol Nurs 30:37–45
9. Pagliano P (2012) The multisensory handbook: a guide for children and adults with sensory learning disabilities. Routledge
10. Jakob A, Collier L (2013) Multisensory environments (MSEs) in dementia care: the role of design—an interdisciplinary research collaboration between design and health care. In: Proceedings of the 2nd European conference on design 4 heal. Sheffield UK, p 135
11. García-Navarro M, Tacoronte M, Sarduy I et al (2000) Influence of early stimulation in cerebral palsy. Rev Neurol 31:716–719

12. Singh NNN, Lancioni GEG, Winton AASW et al (2004) Effects of Snoezelen room, Activities of Daily Living skills training, and Vocational skills training on aggression and self-injury by adults with mental retardation and mental illness. Res Dev Disabil 25:285–293. doi:10.1016/j.ridd.2003.08.003

13. Mauri C, Solanas A, Granollers T (2012) Nonformal interactive therapeutic multisensory environment for people with cerebral palsy. Int J Hum Comput Interact 28:202–2012

14. Lancioni GE, O'Reilly MF, Oliva D, Coppa MM (2001) A microswitch for vocalization responses to foster environmental control in children with multiple disabilities. J Intellect Disabil Res 45:271–275. doi:10.1046/j.1365-2788.2001.00323.x

15. Saunders M, Questad K, Kedziorski T et al (2001) Unprompted mechanical switch use in individuals with severe multiple disabilities: an evaluation of the effects of body position. J Dev Phys Disabil 13:27–39. doi:10.1023/A:1026505332347

16. Lancioni GE, Singh NN, O'Reilly MF, Oliva D (2002) Using a hand-tap response with a vibration microswitch with students with multiple disabilities. Behav Cogn Psychother 30:237–241

17. Shih C, Shih C, Lin K, Chiang M (2009) Assisting people with multiple disabilities and minimal motor behavior to control environmental stimulation through a mouse wheel. Res Dev Disabil 30:1413–1419. doi:10.1016/j.ridd.2009.07.001

18. Shih C, Chang M, Shih C (2010) A new limb movement detector enabling people with multiple disabilities to control environmental stimulation through limb swing with a gyration air mouse. Res Dev Disabil 31:875–880

19. Shih C-H, Shih C-T, Chu C-L (2010) Assisting people with multiple disabilities actively correct abnormal standing posture with a Nintendo Wii balance board through controlling environmental stimulation. Res Dev Disabil 31:936–942. doi:10.1016/j.ridd.2010.03.004

20. Shih C, Chang M, Shih C (2010) A limb action detector enabling people with multiple disabilities to control environmental stimulation through limb action with a Nintendo Wii Remote Controller. Res Dev Disabil 31:1047–1053

21. Lancioni G, O'Reilly M, Singh N et al (2005) A new microswitch to enable a boy with minimal motor behavior to control environmental stimulation with eye blinks. Behav Interv 20:147–153

22. Dahiya RS, Valle M (2013) Robotic tactile sensing technologies and system

23. Spelmezan D, Jacobs M, Hilgers A, Borchers J (2009) Tactile motion instructions for physical activities. In: Proceedings of the SIGCHI Conference on Human Factors in Computing Systems. ACM, New York, NY, USA, pp 2243–2252

24. Morrison A, Manresa-Yee C, Knoche H (2015) Vibrotactile vest and the humming wall: I like the hand down my spine. In: Proceedings of the interacción 2015, Artic. 3. ACM, New York, NY, USA, Vilanova i la Geltrú, Spain, pp 3:1–3:8

25. Morrison A, Knoche H, Manresa-Yee C (2015) Designing a vibrotactile language for a wearable vest. In: HCII2015, design user experience usability users interact. (LNCS 9187), pp 655–666

26. Ghiani G, Leporini B, Paternò F (2008) Vibrotactile feedback as an orientation aid for blind users of mobile guides. In: Proceedings of 10th International Conference on Human-Computer Interaction with Mobile Devices and Services. ACM, New York, NY, USA, pp 431–434

27. Uchiyama H, Covington MA, Potter WD (2008) Vibrotactile Glove Guidance for Semi-autonomous Wheelchair Operations. In: Proceedings of 46th Annual Southeast Regional Conference on XX. ACM, New York, NY, USA, pp 336–339

28. Flores G, Kurniawan S, Manduchi R, Martinson E (2015) Vibrotactile guidance for wayfinding of blind walkers. IEEE Trans Haptics 8:306–316

29. Bahram S, Chakraborty A, Ravindran S, St. Amant R (2013) Intelligent interaction in accessible applications. In: Biswas P, Duarte C, Langdon P, et al (eds) A multimodal End-2-End approach to accessible Computing SE-5. Springer, London, pp 93–117

30. Giudice NA, Palani HP, Brenner E, Kramer KM (2012) Learning non-visual graphical information using a touch-based vibro-audio interface. In: Proceedings of the 14th

international ACM SIGACCESS conference on computers and accessibility. ACM, New York, NY, USA, pp 103–110

31. Krishna S, Bala S, McDaniel T, et al (2010) VibroGlove: an assistive technology aid for conveying facial expressions. In: CHI'10 extended abstracts on human factors in computing systems. ACM, New York, NY, USA, pp 3637–3642

32. Réhman S, Liu L, Li H (2008) Vibrotactile rendering of human emotions on the manifold of facial expressions. J Multimed 3:18–25

33. Yuan B, Folmer E (2008) Blind Hero: Enabling Guitar Hero for the Visually Impaired. In: Proceedings of the 10th international ACM SIGACCESS conference on computers and accessibility. ACM, New York, NY, USA, pp 169–176

34. Morelli T, Foley J, Folmer E (2010) Vi-bowling: A tactile spatial exergame for individuals with visual impairments. In: Proceedings of the 12th international ACM SIGACCESS conference on computers and accessibility. ACM, New York, NY, USA, pp 179–186

35. Jayant C, Acuario C, Johnson W, et al (2010) V-braille: haptic braille perception using a touch-screen and vibration on mobile phones. In: Proceedings of the 12th international ACM SIGACCESS conference on computers and accessibility ACM, New York, NY, USA, pp 295–296

36. Nanayakkara S, Taylor E, Wyse L, Ong SH (2009) An enhanced musical experience for the deaf: design and evaluation of a music display and a haptic chair. In: Proceedings of the SIGCHI conference on human factors in computing systems. ACM, New York, NY, USA, pp 337–346

37. Yao L, Shi Y, Chi H, et al (2010) Music-touch shoes: vibrotactile interface for hearing impaired dancers. In: Proceedings of the fourth international conference dedicated to research in tangible, embedded, and embodied interaction. ACM, New York, NY, USA, pp 275–276

38. Grierson LEM, Zelek J, Lam I et al (2011) Application of a tactile way-finding device to facilitate navigation in persons with dementia. Assist Technol 23:108–115. doi:10.1080/10400435.2011.567375

39. Knudsen L, Morrison A, Andersen H (2011) Design of vibrotactile navigation displays for elderly with memory disorders

40. Manresa-Yee C, Ponsa P, Varona J, Perales FJ (2010) User experience to improve the usability of a vision-based interface. Interact Comput 22:594–605. doi:10.1016/j.intcom.2010.06.004

41. Manresa-Yee C, Ponsa P, Salinas I et al (2014) Observing the use of an input device for rehabilitation purposes. Behav Inf Technol 33:271–282

42. Manresa-Yee C, Varona J, Perales F, Salinas I (2014) Design recommendations for camera-based head-controlled interfaces that replace the mouse for motion-impaired users. Univers Access Inf Soc 13:471–482. doi:10.1007/s10209-013-0326-z

43. Manresa-Yee C, Mas R (2014) Designing an accessible low-cost interactive multi-touch surface. Univers Access Inf Soc 1–11. doi:10.1007/s10209-014-0396-6

44. Varona J, Manresa-Yee C, Perales FJ (2008) Hands-free vision-based interface for computer accessibility. J Netw Comput Appl 31:357–374. doi:10.1016/j.jnca.2008.03.003

45. Manresa Yee C, Muntaner JJ, Arellano D (2013) A motion-based interface to control environmental stimulation for children with severe to profound disabilities. In: CHI'13 extended abstracts on human factors in computing systems. ACM, New York, NY, USA, NY, USA, pp 7–12

46. Bradski GR (1998) Computer vision face tracking for use in a perceptual user interface. Intel Technol. J. Q2

47. Manresa-Yee C, Morrison A, Larsen J, Varona J (2014) A vibrotactile interface to motivate movement for children with severe to profound disabilities. In: Proceedings of the XV international conference on human computer interactions ACM, New York, NY, USA, NY, USA, pp 10:1–10:4

48. Geldard F, Sherrick C (1972) The cutaneous "rabbit": a perceptual illusion. Science 178 (80):178–179

49. McDaniel T, Villanueva D, Krishna S, Panchanathan S (2010) MOVeMENT: A framework for systematically mapping vibrotactile stimulations to fundamental body movements. In: 2010 IEEE International symposium on haptic audio-v. environment games (HAVE), pp 1–6
50. Manresa-Yee C, Morrison A, Muntaner JJ (2015) First insights with a vibrotactile interface for children with multiple disabilities. In: CHI'15 extended abstracts on human factors in computing systems ACM, New York, NY, USA, pp 905–910
51. McDaniel T, Villanueva D, Krishna S, Panchanathan S (2010) MOVeMENT: a framework for systematically mapping vibrotactile stimulations to fundamental body movements. In: Proceedings of the HAVE 2010. IEEE, pp 1–6
52. Spelmezan D, Hilgers A, Borchers J (2009) A language of tactile motion instructions. In: Proceedings of the 11th international conference on human-computer interaction with mobile devices and services. ACM, New York, NY, USA, pp 29:1–29:5
53. Davis AB, Moore MM, Storey VC (2002) Context-aware communication for severely disabled users. SIGCAPH Comput Phys Handicap 106–111. doi:10.1145/960201.957224

Chapter 10
Interactive Furniture: Bi-directional Interaction with a Vibrotactile Wearable Vest in an Urban Space

Ann Morrison, Jack Leegaard, Cristina Manresa-Yee, Walther Jensen and Hendrik Knoche

Abstract In this study we investigate the experience for participants while wearing a vibrotactile vest that interacts with a vibroacoustic architecture *The Humming Wall*, set in an urban space. This public large scale artefact is built to exchange vibrotactile and physiological interactions with a vibrotactile wearable vest. The heart beats and breath rates of the vest wearers are vibroacoustically displayed at *The Humming Wall*. In addition, participants can swipe and knock on *The Humming Wall* and the vest wearer is effectively swiped and knocked upon. We work with overlapping vibrotactile outputs in order that the wearers experience a flow of sensations similar to a touch gesture. The communication advantaged vibroacoustic and vibrotactile as the primary interaction modalities for both vest wearers as well as for a passing public. The participants found the experience favourable and analysis reveals some patterns on the vest and zones at the wall impact relaxation in the form of calming and feel-good sensations, (even therapeutic) as well as activation and warning on the vest. We contribute to this research field by adding a large scale public object and visibly responsive interactive wall that was positively received as the partner responder for the wearers of a vibrotactile vest set in an urban environment. Participants reported calming, therapeutic, feel good sensations in response to the patterns.

A. Morrison (✉) · J. Leegaard · W. Jensen · H. Knoche
Aalborg University, Aalborg, Denmark
e-mail: morrison@create.aau.dk

J. Leegaard
e-mail: jhl@create.aau.dk

W. Jensen
e-mail: bwsj@create.aau.dk

H. Knoche
e-mail: hk@create.aau.dk

C. Manresa-Yee
University of Balearic Islands, Palma, Spain
e-mail: cristina.manresa@uib.es

© Springer International Publishing AG 2017
A.L. Brooks et al. (eds.), *Recent Advances in Technologies for Inclusive Well-Being*, Intelligent Systems Reference Library 119, DOI 10.1007/978-3-319-49879-9_10

183

Keywords Vibroacoustics · Vibrotactile · Wearable computing · *The Humming Wall* · Wearable vest · Breath rate · Heart rate · Assistive technology: well-being

10.1 Introduction

Multiple soft toys and comforters feature in many people's early childhoods. There are numerous examples of electronically enhanced huggie style jacket versions that support remote tactile interaction for children or adults. These focus on haptic and 'hug-like' sensations to calm the wearer. Mueller et al. developed an interface that required patting a soft koala artifact to send hug sensations to an unobtrusive inflatable vest of the partner [1]. Huggy Pajama (for parent and child) also uses inflatable clothing to send and receive touch and hug interactions via a doll interface [2]. Haans et al. stress the importance of combining haptic with visual feedback for those involved in the interactions, so partners actions are visible to each other, in order to enable reciprocity and a sense of proximity with the other [3], or to provide the illusion of interaction through a shared object [4]. Recent work by Tsetserukou et al. works with wearable humanoid robots to augment emotions [5], and the XOXO system provides a haptic interface for mediated intimacy with kiss and hug interactions [6]. All rely on one-to-one interactions and offer the sense of personal and/or intimacy through the interface and interaction.

In contrast, *The Humming Wall* (see Fig. 10.1) is a public object that provides interaction as a shared artifact and addresses issues raised about the invisibility of the experience of others involved in remote interactions (where for the participants, it is not clear or visible what the experience for the others involved in the interaction is). Additionally, the large shared artifact expands the interaction to being available beyond prior works that are largely designed for those in already established intimate relationships [4]. People can easily see actions performed by others on the wall and it is available for both one-to-one interactions and also for a broader

Fig. 10.1 The Humming Wall with a participant using the Vibrotactile Sentate Vest

audience to interact with the wall and with the interactively enabled party (the vibrotactile vest wearer). Acting as an interactive vibroacoustic environment *The Humming Wall* combines haptic, audio, vibration and feedback displaying the interacting partner-people's actions, exposing reciprocal interactions [3], which in turn trigger and enable immediate responsive interactions by others.

In this research, we trial a comfortable wearable vest that acts to calm and relax and also activate, guide and warn the participants. Further to this, many participants reported feeling stroked and emotionally supported. *The Humming Wall* also provided a safe place for passers-by and vest wearers alike. We witnessed children climbing happily all over it, and adults and children alike tapping and swiping different beats and rhythms on the responsive vibroacoustic surface. Many spent time sitting and lying upon it with its vibroacoustic qualities at 42 Hz (within the 40–50 Hz range), known for its therapeutic qualities. While we gather physiological data from the vibrotactile vest wearer capturing heartbeat and breathrate, the dynamic analysis of this data to gauge stress levels is still in early stages of development. The overall long term design goal is to work with a real time reading of stress levels to develop assistive adaptive wearables and responsive urban environments to aide relaxation and improve health and general well-being of individuals and the natural inhabitants of these spaces.

The chapter is structured as follows: after summarising the related work, we describe the site (evolution, natural use, aesthetics) that we addressed in the design of *The Humming Wall* as an interactive urban environment. Next, we explain in detail the rationale for design choices for both the vest and the wall systems and their bi-directional interaction. We then present the field trials, the participants, the data gathered and the results from the multiple evaluation methods. We summarise the findings and enlarge these in a discussion that leads to a statement on the contribution of the work and future directions.

10.2   Related Work

Smart wearables and fitness gadgets are currently part of an exploding fitness market, rife with small wearable technologies that track and respond to your physiological data. We find, for example, smart bras, shirts, connected toothbrushes, glasses, gloves, smartwatches, intelligent jewelry, shoes, vests, pendants, caps, headbands and other responsive clothing [7]. Many use the mobile phone as a proxy communication and visualization manager. We note the successful uptake of devices and products, made possible by leaps in sensor design and refinement and assimilation with new generation integrated circuits and development boards. For example, Infravitals smart watch [8] monitors continuous blood glucose and pulse, blood pressure, ECG—heart health, EEG—nervous system measuring, kidney function working with alerts and triggers and integration with peripheral sensors. In addition, we see advances in textile cabling, conductive elastics, textile connectors and sensors with companies such as Ohmatex [9]—a smart textiles company known for developing a smart sock for measuring astronaut muscle activity for the European Space Agency—integrating smart fabrics and textiles with flexible and stretchable electronics.

Core to these advances is the bringing together of interdisciplinary teams from diverse technology fields of software design, electronic design and manufacture and healthcare. MyWear customised wearing products is developing assistive customised, eco-friendly, portable, low-cost, smart work and sportswear that is assistive for elderly, obese, diabetics and disabled people [10]. Wearable technologies adding stress level readings and integrated fall detection and alarm function which automatically alerts designated contacts that the user is in need of assistance are also reaching the market [11] expanding the arena beyond fitness to include assistive technologies. To date, research on vibrotactile interfaces for disabled users has largely focused on visually-impaired users to assist them in interacting with others and the environment: to enable/enhance navigation [12], to interact with nearby objects [13], to present graphical information non-visually [14], to enrich interpersonal communication [15], to play videogames [16] or to teach choreographed dance [17].

Other research that addresses problems with hearing impairments uses vibrotactile interfaces to translate sounds (music, speech or environmental noises) into physical vibrations [18, 19]. The Emoti-Chair [20] translated sounds, including music, speech, and environmental noises, into physical vibrations using eight discrete audio-tactile channels via a model human cochlea (MHC) acting as a sensory substitution technique used to convert sound into vibrations. The Haptic Chair was designed to enrich the experience of music for the deaf by enhancing sensory input of information via the vibrotactile channel [18] providing sensory input of vibrations via touch. The Music-touch Shoes, embedded with vibrotactile interface, were used by dancers with hearing impairments to feel the rhythm and tempo of music through variable stimulation signals in the soles of the shoes [19]. These shoes counted with vibrotactile points that gave users vibrotactile stimulation according to the rhythm and tempo of the chosen music.

Vibrotactile information can support functionality and motivate users' activities [21]. Spelmezan et al. [22] developed ten patterns as a form of tactile language to assist athletes' snowboarding motor skills. Rosenthal et al. use a Vibrotactile Belt to teach choreographed dance moves through vibrotactile cues [17]. These and much other research motivates and informs the work we report here. A priority though, is to calm and relax people—to work with and adjust their stress levels—as well as to activate, guide and warn the participants while mobile in the environment.

Emotional states and locations are pertinent for developing a vibrotactile language for the body. Nummenmaa et al. had participants draw and map where they thought the emotions anger, fear, joy, happiness, sadness etc. dwelled in the body [23]. Arafsha et al. asked participants to draw where love, joy, surprise, anger, sorrow, and fear lay, and to include vibration, rhythm and warmth as haptic feedback enhancers [24]. These build on soft toys and electronic huggies enhancing feel good states.

In addition, vibroacoustics—music combined with low frequency sound vibration (30–120 Hz) is an expanding research and therapy field working with curative health and comfort effects of vibroacoustic ranges to promote relaxation. Evidence supports 40–50 Hz (the range we use) having a positive influence on pain with 40 Hz recognised as having an essential role in regulating cognitive processes, in particular selective attention Rüütel [25]. However, in most cases a treatment session lasts 20–30 min and requires a series of treatments. The work we address here aims to communicate solely by means of vibroacoustic (at the wall) to vibrotactile (on the vest) sensations as a preliminary investigation into how effective this more singular media modality communication can be—as well as to effect movement—to motivate and activate.

Multiple factors need consideration when designing for the body. Form factor considerations include robustness, long term wear, comfort, accuracy, mobility, a wide array of body shapes and sizes etc. Additionally, the placement, location, spacing, number and types of tactors and sensors; and the arrangement, frequency, tempo, sequencing, overlap, duration, as well as the connection methods used all have an impact on the design of wearable vibrotactile clothing-computing [26, 27]. Karuei et al. found that the wrists and spine are the consistently most sensitive sites, with movement an impacting factor in decreasing detection rate [27]. Morrison et al. found that conducting other activities reduced detection performance of vibrotactile stimuli [28]. Much vibrotactile research works with ensuring each tactor can be individually felt and ensuring distance between vibrators and pauses between vibrations are correct to ensure this (see for example [29–31]), in particular to elicit the effect of movement.

In order to study the paired vibrotactile applications in realistic settings and situations that allow for mobility and responsiveness, we staged the interactive scenarios in an urban environment (Utzon Park). We designed an environment to invite activities for passers-by that were open-ended, meaning there is no 'correct set order in which events must take place. The work is designed to enable a non-narrative, non-predetermined experience where people can set their own pace and essentially 'find their own meaning'. The interaction was embodied and playful with *The Humming Wall* calling for bodily gestures—swiping, knocking, sitting, lying, running, touching type forms of embodied play—in order to interact with the work and others at the site. The interactive furniture invited speculative play, where people—members of the same group and also often strangers—would interact with each other to determine what to do, where to do what and eventually how the wall as a system worked. In many cases people often experimented and feedback their findings to each other, informing new comers (or members of their party) while interacting with the responsive urban architecture-furniture [32–34].

10.3 Addressing the Site

The conversion of Aalborg Harbour front from an industrial port that supported multiple industries (e.g., cement and concrete industries, slaughterhouse, grain and feed, fertilizer factory, shipyard, power station, fishing port etc.) to an inviting community resource, with people-friendly and culture-enabled environments has taken due consideration. The site of Utzon Park (where we installed *The Humming Wall)* stands in the middle harbour between two of Utzon's architectural structures —Utzon Center and student housing—encompassing aesthetics and design principles that needed to be taken into consideration. We designed *The Humming Wall* with use for the natural inhabitants of the area in mind. In addition, we incorporated Utzon architectural concerns, echoed in the organic waving lines of the park paths, by designing a modular, replicating, geometric structure answering to Utzon architectural and aesthetic considerations [35].

10.3.1 Data Collection

In order to better understand the site, address its natural use and the wishes of those invested in development, we conducted interviews and used questionnaires to obtain information on the history of the development and the current use of the site by stakeholders and natural inhabitants.

10.3.2 Stakeholder Interviews

A team from City of Aalborg's Central Harbour working together over a 10 year period found pedestrian and cycle access with heavy roads behind the harbour area meant earlier urban area and central harbour front improvements were still not easy to access from the city (see Fig. 10.2a, b).

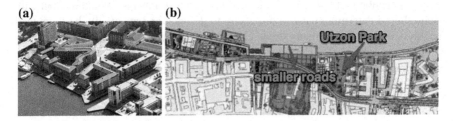

Fig. 10.2 a Earlier heavier development of Western Harbour. **b** Utzon park and surrounding areas with roads pulled back and more park and open areas to the water

(a)   **(b)**

Fig. 10.3 a Harbour front area with common area between sea and land and soft traffic area beside water. **b** Scenic lighting at night highlighting features of harbor front

The central principles that underlie the redesign of the area are:

- Area between sea and land remains as a common area (see Fig. 10.3a)
- Soft traffic pedestrian and bikes on the main harbour front, easy to get from the city to water. The hard traffic double lanes, now single with standing place in middle (see Fig. 10.3a, b)
- View of water throughout the city, even from small streets can see to the water front with few obstructions
- Reminder of the port and its past with industrial style of architecture with rusted artefacts etc.
- Added lighting scenic, so certain parts spotted/highlighted (see Fig. 10.3b)

For Utzon Park in the warmer months, striped deckchairs are supplied for locals and passers by to sit on the grass and enjoy the fjord view, the park space and the warmer climate. Public event bookings include sports, boating, tallships, education and promotion events.

10.3.3   Natural Inhabitant Interviews

The findings from interviews with 91 natural inhabitants and/or regular visitors to central Aalborg show the main attendance is by people passing through the area, in the warmer months only, occasionally and/or for attending events (see Fig. 10.4a). The age range of participants ranges from less than 12, to more than 75 with 58 % between 21 and 27 years of age (see Fig. 10.4b). The area is designed as 'space' in the city and that people are walking through and along the harbour front instead of in the city streets fits the vision well. We looked at ways to assist the Kommune vision to make the park a place where people also come to spend time.

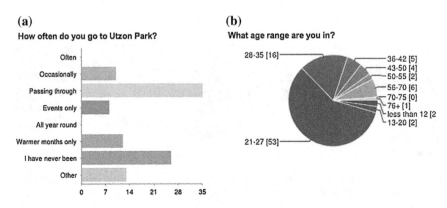

Fig. 10.4 **a** Attendance at Utzon Park with the majority as passing through. **b** Demographics: Age range

10.3.4 Addressing the Aesthetics of Utzon and Utzon Park

Jørn Utzon, the architect perhaps best known for designing the Sydney Opera House, works with inherent modular geometric ordered properties as joint aesthetic and function considerations [35]. We were looking to provide an open-ended, free-play experience for passersby and to reach this goal, we introduced fundamental considerations from Utzon's aesthetic as well as interrogating the natural use of the site. We addressed Utzon's design considerations such as light, colour, geometry, structural expressions and additive geometry with references to nature (and natural form) and harmony and unity. A criterion was to integrate nature into architecture—that is to design a building (in our case interactive furniture) which fits comfortably into its surrounding. Utzon referred to important elements of his work as *Additive Architecture*, [36] comparing this approach to the growth patterns of nature, requiring the design be based on the use of a limited number of repeating elements [35].

In addressing the organic shape of the site with the rounded 'wriggling' pathways through (see Fig. 10.5a), we have echoed the curving shapes in the shape and lines of *The Humming Wall*. We designed *The Humming Wall* to include harmonious modular repeating segments with internal voids large enough for arms to access electronics and cables to enable interactivity, serving both the aesthetics and function well.

The rounded seating and sides protected participants from wind at the fjord and allowed enough room for groups of people to sit and lie down together at the wall and complemented the shapes in the existing seating at the site (see Fig. 10.5b). The curving shapes and the give in the epoxy resin surface made the structure smooth and soft to touch. The bright red colour gave contrast to the otherwise minimal white and grey buildings and green grass colours at the site (see Fig. 10.6).

(a)   (b)

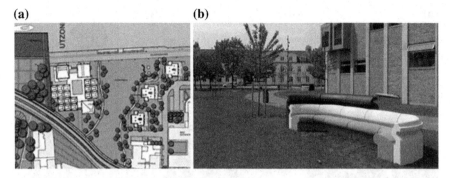

Fig. 10.5 a Birds eye view of Utzon Park shows the organic pathway. **b** Utzon designed seating in the park

Fig. 10.6 The general passing public playing with *The Humming Wall*: set in Utzon Park beside Utzon Center

10.4   The Systems: The Vest, the Patterns, *The Humming Wall*

In this section we describe the technical details of the vest, the vibrotactile patterns and *The Humming Wall*. We also discuss our approach to enabling a human-centered interaction between the systems.

10.4.1   The Vibrotactile Wearable Sensate Vest

The vest has been developed in a user-centered design approach leading to a series of iterative design cycles. The vest is made of two layers. The inner layer comprises an adjustable harness and the outer a padded stretchable vest. The inner adjustable harness, is a 'one-size-fits all' that has capacity for 32 actuators, easily moveable in order to ensure they are correctly placed against the required points for each different shaped body (see Fig. 10.7a, b). The lower harness fits around the legs, ensuring that the harness remains pulled down and holds the actuators in place as

(a)   **(b)**   **(c)**

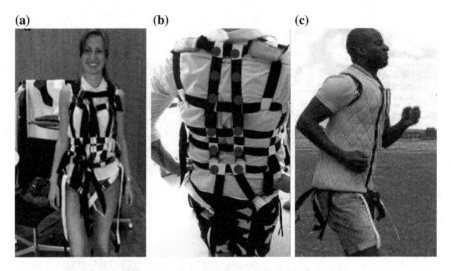

Fig. 10.7 **a** Harness front. **b** Harness back. *Blue dots* indicate actuator placement. **c** Outer shell. Participant raising heart rate to see, hear and feel at the walls' 1st zone

participants move about. The outer shell is a snuggly-fitting, padded layered stretchable vest, in three sizes. It is designed to keep the actuators tightly placed against the bowed-curved areas of the body (such as the lower back and chest) to ensure the vibrations are evenly felt all over (Fig. 10.7c).

10.4.2 The Vest as a System (The Vest Design Considerations)

Two custom-made electronic boards drive the wearable system. One board controls communication, the other powers and controls the actuators. The Zephyr Bioharness 3 [37] is integrated into the system and worn against the skin, to read the participant's physiological states: R-R interval (heart rate), breath rate and activity, which we used for analysis.

The positions of the actuators and combinations of movements for instigating these states were designed in close collaboration with a Kinesiologist-Neurophysiologist. Informed by this collaboration and prior research, we placed the actuators with a total of 29 actuators. Activating these actuators in correct patterns allowed the potential to give the participant different haptic synesthesia sensations such as sense of movement, shiver and states of activation and/or calming as well as navigational cues.

10.4.2.1 Harness Design

The harness is made of strong thick elastic strips woven into a grid-like structure that can easily be aligned to the correct positions on the body. Each connection between a horizontal and vertical elastic strip is able to be moved in all four directions, allowing for more accurate placement of the connected actuator. Some actuators are sewn into small velcro pockets which can slide along or up and down on these elastic strips, while others are sewn into the intersections. The actuators are moveable to a certain degree in their respective pockets and the whole woven grid can be made smaller or larger, while the small pockets still keep the actuators close to the body. The interconnecting pockets also stabilize the overall structure of the harness.

The communication and actuator boards are fastened via velcro pockets specifically designed to allow wire connectors to be accessed at all times. These are attached to the elastic strips which makes it possible to place these anywhere on the grid structure, with enough space between the adjacent interconnector pockets and distributing the weight evenly on the body. To make sure the harness is adjusted to the body, if the wearer bends over or twists their upper body, we developed a leg harness system resembling a lightweight climbing gear harness. This system ensures that the lower part of the harness will return back to the original position after a bend or twist of the upper body. The harness was designed so the left front side of the harness is pulled apart (and connected again) by a mixture of plastic clip fasteners and square metal D rings. All clips/connections had excess lengths of bands for lengthening or shortening according to the body size.

10.4.2.2 The Vest: Technical Description

We chose Bluetooth as the communication platform as it is easy to setup, control and has a low power mode. The communication board, is the central communication point in the system. It receives commands over Bluetooth from a custom-made Android application, relays physiological data from the BioHarness and controls all actuators via transmitting I2C commands to the motor board. We developed a custom-made Android application for easy remote actuator control, to enable timed activation in different lengths and intensities.

The custom-made communication board is based on the Arduino Mega (Atmega2560P) chip, and has two Roving networks RN42 Bluetooth modules integrated with built-in power control connected via UART. One Bluetooth module is a slave device and handles all incoming communication with the Android application, while the other acts as a master device communicating with the BioHarness. The advantage of two Bluetooth modules is to reduce software complexity of communication at the cost of higher power usage. The chosen actuator, is a generic eccentric rotating-mass (ERM) pancake shaped Pico Vibe 312-101 vibration motor from Precision Microdrive [38]. The motors have a small footprint, light weight, easy to control and inexpensive compared to alternative solutions. For stability and robustness, the actuator is glued to a small custom-designed PCB. The frequencies for the actuator are from 110 Hz up to 250 Hz at maximum amplitude. A common power source is used for both drivers to insure a fixed common maximum amplitude on all actuators. The motor board is powered by a LiPo battery, running 3.7 V capable of running 10 actuators continuously at maximum power.

10.4.2.3 Software

The firmware utilizes an open-ended communication scheme and supports both dynamic real-time creation and execution of patterns. The Bluetooth devices use secure simple pairing (SSP) and serial port profile (SPP) for communication. We developed an event based control system for the actuators to generate tactile patterns with differing duration, overlap, amplitude and actuator activations per phase. A pattern consists of one or more phases. Each phase has individual duration, overlap and amplitude.

10.4.2.4 Android Application

To facilitate ease of use in multiple Android applications, we developed an Android API specifically for interfacing with the haptic vest. The API supports transmission of command packets to start and stop hard coded patterns as well as an application for designing patterns. With this application we can easily add and remove phases, set the number of active actuators in each phase and adjust the duration, overlap and amplitude. When a pattern is ready to be used, it can be exported from Android and added to the firmware of the vest after which it can be played back using the normal command packets in the API.

10.4.3 The Vibrotactile Patterns

We examined prior work on responsive audio-vibrotactile environments. We needed the vibrotactile vest to have a larger environment to be in response-interaction with. MEDIATE environments integrated sonic, visual, and tactile interfaces in both sensing and feedback for education with autistic children and provided strong design guidelines, customised to the needs of each specific user for learning purposes [39]. In the patterns we developed, we deliberately overlap transitions between 0 and 400 ms and vary duration to enable a smooth flowing sensation, rather than short bursts as found in saltation effects [29]. Emulating natural touch was the effect we were working towards. We developed a range of vibrotactile patterns in the vest to induce five sensation types: Calming; FeelGood; Activation; Navigation such as full body turning; and Warning prompts, to slow down or stop. In addition, the vibroacoustic *Humming Wall* reacts to human touch—knock, swipe etc.—and conveys these both at the wall and to the vest and responds to and displays the heartbeat and breath rate of the vest wearer.

10.4.3.1   The Sensate Vest Design

We worked on the positioning and necessary combinations of actuators and patterns onto responsive points and zones of the body with an experienced Kinesiologist-Neurophysiologist. We aimed to emulate the hands on work a Kinesiologist does in activating or calming inactive or overactive sequential points of the body. Kinesiology works with an understanding of the body as a set of rhizome like structures acting as communication tracks between the larger organs. The touch can be calming but may also be probing or jabbing running in sequences of jiggle, pause, jiggle (similar to saltation effect) but largely running in longer duration, working on one area at a time and with longer pauses before moving to another area-system.

The patterns fall into five categories (coloured on Fig. 10.8):

(1) Blue: Calming and/or Feel Good (Back and Waist);
(2) Feel Good (Dark Blue, Waist);
(3) Activating (Red, Front and Back);
(4) Navigation (Green, whole body) and
(5) Warning (Yellow, mid front).

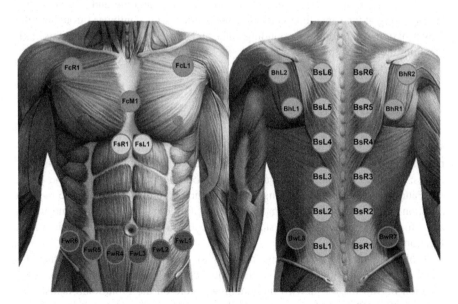

Fig. 10.8 Actuator locations and naming scheme: 10 actuators around the waist, 6 down each side of vertebra (including 2 from waist), 2 on each lower shoulder, 2 on the solar plexus, 1right below each collarbone and 1 in the middle of the chest

The actuators operate in overlapping patterns in order to provide various haptic synesthesia sensations such as sense of movement, a body shiver, states of activation and/or calming as well as providing navigational whole body-turning cues. For example, actions such as: (1) calming-comforting, stroking the back to calm or comfort a person (see Fig. 10.8); (2) guidance-navigation, placing hands on nape of back and shoulder and turning at the hip as if to support and guide an elder; and/or (3) warning, stopping the body with pulses to the solar plexus—acting as if a warning. Patterns on the front of the body include Shiver and Tarzan sequences with for example, pulses in the top two actuators under the collarbones in rapid succession, then longer duration (appearing as stronger when felt over time) pulses to the midpoint actuator emulating 'The Tarzan Effect'—Tarzan thumps his chest to raise his energy levels before action. Shiver includes shorter versions of these same patterns as well as patterns up and down the back.

In the patterns we use overlapping transitions between 0 to 400 ms (with most running between 200 and 350 ms) and vary duration from 175 to 500 ms to enable a smooth flowing sensation between pulses, rather than shorter saltation bursts to elicit muscle movement [29]. So, for example, for Calming Sensation located on the Back (on the muscles on both sides of the spine), the overall length of pattern is 1225 ms, with an individual duration for each actuator of 350 ms with an overlap between the actuator sensations of 175 ms, and an amplitude of 3500 (ampl). This means BackSideLeft6 vibrator is still operating when BackSideLeft5 actuator kicks in (see Fig. 10.8). As the pulsations move down both sides of the spine, this is repeated at the same time on the right side of the spine at BackSideRight6 (BsR6) and BsR65 with an overlap where all 4 vibrators (the outgoing and the incoming) are operating at the same time in the 175 ms overlap. This continues throughout the Down pattern cycle on corresponding right and left sides actuator locations until reaching the end of the pattern at BackSideLeft1 (& BackSideRight1).

10.4.4   Patterns System Details

We describe the qualities and placement on the body for the types of sensate patterns in Table 10.1. In Table 10.2, we present a selection of each of the categories: calming, activating, navigation, feel good and warning patterns and detail the Pattern structure, the total length, the individual activation length, the overlap duration of all consecutive activations, the amplitude of the activation and which activators are activated at each step. See Fig. 10.8, for naming conventions and locations.

Table 10.2 displays the detail and logic of the different sensations that are enabled with each pattern category and sequence. For example, for Stop we see a duration of 900 ms for the two individual actuators on the FrontSolarPlexusRight and FrontSolarPlexusLeft, activating at the same time. This is a warning pattern, designed to stop the person (or give them the sense of being pushed back) and the pattern with the longest ms duration. In contrast, the WaistLeftToRight pattern runs

Table 10.1 Patterns: the category of sensation, the type of pattern and body placement

Sensation	The pattern	Placement
Calming	**Down** A pattern going down the back, in a succession of overlays that will act as if water running down the spine to calm the person	Back (spine)
	Up A pattern going up the back. For some this worked as an activation, for others as a calming feeling	Back (spine)
Feel Good	**Waist Left to right and Right to left** A pattern starting at one side of the waist and going to the other. Produces a feeling of happiness or butterflies in the stomach.	Waist (around the waist)
	Mid front to back A pattern starting in the middle front of the waist and going around both sides at the same time to end at the middle back	Waist (around the waist)
Activating	**Shiver** A pattern going first up the front and then down the back, meant to give a shivering experience	Front (top) and Back (spine)
	Tarzan A pattern providing long and strong series of vibrations on the upper part of the chest and shoulder	Front (top)
Navigation	**Turn left and right** A pattern with multiple activation zones, emulating touching and steering a person on the middle of the back, the shoulder and the opposite front of waist to assist turning the person in either a right or left direction	Back (side, top) and Waist
Warning/information	**Stop** A fast pattern placed on two actuators on the solar plexus (under rib cage) to act as a warning to slow down or stop	Front (mid top)

Table 10.2 A selection of patterns. Pattern structure, total length (len), individual activation length (act.), overlap of consecutive activations (overl.), amplitude of activation (ampl.), and activators activated per step (cf. Fig. 10.8)

Actions	Len.	Act.	Ovel.	Ampl.	Pattern sequences run left to right
Down	1225	350	175	3500s	(BsL6, BsR6), (BsL5, BsR5), (BsL4, BsR4), (BsL3, BsR3), (BsL2, BsR2), (BsL1, BsR1)
Tarzan	2900	500	−100	4000	2x(FcR, FcL), 3xFcM
Turn Left	2500	500	0	4000	Bw7, (BhR1, BhR2), (Fw3, Fw2), (Fw2, Fw1), (Fw1, Bw8)
Waist left to right	2100	600	300	4000	Fw1, (Fw2, Bw8), (BsL1, Fw3), (BsR1, Fw4), (Fw5, Bw7), Fw6
Stop	900	900	0	4000	(FspR, FspL)

at a duration of 600 ms, with overlaps between the actuators of 300 ms (running around the waist), meaning at 300 ms into e.g., FrontWaist1, the two next actuators FrontWaist2 and BackWaist8 kick in, and 300 ms later FrontWaistI drops out but BackSideLeft1 and FrontWaist3 kick in. At no time (aside from the beginning and end actuators) are the actuators running solo. The sequence acts as a generalized sensation moving slowly around the waist. The whole cycle lasts for 2100 ms.

10.4.5   The Humming Wall

The Humming Wall contains seating, playing and lying on areas. The overall structure of the wall is curved, which also provides cover from wind blowing at the harbour front. We designed the 12 m wide, 3.5 m deep and 2.7 m high wall with commissioned foam structures. The wall is made from 11 segments of EPS150 foam blocks. Each segment is coated with PolyUrea for strength and durability. Part of the wall is fitted with warm-and cold-white lights, used for ambient and indicative visual effects. An array of tactile transducers, placed in the hollowed out sections of the wall generate different haptic sensations, audio cues and haptic feedback.

Each segment has its own 80 W Bass-Shaker mounted with a frequency range suitable for haptic and audio feedback and amplified so that each segment functions as a vibrating speaker. *The Humming Wall* has its own baseline vibration and audible hum (42 Hz, modulated by ¼ Hz). The breathing sound is relayed at a frequency combination of 40 and 45 Hz. The heartbeat sound effects (70 ms duration) operate at a base frequency of circa 70 Hz. A combination of piezo- and capacitive electronics to read and analyse touches, enables the differentiation of knocking (tapping) and swiping gestures.

The overall vision for *The Humming Wall* is a multimodal interactive wall that increases the experiential value in urban areas where citizens can play on the wall as well as relax. All elements needed to be robust enough to support rough handling, and the electronics need to be safeguarded from weather. Feedback in the system needs to be fast enough to not stifle playfulness requiring low latency responses for knocking and swiping. To increase the incentive for more persons engaging in playing with the wall we developed with multi-use in mind, that is the system needs to be able to alter the vibration and/or audio being played back when multiple players were present and interacting with the wall.

In addition, to accommodate the electronic systems the internal voids were designed wide enough to mount the electronics and cables and able to be aligned to adjacent segments so that cables and wires can be pulled through the wall from one end to the other. The entire system is run by an array of systems responsible for the audio, vibration and interactivity modalities.

10.4.5.1   The Complete System

The Humming Wall consists of knock and swipe as interactive technologies, audio and LED output, and to control these systems, we have a main Server, an audio server, one DMX controller and a set of Raspberry PI. All systems controlling the wall are connected using Ethernet in a LAN configuration, with switches and WIFI routers internally in the wall, connected to the external servers. Because the vest uses Bluetooth communication, a proxy mobile device is used as an access point to connect via WLAN to the main server.

Swipe: We devised a simple 'touch'/swiping interface, that would let people interact with the wall other than knocking. The swipe detection is divided into 4 overlapping sub matrices each with the possibility of going either up, down, left or right. When a swipe is determined, the direction is detected and transmitted to the main server. The main server then triggers the sound, light and during trials triggers a pattern on the vest.

Lighting: To draw the attention of passersby, we added 28 DMX pixel lights placed in two lines as a visual effect to one part of the wall. The light renderer uses simple algorithms to create a slow glowing ember effect with random occasional flickering. In addition, we developed a blinking algorithm so that when people were knocking on a particular segment, the lighting for this segment would light up and slowly fade out again.

Audio: We designed a semi-segmented system which uses two "T.C. Electronic Studio Konnekt 48" sound cards, one MACbook pro running MaxMSP as audio server. All segments were mounted with an 80 W Bass-Shaker making the entire segment function as a speaker/vibrator with a frequency range of 15–80 Hz, suitable for haptic feedback and low frequency audio feedback. To attract passersby, *The Humming Wall* had its own baseline vibration and audible hum (42 Hz, modulated by ¼ Hz). During the trials, the audio was coupled with physiological readings in different zones. For example, breathing rate was relayed at a frequency combination of 40 Hz and 45 Hz. The measured heartbeat was played back as a sound effect (70 ms duration) operating at a base frequency of circa 70 Hz. Because the Bio Harness' breath-per-minute is based on a sample interval from the last minute, there were delays from 20 to 60 s for a person holding their breath.

Knock: To detect knocking on a segment, we used a series of piezo-sensors attached inside each segment coupled with a MaxMSP patch running on the audio server, to differentiate and filter knocking on the wall. When a knock is detected, the placement (segment number) is coded and transmitted to the main server. Immediately after receiving the knock, a sound is triggered on the audio server, and a random actuator is activated in the corresponding zone on the vest.

10.4.6 Bi-directional Wall and Vest Interaction Administration

All communication goes through the main server and all activity to and from the main server is logged. This includes physiological feedback from the users, swipes, knocks, wall state changes, light activity and all pattern activation triggered by the system. For the wall and vest to interact with each other, we developed an android communication application so we could connect to the vest using Bluetooth and connect to *The Humming Wall*'s WiFi at the same time. The application converted any messages from the wall to vest commands and converted physiological readings from vest to messages for the wall. As the Android device was located on the wearer and is connected to the wall's WiFi it was possible to move around in a much larger area while still being able to receive and send messages between vest and wall. To control the trial and swap between the different interactive zones, we developed an administration application for the experiment facilitator, and to activate different pre-recorded patterns on the vest.

10.5 Field Trials: Adding People

We ran trials for 5 weeks at Utzon Park. Initial pilot testing showed it worked better to have pairs experience the wall-vest interactions (see Results: Public/private for details), so we asked participants to invite another person to take part with them. The duration of the trials varied from between 1.5 to 4 h per pair, averaging 2.5 h. We ran the trials in four phases (see Table 10.3) and one researcher guided/assisted in all phases and one videoed all activity phases 2–4:

10.5.1 The Activities

We sectioned the wall into zones of discrete activities for the trials, either (1) calming or (2) activating, consisting of:

(1) Physiological data from the vest conveyed to the wall.

    (a) Heartbeat—displayed as audio, vibration and light (at the same beat) in the wall (Fig. 10.9a).

    (b) Breath rate—displayed as audio and vibration in the wall (Fig. 10.9e).

    Both these zones held seating areas.

(2) Actions at the wall were vibroacoustically displayed at the wall and also transferred to and felt in the vest.

Table 10.3 The Four Phases of the Experiment Interacting with The Humming Wall

Phase#	Phase type	Activity	Time
1	Fitting	• Filled in consent and demographic info • Fitted the BioHarness, harness and vest • Tested for accurate actuator placement, private environment	15–35 min
2	Training	• Participants walked up and down in the park • Experienced the 10 vibration patterns X 3 • Gave talk aloud responses while moving • While stationary articulated response to a repeat of each individual pattern • Same as above but walked slowly for those related to motion	10–15 min
3	Interaction	• Participants moved clockwise around the wall • Participants & pair interacted with 5 zones • Instructed on what to do arrival at each zone (e.g., knock on these 3 panels; swipe these 3 panels; sit & breathe). How they interacted, breathe, knock or swipe—fast, slow, or with rhythms, was left to them. • Repeat visit to 2 physiology zones—1 displaying heartbeat, 1 breath rate • Other three zones respond to gestures. • Total of 7 stops	20–90 min
4	Evaluation	• Removed vest and BioHarness • Filled in post-trial questionnaires • Semi-structured interviews, • While pair began the fitting phase	15–40 min

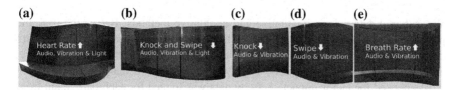

(a)   **(b)**   **(c)**   **(d)**   **(e)**

Fig. 10.9 **a** Heartbeat: Vest transmits to Wall; Audio, Vibration and Light echo vest wearers heartbeat at wall. **b** Knock and Swipe: Wall transmits to Vest; Audio, Vibration and Light respond to knock and swipe. **c** Knock: Wall to Vest. **d** Swipe: Wall to Vest. **e** Breath Rate: Vest to wall

(c) *Knock*—3 panels responded to knocking with audio and vibration in the wall and knocks felt in individual actuators of the vest. Right panel knocks transferred to the right side of the body (front & waist); Middle panel transferred to the back; Left panel transferred to left side of the body (front & waist) (Fig. 10.9c).

(d) Swipe—3 panels that responded to up, down, left and right gestures. These played as audio and vibration at the wall with a corresponding gesture played on the vest—e.g., swiped up or down the back and swiped across the waist—left to right or right to left—depending on direction of gesture (Fig. 10.9d).

(e) Knock & Swipe—4 panels that responded to knock and swipe at the wall and played on the vest. The swipe gestures ran along the 4 sections with audio, vibration and lights. The knocks were relayed alternatively to all actuators and felt all over the body. Simultaneous knocking on 2 panels activated 2 actuators, knocking on three activated three (Fig. 10.9b)

10.5.2 The Participants

We enlisted 39 volunteers with ages ranging from 12 to 65 years (average age 39), 20 females and 19 males. 19 participated in mixed gender groups, 11 in female/female and 9 in male/male (uneven numbers where only one of the participants in one pair wore the vest). Most paired with close or good friends (11), their partners (10), colleagues (10), family members (5), or social friends or acquaintances (3). 22 self-reported basic or average IT skills with 17 advanced or above. 28 averaged over 20 h on a computer each week and 12 played a musical instrument regularly. 31 had tertiary level qualifications. 14 were knowledgeable about wearable technology.

10.5.3 Data Collection

We gathered data using quantitative and qualitative methods. Before the trial, the participants filled in informed consent forms and a demographic questionnaire asking about fitness levels and experience with IT, vibrotactile technology, embodied interaction, large public displays and playing musical instruments to be cross checked for factor impact. Activity data was logged for each participant from the actuators and also from the BioHarness, for heartbeat and breath rate. The activity on the wall was also registered. This included logging knocking and swiping activities (the frequency and direction), identifying at which segment the activity took place as well as the time spent in each zone. Each pair of participants was accompanied throughout the trial by two researchers, one guiding and one videoing. After completing the required activities around the wall, participants completed shortened adapted versions of questions from MEC Spatial Presence Questionnaire [40], Flow State Scale [41] and Intrinsic Motivation Inventory [42] to gauge reactions to the sensations in the vest, bi-directional interaction between the vest and the wall, and the interactions with *The Humming Wall*. The questionnaires comprised 21 five-level Likert-scale items to analyse and cross-check users' perceptions. For Presence, we measured concentration, errors, activated thinking, and imagining space. For IMI, we measured interest/enjoyment, perceived competence, pressure/tension, and effort/importance. For flow, we measured challenge-skills balance, goals, autotelic, concentration on task, and sense of control. The overall

Table 10.4 Selected questionnaire items: IMI items (1, 2, 3, 14, 16, 17, 19, 21); FSS items (1, 4, 6, 15); MEC-SPQ items (1, 7). Q4 & 19 were negative questions and reversed

ID	Verified question	Categories	Question	Fig category
Q1.	IMI. FSS. MEC-SPQ	Effort, concentration, concentration	I concentrated on the sensations in the vest	(concentration)
Q2.	IMI.	Interest/enjoyment	I found doing the tasks asked of me in the trial pleasant (e.g., knocking on the wall, hearing breath)	(tasks pleasant)
Q3.	IMI.	Perceived competence	I found some sensations made more sense than others	(made sense)
Q4.	FSS. Reversed	Sense of control	I found some of the vest vibration sensations uncomfortable	(vest vibration uncomfortable)
Q6.	FSS	Sense of control	I felt I could be active in my surrounding environment (move around etc.)	(active)
Q7.	MEC-SPQ	Activated thinking imagining space	I thought about whether these kinds of sensations could be useful to me for other activities	(useful)
Q14	FSS IMI	Goals perceived competence	I understood how to do the tasks (e.g., knocking on the wall, swiping the wall etc.) task	(understanding)
Q15	FSS	Autotelic	I was not as aware of time passing (or of my surrounding environment) outside of the trial, as I would usually be	(lost sense of time)
Q16	IMI	Interest/enjoyment	I enjoyed interacting with the wall	(wall enjoyable)
Q17.	IMI. FSS	Percseived competence Challenge-skills balance	The vest vibrations made it easier to understand the tasks and responses	(vest vibrations helped understanding)
Q19	IMI R	Pressure/tension	I felt nervous while doing the tasks	(nervous at tasks)
Q21.	IMI.	Effort/importance pressure/tension	It was important for me to do well at interacting at the wall	(important do well)

experience was measured through 10 semantic differential scale items (See Table 10.4). Lastly, each participant described their experience, highlighting chosen aspects in semi-structured recorded interviews.

10.5.4   Data Processing and Analysis

Data post-processing tasks included transcribing user interviews and translating them into English for further analysis. Videos were cut into manageable chunks and coded focusing on where activities and interactions occurred. We transcribed the talk aloud responses of the participants to each vibration and coded them numerically. For each zone we measured the amount of time the participants spent, coded to what degree they explored the interrelation of knocking and swiping on the wall and corresponding sensations for the vest wearer. From the translated transcribed debrief interviews, we counted word frequencies, added all sensation descriptions, and those responses that received more than 10 mentions were grouped into meta-categories, and summarized in Fig. 10.12.

We used 5 level Likert-type scales (from completely disagree to completely agree) to measure items adopted from IMI, FSS and MEC-SPQ. The levels of agreement were mapped to values from 1 to 5. The semantic differential scale items featured 5 numerical levels with the differential labels at the extreme values (e.g., Impersonal 1, 2, 3, 4, 5 Personal). We compared demographic, fitness levels and prior knowledge factors against the data to assess impacting factors.

For analysing the physiological data, we prepared the recorded heart rate data by a 2-stage filtering in Matlab with an upper limit at 240 beats per minute (bpm) with a lower limit of 250 ms R-wave to R-wave, removing outliers and discarding noisy data samples. We then employed a median value of a circular FIFO buffer, and discarded those outside a 40 % deviation of this. The resulting IBIs were used to analyse how active the participants were and in time domain based heart rate variability analysis to assess their degree of relaxation. Heart rate variability (HRV) has been employed as a measure of mental load or effort [43] or stress [44] with higher variability indicating lower effort or relaxation. For HRV we relied on the measure of pNN10 [45]. We compared the participants' heart rates (bpm) and heart rate variability (pNN10) from the period during the pattern training phase and compared their physiological values to a baseline value from right before they experienced that pattern. We used t-tests to compare physiological responses (before to during pattern exposure), excluding 3 participants data with poor readings.

10.6   Results

In this section we report on findings from questionnaires, system and BioHarness logging, video footage and semi-structured interviews.

10.6.1   Overall Experience

We found the participants joined in the spirit of play well at *The Humming Wall*, and were generally interested in uncovering what the vest patterns, the interactions at the wall and between the wall and the vest might mean.

10.6.1.1   Self-reporting

According to our questionnaire items (see Table 10.4 and Fig. 10.10 for a presented subset) users enjoyed interacting with the vest and *The Humming Wall*. They understood what was required, concentrated, found the activities and tasks made sense and the experience to be enjoyable and pleasant. They reported little or no discomfort associated with the vibrations from the vest or wall. Further, most users felt that they lost track of time during the trial so they were immersed in the activities and unaware of the passing of time.

In Table 10.4, we see the subset of the questions where significant differences were found. We identify their category in the validated questionnaires scales and the category they are presented at (last column) as shown in Table 10.4 and represented numerically and visually in Fig. 10.10.

Most responses show that users enjoyed interacting with *The Humming Wall*. Participants were relaxed when carrying out the tasks (Q19) and they found it pleasant to complete the different activities at the wall (Q2, Q16). They reported no discomfort associated with vibrations from the wall (Q4). Most felt they lost track of time during the trial (Q15). Responding to vest questions, users reported little discomfort associated with the vibrations (Q4), felt it was possible to be active in the surroundings (Q6) and the vest vibrations made it easier to understand the tasks and responses (Q17). In general users stated that they were concentrating on the sensations in the vest (Q1) and thought whether these sensations could be useful for other activities (Q7). Most participants found that some sensations held greater meaning than others (Q3).

Results summarising the overall experience with the vest and the wall are shown in Fig. 10.11. Participants reported to be relaxed and calmed even though they

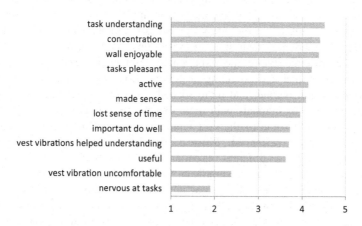

Fig. 10.10 Selected questionnaire items from Likert scale items (mapped 1 to 5, large numbers indicating higher agreement)

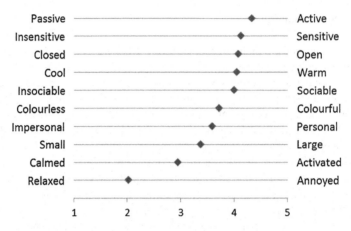

Fig. 10.11 Semantic differential values on overall experience

reported to be active as well. Further, users felt that it was a personal experience, but at the same time, they felt sociable at the installation.

10.6.2 Cross Comparing with Participants Personal Info

When crossing results of the questionnaires with the demographic data using non-parametric Mann-Whitney tests we found the following statistically significant differences. Males agreed significantly more than females that some sensations made more sense than others but enjoyed less learning how to do the tasks suggested by the sensations. They felt less warm and personal than the women. Participants who exercised 2 or more times per week reported significantly warmer and more relaxed experiences than those exercising less. Participants with average and lower IT knowledge found the experience significantly: easier to respond to vibrations, found their understanding of the sensations improving more as the trial progressed and felt warmer and more sociable than the participants with high IT knowledge. Those participants with more experience at large displays agreed significantly more that some of the vest vibrations were uncomfortable when compared to the less experienced. Musical participants who enjoyed playing an instrument reported a different experience from participants who did not. Non-instrument players had a significantly more sociable and open experience. Those less knowledgeable with vibrotactile technology enjoyed learning how to do the tasks suggested by the sensation significantly more and had a more personal experience than the vibro-technology savvy ones. Those knowledgeable in wearable technology reported a significantly more relaxed experience compared to those with little or no knowledge in this area.

10.6.3   Semi-structured Interviews

The semi-structured interviews included questions such as '*How did you find: the sensations in the vest; interacting with the wall?*' '*Did you prefer some sensations to others?*' in order to get participants expanding on their experiences. The most common responses, counted and organised in meta-categories, are depicted in Fig. 10.12. Many participants described the interactions as fun, pleasurable, unusual or new and different. People referred often to feeling and identified parts of the body: Back, Waist Front, Chest and Body per se as well as to Heart or heartbeat and Breath. They also referred often to movement, and/or Different-Confusing-New–Surprise and/or Calm-Relaxing-Natural.

10.6.4   The Humming Wall, the Zones, Physiology Readings and Activity

Participants took up suggestions to do 10 star jumps (see Fig. 10.13 Left A, B) to get their heart beat up (with many trying to maximize to their top heart rate range, see Fig. 10.13 Left C) and then settling back to 'feel-hear' their heartbeat at the wall slowing. They also took their time leaning back in the slower heartbeat and

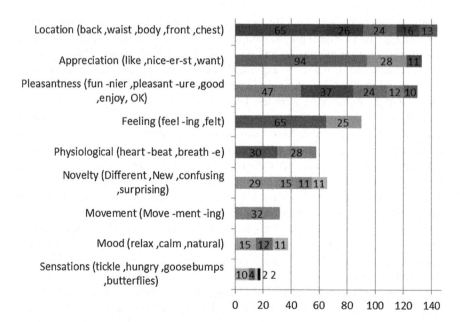

Fig. 10.12 Word counts from debrief interviews structured in meta-categories with the unprompted actual spoken terms in parentheses, and displayed colour coded

(a)   **(b)**   **(c)**

Fig. 10.13 *Left A, B* Participants do star jumps to get their heart rate up. *C.* Running on the spot in an attempt to maximize their heart rate. Right. Basic Instructions in each section included, e.g., 'knock on these three', 'swipe these' in order to ensure participants could found their own meaning with each task

(a)   **(b)**   **(c)**   **(d)**

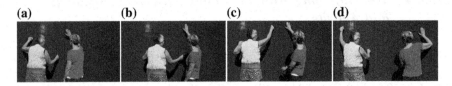

Fig. 10.14 Participants knocked at different segments and heights on the wall to determine if there was direct a mapping between areas on the wall to areas on the body

breathing zones. They voluntarily (without suggestion beyond very basic instruction (see Fig. 10.13 Right) played their own rhythms and patterns in the knock and swipe areas, or were running up and down (see Fig. 10.14), swiping large or small circles at once and/or generally experimenting and playing together at the wall.

We found that participants spent more time in the physiology zones when they advanced at their own pace, were less knowledgeable about vibrotactile technology and if they had tried to find out the mapping of the knocking on the wall to where it landed on the vest. Together these three factors accounted for 58 % of the variance in time spent in the physiology zones. Participants spent more time in the knocking and swiping zones, the stronger their relationship was with their pair and if they tried to find out the mapping of the knocking to the vest (see Fig. 10.14).

Together these factors accounted for 32 % of the variance in time spent in these zones. Twenty-one out of 39 (54 %) participants tried to map knocking on the wall to the same area on the vest wearer's body. That is they assumed a direct mapping between body and wall. '*I thought I could find this spot on the wall relates to this spot on the body*' and '*I thought if I knocked on upper part of wall I would feel it in upper region on body*'.

All people participated with the main swipe movements: up, down, left and right. However, people also experimented. Twenty-four out of 39 (62 %) participants made waving-swipe-circular movements with one or two hands, as if

Fig. 10.15 Pairs swipes and knocks wall and sensations are transmitted to the vest wearer (swiped around waist and knocking on vest), who responds to the sensations

massaging a back while swiping at the wall. Closer pairs often knocked and swiped in patterns and rhythms, playing musical melodies on their partner (Fig. 10.15).

For the physiological analysis, we found the participants' heart rates went up in the more active zones and down in the relaxing zones, whereas HRV moved in the opposite direction (as to be expected). Given the different self-reported experiences depending on the participants' fitness levels, we ran post hoc tests, but fitness levels had no significant effect on the degree of activity carried out at the wall (measured by their knocks and swipes per minute), nor on their HRV readings. So, in contrast to the self-reported results, according to their physiological readings, the fitter (or more regularly active) people were not more relaxed than the low exercisers. There was no significant correlation between how much people exercised and how much their HRV changed from their personal baseline during the different zones. Nor did we find evidence that the frequency of activations from knocks (single) and from swipes (patterns) had an effect on our participants' heart rate.

10.6.5   What People Said: An Overview

From talking aloud while first training with the patterns, while talking and interacting with the wall and moving about generally and from the semi-structured interviews, we found a variety of spoken reactions overall. Participants often described the interactions as interesting, fun, unusual or new and different. For example, participants described the experience as '*Funny - different –a new experience*'; '*I like to try new things, so it was new to me*'. Others took some time to adjust before they felt comfortable: '*In the beginning it was a new feeling and it was stressful… Later I got used to it*' and '*a bit scary, intimidating at start, but then more natural when I understood how it felt*'. Some tried to compare the sensations to other experiences: '*trying to make sense of it, tried to connect it to other experiences, front vibrations felt like standing in room with loud music, connection vibrations going one side to another like being in the ocean*'. And discussed: '*all immersed, when all over body—could feel like me vibrating all over, especially ones on stomach*'.

(a)   (b)

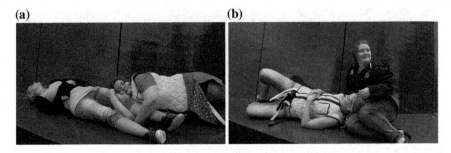

Fig. 10.16 A case of close friends. Vest wearer X lies on pair Y while relaxing in Heartbeat zone during 2nd visit. Vest wearer Y lies on pair X at her turn

For many it was very personal: '*it was much more responsive and personal than I thought it would be*'. The wall appeared to be a safe place where people felt protected and even though in a public space, it was somehow private. Many found the wall helped them relax and requests such as: '*want one in my garden*' were common with participants often referring to the wall as: '*therapeutic*' or '*good to help take care of health*'. One participant referred to the wall as '*the mother*', '*feels like she cares for you*' and another '*the wall was much more connected and it was like stroking a pet*', while most seemed to relax well on it (see Fig. 10.16).

10.6.6 Vibration Patterns

From the training session, where participants walked up and down in the park and experienced the ten vibration patterns, 3 times for each pattern while giving talk aloud responses and then responded to a repeat of each pattern separately again, we found participants reported little discomfort. However, users did interpret some sensations to hold greater meaning than others.

The talk aloud responses to the 10 patterns were coded numerically with a response to each individual pattern judged as either negative (-1), neutral (0) or positive (1). We analysed these in a non-parametric Friedman test rendering a Chi-square value of 51.9, which was significant ($p < 0.001$). Post hoc pairwise comparisons showed that Up, Down, WaistLeftToRight and WaistRightToLeft were all significantly more positively evaluated than every other pattern. Tarzan and Stop were evaluated more negatively, significantly worse than all other patterns with non-significant differences between Stop, TurnRight and Shiver (see Fig. 10.17). Important to note here is that Shiver has top front patterns followed by a Down the back sequence, so this is a combination of upper front and back sensations.

We asked questions such as '*Do the patterns suggest anything to you?*' Responses named sensations as tickling, butterflies, hunger or goose bumps, for example: '*some feeling was like butterflies in the stomach.*' and '*The tickling on the*

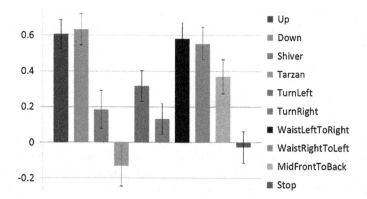

Fig. 10.17 Means of coded responses to the vibration patterns with error bars (standard error of the mean)

side... gave me goose bumps from the outside—like you didn't have the sensation but got the goose bumps anyway' [Waist, FeelGood] or—*'Solar plexus, not so bad once got used to it—at first didn't like it'... 'it could be a warning of don't do that'* [Front, Warning-Stop].

For some the vibrations on the front of the body (the activating *Tarzan* [Front, Tarzan] and *Shiver* [Front & Back, Shiver] or Warning *Stop* patterns [Front, Stop]*)* were the most testing: *'Front—it was not uncomfortable but not as pleasant—the ones on the back are more natural'*. For others: *'The feeling in the chest was best, feel it better also at lower back—very nice'*, while many said similar about the back and waist. *'Up and down the back I liked the most, none were unpleasant.' 'Those on the back were pleasant... nothing was uncomfortable... it was all very gentle'* [Back, Calming] and *'Preferred up and down—lifting you up with up swipe...Up more than down, felt more natural for me* [Back, Up]*'*. Yet there was differences in interpretations and preferences between elements of the patterns: *'Solar plexus* [Front, Warning], *not so bad once got used to it—at first didn't like it—was heavier —then saw it is a warning of don't do that'*, whereas for another *'Front—not as pleasant—not as natural—vibrating there'*.: Some also suggested certain movements took longer to learn: *'turn left or right, took time to get more used to them— needed to do again a few times at first—and was harder to figure out at the beginning, but become almost normal quickly'*. Sensations in more than just one area of the body at the same time often confused people.

Some participants likened the sensation to a real touch: *'Putting up the same emotion as touch—bringing out the same sense of well-being or happiness as when somebody touches or pats or hugs you... the hug* [Front & Back, Shiver] *was very surprising'*. Many were more comfortable with familiar sensations: *'I like the one on my spine the most... it resembles what people do... when they stroke them on the back or something. So I think it was the most comfortable one'... or 'felt comforting, like there, there...'* [Back, Up, Down]. For some their responsiveness increased over time and as the body became more used to the sensations: *'More body got used to listening to the sensation, better it picked up on them.' 'become part of you—sensation quickly become natural, kept my mind occupied when I got more physical'.*

10.6.7 Summary of Findings

Overall: The findings for interacting with the vibrotactile vest and *The Humming Wall* systems were very positive. Participants' self-reported high levels of concentration, they enjoyed the vest, interacting with the wall and the interactions between the wall and the vest, and found the experience enjoyable. In addition, they were motivated but relaxed, they understood what was required of them, were activated and did not notice time passing.

Crossing demographic information with the questionnaires, we found that women reported to have a more personal and warm experience than men. Those who were physically fit, and particularly with less IT knowledge had the best experiences—the latter even understanding the sensations better as the trial progressed. Further, participants less familiar with vibrotactile technology had a more personal experience and enjoyed learning how to do the tasks suggested by the sensations more than experts in this domain.

We found some trends for patterns changing the participants' heart rates but these were not significant. Similarly our analysis did not yield any statistical differences in the participants HRV between the pNN10 scores when comparing during exposure to different patterns.

10.6.7.1 The Vest

In relation to the patterns and the sensations they elicited from the vest wearer, we found, particularly for males, that some sensations made more sense than others, and this was confirmed across genders in the video logging, the semi-structured interviews and to some degree in the physiological responses. Participants responded more positively to Up, Down, on the Waist Right to Left and for Waist Left to Right patterns, (calming and/or feel good patterns) than to e.g., Shiver, Tarzan (activating) or Turn Left (navigation) and Stop (warning).

From the interviews and physiological responses, we found mixed reactions to the patterns—some requested more activating (the chest—Tarzan or Shiver), but most preferred the calming sensations (waist and back). These were considered more comfortable, more like how it usually is to be stroked. That participants found the calming and feel good patterns more pleasant and more comfortable than the activating and warning patterns is to be expected and a strong result. That the activating and navigating and warning patterns were less comfortable reflects the use we want them to be put to—to motivate action and/or reaction, not to soothe or calm.

10.6.7.2   The Wall

Most participants experimented with mapping the knocking to the vest wearer's body and with different types of swipes—emulating massaging a body, assuming a correlation between the wall and the body. Comments on the intimate-or personal nature of the wall and the connection with the partner via the activity on the wall having an effect for both were reflective of a novel experience. These bi-directional interactions between the vest and the wall mediated an innovative form of physical and personal connection between people. The differentiation between the heartbeat and breath rate zones compared to the more general experience at the knock and swipe zones is indicative of the more intimate nature of 'sitting in the vibroacoustics of the physiological responses of another'—different to watching-hearing it as was the experience for passersby, or those not seated in the vibroacoustic action of the physiology zones.

There was an obvious connection to the wall, to a sense of well-being at the wall and an ease in an experience where regardless, their physiology was made publically hearable and visible in an urban space. In addition, a passing public quickly added using *The Humming Wall* as part of the public furniture available at the park, along with Utzon seating, portable deck chairs and the grass.

10.6.7.3   Emergent Themes

We found evidence of several themes emerging from the analyses:

(1) *Language:* a suitable language to discuss often novel sensations.

We noted that in the training and interview discussions, many had difficulty attempting to describe an unfamiliar pattern or new sensation. We found most struggled to find words to describe the sensations and convey their impressions. Others expressed their uncertainty: *'haven't felt before, yes single vibrations from mobile, but not all over body and flowing, so not sure how to describe what it does… is new'.*

(2) *Therapeutic:* Participants reported both the vest and wall sensations as therapeutic.

Additionally, we noted that many found the vibration patterns and the wall vibroacoustics to be therapeutic and/or pleasurable and sighed, smiled and laughed, giving the overall impression of enjoyment in spending their time there. Several mentioned previously sore body areas… *'had an operation… so at first… not used to being touched there any more… got used to it… after a while it felt very safe and good to be touched there again.'*

(3) *Public/private*: participants found a mixture of public and private connection-communication-display elements in the experience.

Some participants described the experience as personal-intimate, connecting with the other in the pair, as well as social. *'Because what she* [partner] *did on the wall, I could feel on my body and it was both intimate and social… she also got a response from sounds* [at the wall] *and I got a response* [initiated] *from her.'* and *'it was kind of personal somehow, like when a person is rubbing a hand down the spine… I think it was a social interaction especially when* [partner] *was stroking the wall and I was being stroked'*. One described it *'like a social furniture… you get a connection with the person you are interacting with. It comes close to your body, so it's a good way to communicate and play'*. Often there was differentiation between the heartbeat and breath rate zones and the knock and swipe experiences: *'heartbeat and breath is too personal and very close… nicer with somebody you really know'*.

10.7   Discussion

In this section we discuss the patterns that were more easily understood and those more difficult, the therapeutic qualities from both the vest and the wall, intimacy, connection, novelty and finding a precise enough language to define the sensations. We then outline the considerations that need to be taken into account when designing for wearables and for well-being environments and the types of users and scenarios these could impact positively.

Looking at the findings for the more successful calming and feel good patterns, we can see that the feel good factor—fun, pleasure, enjoyment, comfort—was not so surprising given existing evidence on stomach vibrations making people feel happy or in love [27, 46] and touch sensations being generally good for well-being. Regardless, that the participants found the calming and feel good patterns more pleasant and more comfortable than the activating and warning patterns is a strong result.

The activating and full body navigation patterns were less clear to many participants. Where more than one location vibrated at the same time, participants found it difficult to determine which vibration to pay attention to. Other studies have similar outcomes [27, 30, 31, 46], and patterns for provoking movement and activation need more consideration to accommodate the diverse target groups—fit and fragile—that this work could be useful for. Additionally, a pattern that activates may need to be less pleasant than a relaxing one. A pattern that turns the body may not be as calming or feel good or comfortable. The pattern is requiring the body move from its current state of inertia (or comfort) to somewhere else, to go at a faster pace or to stop. That the activating and navigating and warning patterns were less comfortable reflects the use we want them to be put to—to motivate action and/or reaction, not to soothe or calm. Certainly that the stop (warning) was the least 'enjoyed/comfortable' reads as a strong and sensible response to a pattern that would be implemented, for example, to stop a visually or aurally impaired person from crossing a busy road, or other such 'emergency' scenario.

That some found the vibrations in the vest (and the wall) to be therapeutic was initially surprising, however there are several factors that can be taken into consideration. We discuss the wall and the vest separately as they provide different experiences (and responses).

For the wall, given the frequency range of the base humming (42 Hz), some feel good factor or sense of well-being was inevitable. The usual vibroacoustic treatment session lasts 20 min (40–50 Hz) [25] with the participants lying still on a vibrating table within the 40–50 Hz range. In contrast, our participants spent on average a total of 13 min, either lying or sitting on the 42 Hz vibroacoustics humming and vibration range. That they were moving around the wall within the vibroacoustic range for a longer period of time (not directly sitting-lying on but in earshot and vibration range), may also have impacted. Regardless, they spent a considerable time lying or sitting upon the wall vibrating within the therapeutic range—most often with a friend, so also in comfortable (potentially comforting, or at the very least relaxing-recreational) circumstances.

For the vest, the vibrotactile 'touches' of the vest and the vibroacoustics of the wall provided mediated touch-vibrations—perhaps even more comfortable for some than a more direct and therefore personal human touch. This may make the touch less confronting—less personal, the element of judgment of the body is removed (particularly for those with e.g., surgery or other conditions that usually prohibit touch), so the mediated touch sensation becomes safe, objective, mechanical (akin to ELIZA, a keyword scripted simulation of a Rogerian psychotherapist), that patients were most comfortable confessing their all too [47]. Additionally the mediated touch (touch is performed by actuators vibrating closely on the body controlled by either the researcher or the pair touching the wall) is private in that it returns no feedback about the body it touches—no communication about skin quality, scarring etc. to the bearer of the mediated touch, beyond what the person themselves portrays in body and/or facial, gestures and/or verbal responses.

Regardless, there were comments on the surprisingly personal nature of interaction with the wall and in particular depending on the connection with the partner through the activities at the wall. The bi-directional interactions between the vest wearer and their pair at the wall mediated an innovative form of physical and personal connection between people. Differentiation of the heartbeat and breath rate zones (when compared to the more general responses at the knock and swipe zones) indicated the more intimate nature of 'sitting in the vibroacoustically delivered physiological responses of another'—a somewhat novel experience and the connection with another, of beneficial consequence.

Our evaluations identified that a key element missing in exploring tactile experience is a sensible vocabulary for verbalising the multiple nuances of the sensations. These are new sensations, there is no current set of terms that easily transfer to precisely define the experience these novel sensations deliver. Accurate terminology is not available for people to be able to easily describe their experiences [48]. For this study, we referred to the sensations as patterns but did not identify each one to the participants in order to avoid influencing their perception-experience. We did because this is still early work with these kinds of full body sensations. We wanted to find out their unprompted, uninformed responses. For ongoing studies, we can build from appropriate terms in existing language sets [46] and add the sensate terms identified by participants from this study—hungry, ants, goosebumps, butterflies, tickling, exciting and/or crawling— as well as specifying location on parts of body and types of movement and mood-determiners (relaxed, activated, calm, excited, overwhelmed, agitated, excited etc.) as initial steps. The training phase would begin familiarising participants with relevant terminology to assist in building an expansive vocabulary.

Interestingly, the usual findings with novelty of use, meaning users lose interest once they have learnt the new technology [49] are somewhat nuanced here. Those knowledgeable in wearable technology reported a significantly more relaxed experience compared to those with little or no knowledge in this area, which is to be expected, but they did not lose interest. However, the participants less familiar with vibrotactile technology had a more personal experience and enjoyed learning how to do the tasks suggested by the sensations more than experts in this domain.

It is important to know the particular use needs of the individuals who may benefit from assistive wearable technology—particularly as in this case with a wearable vest that works on the torso. Advantages when dealing with the torso region means there can be a wide range of potential benefits for a variety of use cases and to suit user-needs. We see potential for adapting, for example, calming-activating patterns to assist with autistic and/or spastic paraplegia needs. The vest is designed to hug, to stroke up and down the spine, to stroke around the waist and to assist with comfort, release and feeling safe. By contrast, the navigation and warning patterns may be best suited to hearing and/or audio-impaired users or those with memory disorders, who are still mobile in the environment and require low-cognitive load assistance to reach their destination safely. Future studies would need to include designing with those specific use cases and needs of individuals in mind.

It is equally important when designing well-being and recreational facilities in an urban environment to access the necessities for the community to ensure that the local natural inhabitants of the space are consulted and that their needs are addressed in the design. It is also important to understand and align with the stakeholders' interests, and to advance these with the insight gained from understanding the everyday (and event-based) use from the natural inhabitants to the site. This ensures that the vision of all the stakeholders—natural users and investors in the site—can be improved upon and gives the site itself some agency in the lives of the users (who also hold some ownership, some investment into how the site is used). There is scope to design smaller versions of furniture and architecture for assistive living (home size and urban scale), for therapeutic settings, and for communities and urban environments. The vibroacoustic components could be embedded in the seating or the architecture surrounding the local community square or similar, providing experience similar to those found in e.g., a Snoelezen room environment, as part of the urban seating environment and making a significant contribution to health and well-being to the larger local population.

The plethora of small devices reading our physiological data need somewhere-something to talk to and our public spaces may be the place to inform natural inhabitants and raise health awareness to new levels and enhance general quality of life. The interactive environment could allow the coupling of peoples personal devices (enabling a vibroacoustic display of their physiological readings) via blue tooth communication with the vibroacoustic interactive seating/architecture. This would require managing multiuse in public spaces [49]. However, the major contribution would be rehabilitation for those with either purpose fitted wearables and/or requiring vibroacoustic therapy who are mobile enough to leave their homes and seek company in a relaxed urban environment. Such an environment would ensure the healthiness of daily interaction with others (outside of family and healthcare workers) as well as meeting their specific reha-bilitation needs. This could change the face of assistive care and rehabilitation and build a community of use and support (scheduled with times for the Kommune healthcare workers to be on-hand) within the rehabilitative urban environment: the city square revisited!

10.8 Conclusion and Future Work

Overall we had overwhelmingly positive responses from vest wearers with their partners from interacting with the vibroacoustic *Humming Wall.* We found that communication privileging vibroacoustic and vibrotactile as the primary interaction modalities for the vest wearers and their pair was successful.

Ongoing work with the vest includes refining the successful calming and feel-good patterns and continuing investigation and development of the activating, full body navigation and warning arrangements. Longer repeating pattern sequences working with one body location at a time seems a way to progress. Varying the duration and overlap length ratios proportionally with shorter ratios used for acti-vating and longer for calming shows potential for future investigation from several further lab studies. In addition, responses to therapeutic benefits from the vest require more consideration. Targeting specific use groups could prove useful and testing with blind-folded users with several of the improved navigation patterns is in progress. Naming patterns and beginning foundational work with verbalising haptic sensations is also called for. Future design considerations for the wall include mapping zones more directly to the morphology of the body and continuing work within the 40–50 Hz vibroacoustic frequency range to address therapeutic responses in more depth.

Many passers-by interacted with *The Humming Wall,* demonstrating potential for future interactive furniture and embedded vibroacoustics in interactive archi-tecture in public spaces. Future analysis of video footage of the passing public is still pending to gauge how successful communication that privileges vibroacoustic and vibrotactile as the primary interaction modalities can be for a passing (or sitting-standing-interacting) public.

In this study, we contribute successful vibroacoustic and vibrotactile communication and interaction between vest wearers and people at the wall with a public vibroacoustic artifact. In addition, we identify successful calming, feel good and warning patterns for the torso (the duration, overlap, rhythm and placement on the body) that hold potential for use to augment and improve mobility and quality of life for those with multiple disabilities This is foundational work demonstrating the promise of vibrotactile and vibroacoustic as communication mediums with potential for therapeutic and assistive well-being.

Acknowledgements We thank the volunteers for the field trials, the stakeholders, colleagues and the many participants who gave freely of their information and time, our Kinesiologist Bettina Eriksen, Odico Robotics, RealDania, Det Obeleske Familie-fond and Aalborg Kommune for support. This work is supported by the EU funded project CultAR (FP7-ICT-2011-9 601139), the FUEIB agreement 3194 and mobility grant CAS12/00199, Programa José Castillejo granted by the Spanish MECD.

References

1. Mueller F "Floyd," Vetere F, Gibbs MR et al (2005) Hug over a Distance. In: CHI '05 extended abstracts on human factors in computing systems. ACM, New York, NY, USA, pp 1673–1676
2. Teh JKS, Cheok AD, Peiris RL et al (2008) Huggy Pajama: a mobile parent and child hugging communication system. In: Proceedings of the 7th international conference on Interaction design and children. ACM, New York, NY, USA, pp 250–257
3. Haans A, IJsselsteijn W (2006) Mediated social touch: a review of current research and future directions. Virtual Real 9:149–159. doi:10.1007/s10055-005-0014-2
4. Brave S, Dahley A (1997) inTouch: a medium for haptic interpersonal communication. In: CHI '97 extended abstracts on human factors in computing systems ACM, New York, NY, USA, pp 363–364
5. Tsetserukou D, Neviarouskaya A (2010) World's first wearable humanoid robot that augments our emotions. In: Proceedings of the 1st augmented human international conference. ACM, New York, NY, USA, pp 8:1–8:10
6. Samani H, Teh J, Saadatian E, Nakatsu R (2013) XOXO: haptic interface for mediated intimacy. In: IEEE international symposium on next-generation electron. (ISNE), 2013, pp 256–259
7. Chan M, Estève D, Fourniols J-Y et al (2012) Smart wearable systems: current status and future challenges. Artif Intell Med 56:137–156. doi:10.1016/j.artmed.2012.09.003
8. InfaVitals Innovative Wearable Technologies. http://infravitals.com/. Accessed 20 May 2008
9. Ohmatex. http://www.ohmatex.dk. Accessed 1 Aug 2015
10. My Wear. http://www.mywearproject.info/. Accessed 1 Aug 2015
11. Ambiotex. http://www.ambiotex.com/. Accessed 1 Aug 2015
12. Ghiani G, Leporini B, Paternò F (2008) Vibrotactile feedback as an orientation aid for blind users of mobile guides. In: Proceedings of the 10th international conference on human-computer interaction with mobile devices and services. ACM, New York, NY, USA, pp 431–434

13. Bahram S, Chakraborty A, Ravindran S, St. Amant R (2013) Intelligent interaction in accessible applications. In: Biswas P, Duarte C, Langdon P et al (eds) A multimodal end-2-end approach to access. Comput. SE-5. Springer London, pp 93–117

14. Giudice NA, Palani HP, Brenner E, Kramer KM (2012) Learning non-visual graphical information using a touch-based vibro-audio interface. In: Proceedings of the 14th International ACM SIGACCESS conference on computers and accessibility. ACM, New York, NY, USA, pp 103–110

15. Krishna S, Bala S, McDaniel T et al (2010) VibroGlove: an assistive technology aid for conveying facial expressions. In: CHI '10 extended abstracts on human factors in computing systems. ACM, New York, NY, USA, pp 3637–3642

16. Yuan B, Folmer E (2008) Blind hero: enabling guitar hero for the visually impaired. In: Proceedings of the 10th international ACM SIGACCESS conference on computers and accessibility. ACM, New York, NY, USA, pp 169–176

17. Rosenthal J, Edwards N, Villanueva D et al (2011) Design, implementation, and case study of a pragmatic vibrotactile belt. Instrum Meas IEEE Trans 60:114–125. doi:10.1109/TIM.2010. 2065830

18. Nanayakkara S, Taylor E, Wyse L, Ong SH (2009) An enhanced musical experience for the deaf: design and evaluation of a music display and a haptic chair. In: Proceedings of the SIGCHI conference on human factors in computing systems. ACM, New York, NY, USA, pp 337–346

19. Yao L, Shi Y, Chi H et al (2010) Music-touch shoes: vibrotactile interface for hearing impaired dancers. In: Proceedings of the fourth international conference on tangible, embedded, and embodied interaction. ACM, New York, NY, USA, pp 275–276

20. Karam M, Branje C, Nespoli G et al (2010) The emoti-chair: an interactive tactile music exhibit. In: CHI '10 extended abstracts on human factors in computing systems. ACM, New York, NY, USA, pp 3069–3074

21. Manresa-Yee C, Morrison A, Muntaner JJ (2015) First insights with a vibrotactile interface for children with multiple disabilities. In: CHI '15 '10 extended abstracts on human factors in computing systems. ACM, New York, NY, USA, pp 905–910

22. Spelmezan D (2011) A language of tactile motion instructions. PhD thesis. RWTH Aachen

23. Nummenmaa L, Glerean E, Hari R, Hietanen JJK (2014) Bodily maps of emotions. Proc Natl Acad Sci 111:646–651

24. Arafsha F, Alam KM, El Saddik A (2012) EmoJacket: consumer centric wearable affective jacket to enhance emotional immersion. In: 2012 international conference on innovations in information technology (IIT), pp 350–355

25. Rüütel E (2002) The psychophysiological effects of music and vibroacoustic stimulation. Nord J Music Ther 11:16–26. doi:10.1080/08098130209478039

26. Gemperle F, Hirsch T, Goode A et al (2003) Wearable vibro-tactile display. Report CMU Wearable Group, Carnegie Mellon University

27. Karuei I, MacLean KE, Foley-Fisher Z et al (2011) Detecting vibrations across the body in mobile contexts. In: Proceedings of the SIGCHI conference on human factors in computing systems. ACM, New York, NY, USA, pp 3267–3276

28. Morrison A, Knudsen L, Andersen HJ (2012) Urban vibrations: Sensitivities in the field with a broad demographic. In: Proceedings of the ISWC 2012. IEEE, pp 76–79

29. Geldard F, Sherrick C (1972) The cutaneous "rabbit": a perceptual illusion. Science 178 (80):178–179

30. McDaniel T, Villanueva D, Krishna S, Panchanathan S (2010) MOVeMENT: A framework for systematically mapping vibrotactile stimulations to fundamental body movements. In: Proceedings of the HAVE 2010. IEEE, pp 1–6

31. Spelmezan D, Jacobs M, Hilgers A, Borchers J (2009) Tactile motion instructions for physical activities. In: Proceedings of the SIGCHI conference on human factors in computing systems. ACM, New York, NY, USA, pp 2243–2252

32. Bullivant L (2006) Responsive environments: architecture, art and design. V&A, London

33. Morrison A, Viller S, Mitchell P (2011) Building sensitising terms to understand free-play in open-ended interactive art environments. In: Proceedings of the SIGCHI conference on human factors in computing systems. ACM, New York, NY, USA, pp 2335–2344
34. Shepard M (2011) Sentient city: ubiquitous computing, architecture, and the future of urban space. The MIT press
35. Weston R (2002) Utzon: inspiration, vision. Architecture, Edition Blondal
36. Weston R (2009) Jørn Utzon logbook, vol V: additive architecture. Edition Blondal
37. Zephyr Bioharness 3. http://www.zephyranywhere.com. Accessed 1 Aug 2015
38. Precision Microdrives—Micro DC Motors. http://www.precisionmicrodrives.com/. Accessed 1 Aug 2015
39. Pares N, Carreras A, Durany J et al (2004) MEDIATE: An interactive multisensory environment for children with severe autism and no verbal communication. In: Proceedings of the third international work virtual rehabilitation
40. Vorderer P, Wirth W, Gouveia F et al (2004) MEC spatial presence questionnaire (MEC-SPQ): Short documentation and instructions for application
41. Jackson S, Marsh H (1996) Development and validation of a scale to measure optimal experience: the Flow State Scale. J Sport Exerc Psychol 18:17–35
42. Deci EL, Ryan RM (2000) The "What" and "Why" of goal pursuits: human needs and the self-determination of behavior. Psychol Inq 11:227–268. doi:10.1207/S15327965PLI1104_01
43. Rowe DW, Sibert J, Irwin D (1998) Heart rate variability: indicator of user state as an aid to human-computer interaction. In: Proceedings of the SIGCHI conference on human factors in computing systems. ACM Press/Addison-Wesley Publishing Co., New York, NY, USA, pp 480–487
44. Hall M, Vasko R, Buysse D et al (2004) Acute stress affects heart rate variability during sleep. Psychosom Med 66:56–62
45. Mietus JE, Peng C-K, Henry I et al (2002) The pNNx files: re-examining a widely used heart rate variability measure. Heart 88:378–380
46. Obrist M, Seah SA, Subramanian S (2013) Talking about tactile experiences. In: Proceedings of the SIGCHI conference on human factors in computing systems. ACM, New York, NY, USA, pp 1659–1668
47. Weizenbaum J (1966) ELIZA-A computer program for the study of natural language communication between man and machine. Commun ACM 9:36–45. doi:10.1145/365153.365168
48. Moussette C (2012) Simple haptics: sketching perspectives for the design of haptic interactions. Dissertation: Umeå University, 2012. Umeå, Sweden. ISBN 9789174594843
49. Morrison A, Salovaara A (2008) Sustaining engagement at a public urban display. Situated large displays work. OzCHI

Chapter 11
The Acceptance, Challenges, and Future Applications of Wearable Technology and Virtual Reality to Support People with Autism Spectrum Disorders

Nigel Newbutt, Connie Sung, Hung Jen Kuo and Michael J. Leahy

Abstract This chapter provides a brief review of how virtual reality technologies (VRTs) have been used within research-contexts to support people on the autism spectrum. One area of innovation and research that has evolved since the mid-1990s to aid people with an ASD is that of the use of virtual reality technology (VRT). Research has focused on helping develop personal, social, functional, and pre-vocational/vocational skills. This chapter will provide a review of literature in this field in addition to distilling the key affordances, issues and challenges identified in this evolving area of research. This review will focus on the potentially useful application of VRTs to train and support people with an ASD in developing life-skills (i.e., social skills, job skills, independent living skills) and where there has been successful implementation in applied contexts. The chapter will then describe a project the authors undertook, that sought to ask questions surrounding the role of head-mounted displays and the impact these might have for this population. As more affordable and accessible wearable devices (e.g., Oculus Rift™) are commercially available, we suggest that questions surrounding the acceptability and practicality quickly need to be addressed if we are to develop a sustainable line of inquiry surrounding HMDs and VRTs for this specific population. Therefore this chapter will report, in a rich, descriptive and illustrative manner, the process we engaged to work with a range of participants with ASD (here range refers to low-high functioning ASD with a wide range of IQ) to assess and measure the acceptability of, and experiences within, HMD VRTs (immersion, presence, ecological validity, and any negative effects). We will discuss, in some detail, the ethical approaches the research team took in using HMDs with this population (as this is sometimes a neglected aspect of this type of research), and suggest some guidance for future scholars/practitioners working in this field. Finally, based on

N. Newbutt (✉)
The University of the West of England, Bristol, UK
e-mail: Nigel.Newbutt@uwe.ac.uk

C. Sung · H.J. Kuo · M.J. Leahy
Michigan State University, East Lansing, USA

© Springer International Publishing AG 2017
A.L. Brooks et al. (eds.), *Recent Advances in Technologies for Inclusive Well-Being*, Intelligent Systems Reference Library 119,
DOI 10.1007/978-3-319-49879-9_11

our preliminary study, this chapter will conclude with how we think HMDs and VRTs might be used in the future to help enable ASD populations address some of the challenges faced on a daily basis. Here we are primarily concerned with the challenges, but also the contexts in which this technology can be applied, with specific focus on moving research from labs to real-life (and potentially beneficial) contexts.

11.1 Introduction and Background: The Promise and Potential of Virtual Reality Applied Within Autism and Developmental Contexts

Virtual reality applied to persons with a developmental disability has been long investigated and researched. Since the early 1970s multimedia applications have been used as tools to help people with autism overcome several difficulties (i.e. applying attention, reading facial expressions, social skills development). This area of research coincide with the increase of multimedia computing in the area of educational support for children with autism and has since been the focus of much research; from the early work of Colby [7] and Heimman et al. [11] to the work of Silver and Oakes [34]. This published work examined specific skills that multimedia could address in the educational development of children with autism.

Looking back to the early-mid 1990s Strickland et al. [38] presented an early study that assessed the effectiveness of Virtual Reality (VR) as a learning tool to engage children with autism; their study was primarily designed to determine if children with autism would tolerate VR equipment and respond to a computer-generated world. They considered the differences between VR and computer programs; the level of interaction with computer-generated images; and independence in determining motion and objects in a VR world as a way to present real-life experiences. The use of VR for children with autism was considered, based on: sensory problems, lack of generalisation, visual thought patterns, individualised treatment and responsiveness to computer technology. The aim of the study was to help children with autism learn how to cross a road safely. They used VR helmets to immerse the users in a 3D environment, so that users could identify cars, the colours of objects and how they were moving. The children were presented with various scenarios to determine generalisation, and were asked to "walk" into the scene and interact with signs. Findings gained from the study suggested that the two partic-ipants (with ASD) were able to interact with the environments successfully (ac-cepting the virtual helmet, tracking in-world objects; moved their bodies and heads, located in-world objects and moved towards them). In addition the study suggested that the participants appeared to become immersed (labelling in-world objects and moving their bodies interactively).

Building on the work of Strickland et al. [38], Charitos et al. [3] designed a virtual environment that was controlled by virtual-reality input devices to aid the organisational skills of people with autism. The study aimed to teach social skills through the provision of a virtual interface for people with autism to navigate through. Charitos et al. provided a series of reasons why computer-based systems are well suited for people with autism, based on the work of Murray [20]. Examples cited by Murray included being able to set clear boundaries and controlling the stimuli (through a step by step process). These both allowed for greater control and focus of material in the presentation and learning. Another advantaged cited by Murray links to joint attention and restrictive context. These both, in some ways, play to the strengths of people on the autism spectrum in that focus of interest can be taken into account in addition to restricting other sources of information. Overarching these aspects, Murray suggests that safety, flexibility and adaptability can all be instilled along with a sense of prediction. Again these aspects play to the strengths of people on the autism spectrum.

Other studies have found that the VR can be utilized by individuals with autism as a learning tool [38, 9], to teach independent living skills [18, 38], to teach safety skills [13, 32], to hold their interest [6, 20], monitor eye gaze [16], aid following directions [30], and interpret emotions of avatars accurately [4, 19]. VRTs can help to provide an individual a safe space to test social situations and responses to these situations, and develop confidence and a greater awareness [24]. The person does not have to be exposed to the real stressor, which is usually a barrier to entering treatment; the person can disengage immediately from the situation should it become unbearable, and the social environment and conditions can be structured and individualized to each case situation.

In addition, Kandalaft et al. [14] then created an engaging Virtual Reality Social Cognition Training intervention to help young adults with autism work on social skills. Eight young adults diagnosed with high-functioning autism completed 10 sessions across five weeks. It uses brain imaging and brain wave monitoring, and essentially puts individuals in situations like job interviews or blind dates using avatars. They work on reading social cues and expressing socially acceptable behaviour. The study found that after completing the intervention program, participants' brain scans showed increased activity in areas of the brain tied to social understanding. There were also significant increases on social cognitive measures of theory of mind and emotion recognition, as well as in real life social and occupational functioning. Such findings suggest that the virtual reality platform is a promising tool for improving social skills, cognition, and functioning in autism [14].

Through such studies an argument has been developed for the use of virtual environments in providing a unique affordance for users with ASDs. Scholars such as Strickland et al. [38], Rutten et al. [31], Parsons et al. [26], [23], Wallace et al. [41] and [4] have each explored the role of immersion, realism, engagement and learning within virtual environments. Building on this, Kandalaft et al. [14], provided a specific example of the role virtual worlds can play in social-cognition training. In a recent overview, Wilson and Soranzo [45] provide a compelling argument for the potential of VR used in psychology stating that: "the proliferation of available virtual reality (VR) tools has seen increased use in experimental psychology settings over the last twenty years". In listing several advantages of VR in this setting, the authors suggest that the following points provide a clear rationale for using and continuing to develop VR for use in psychology, as it can provide:

- Greater control over stimulus presentation;
- Variety in response options;
- Presentation of stimuli in three dimensions;
- The creation of complex scenarios;
- The generation of varying levels and combinations of multimodal sensory input potentially allowing audio, haptic, olfactory, and motion to be experienced simultaneously to the graphically rendered environment or objects;
- The possibility for participants to respond in a more ecologically valid manner.

In addition, Riva et al. [29] provide several other advantages for using VR in mental health related fields:

- Cost effectiveness;
- Control of the environment to help assure patients;
- Feel a situation (presence) as real;
- Possibility of person adventures;
- Some reports of high acceptance (although in neuro-typical populations);
- Privacy and confidentiality.

The next section of this chapter seeks to outline some of the above affordances in further detail, providing some examples of how VRTs have been, used by ASD populations based on some key characteristics that VR provides.

11.2 Application of Virtual Reality in Autism

VR provides opportunities to practice dynamic and real-life social interactions, which has been used previously and shown to be an effective intervention tool for people with autism. Its utility is likely due to several unique characteristics. Here are some major reasons.

11.2.1 Interactive Nature Increases Users' Motivation

VR is an interactive, experiential medium that users become directly engaged with the effects of the VR experience in a consistent manner. Because of its interactive nature, VR can increase motivation by making the experience fun. Individuals will be much more motivated to learn tasks when presented with an engaging virtual reality video game environment than simply role playing.

11.2.2 Provide Safe and Controlled Environment

VR represents real-life experiences in a safe, controllable manner that allow for repeated practice and exposure. It is a unique setting where individuals can explore and act without feeling threatened or frightened of real- world consequences, or they can make mistakes without fear of dangerous, real, or humiliating conse-quences. This is particularly important for the autism population in learning tasks. VR can also provide naturalistic environments with unlimited social scenarios and has been shown to replicate social conditions effectively [41].

11.2.3 Provide Repeated Practices and Immediate Feedback to Facilitate Learning

The computer-mediated role-play might present a vital opportunity for individuals to experience different perspectives, which, in turn, might nurture more general skills in mental simulation. Responses to different scenarios can be practiced before, during or after being taught. Tasks can also be repeatedly presented and practiced in a consistent way without the fatigue; an issue sometimes associated with task repetition by human instructors [8]. The ease of repetition of the task could facilitate rote learning of social rules in a specific context before moving on to allow practice of the skill in a different context. VR also offers immediate, real-time feedback about performance and can be tailored to each individual and monitored to test his or her ability to perform certain tasks over time based on progress [2].

11.2.4 Allow Individualized Customization

VR can be adapted to suit individual needs [40]. As individuals begin to show improvement, the tasks can be made increasingly more complex and/or difficult, creating challenge and potentially continual learning. For example, a café envi-ronment—in which the user has to order some food and find a place to sit down—could start off with lots of empty seats, but become increasingly populated and busy as the user moves through the program. A railway carriage scenario could work in a similar way, with increasingly fewer choices of empty seats. An increase in com-plexity could also work with a prompt-fading sequence, in which the user is initially provided with numerous prompts when beginning the task on a more difficult level, before the prompts fade gradually over time. Text boxes, verbal instructions and flashing red areas of interaction (e.g. the lock on a bathroom door), have been used successfully as prompts in previous training packages for people with learning disabilities [1].

11.2.5 Permit Manipulations of Stimuli and Distractions

VR allows people with autism to explore environments without the distracting or restricting presence of other actors. The amount, type and level of particular inputs (e.g. visual, auditory) can be controlled directly, allowing basic skills and tasks to be completed in the absence of competing and distracting cues [44]. Virtual reality can also simulate many situations that may otherwise be difficult to control or simulate in real life, such as a fire emergency or street crossing, resulting in an ecologically valid and dynamic assessment and training. While controlling for outside distractions, VR can systematically introduce distractions that can be manipulated in ways that the real world cannot [43]. For example, VR can convey rules and abstract concepts without the use of language or symbols for individuals with autism with little or no grasp of language. The possibility of developing different VEs designed to mimic a variety of social situations is a particular strength of this approach.

11.2.6 Hierarchical Learning Approach to Promote Generalization of Skills

The possibility of scenarios differing slightly each time that the user encounters them could promote a more flexible style of responding. For example, behaviour demonstrated on a previous occasion (e.g., walking to a particular point at the restaurant to order a meal) may need to be changed slightly on a subsequent trial because of a small change in the environment (the same route to the counter cannot be taken because people are standing in the way). Thus, the user has to think of different ways of solving the same problem. This hierarchical approach to developing social skills within a specific environment could then be repeated in a new environment, ensuring that the same skill could be practiced across different contexts. This aspect of task presentation could improve the chances of generalizing skills across contexts. In addition, the inherent properties of VEs may facilitate the crucial transfer of understanding from the virtual to the real world because of the shared features between virtual and real environments, in the form of realistic images and scenarios.

These taken as a whole, and although the above affordances have been applied to a range of users, participants and groups, they highlight some key characteristics and possibilities of VR in health-related areas. Back to considering the role of VR applied to people with autism, the field is less well furnished with studies and data to support the efficacy therein. However, several studies have shown that the field is well poised to take advantage of new technologies; specifically head-mounted displays and VEs [27]. Within this specific field (autism and HMDs) very few, if any, studies have considered even initial research questions around acceptance. In this chapter we suggest that questions surrounding the acceptance (wearable), experience (presence) and any negative effects (dizziness, feeling sick, displacement) are paramount and in urgent need of inquiry.

In summary, the evidence-base to date is still in the early stages of providing substantial answers to the role of virtual realty used by people with an ASD [27], but we are moving closer to supporting claims that their might be a potential benefit to using this technology for this group of users [22]. However, many questions still remain. Within this chapter we aim to address some of the fundamental questions around head-mounted display technologies, and present a case study to help understand users' experiences.

11.3 Key Questions that Emerge from Historical and Recent Literature

There are several reasons to be excited about the potential of VRTs applied to the field of ASD. Parsons and Cobb [27] provide a clear overview of this, as do the studies of Wallace et al. [41] and Wallace et al. [42]. However, there is still very little known about how HMD technology can or should be used with ASD groups. Key questions in this domain relate to:

- Acceptance: Will people with autism be willing to make use of wearable technology?
- Ethical and dosage issues: Will and can people with autism experience increased negative effectives using HMDs and VE therein?
- Can HMDs and VEs provide a successful and appropriate form of skills training (i.e. social skills, confidence, job interview [35, 36], safety, etc.)? If so which ones and why?
- In what settings can HMD VRTs be most effective?
- What are the views and opinions of people with autism in using these technologies?

There is a great need to further consider VRTs and especially HMD technologies as while on the one hand they provide an evolution of VEs and the next stage of immersion, on the other hand they also provide the need to essentially re-assess the state-of-the-art and start to consider some of the key questions above. More than anything, the most fundamental and basic questions (related to acceptance) are in urgent need of investigation to enable a basis on which to establish a need to further investigate these immerging technologies.

In sum, while some evidence has been developed and presented, this has mainly been in lab-based settings and the role of users is often over-looked. There is a great need to seek ways to better involve end users in this research endeavour, but in addition, ways that this exciting technology can be applied and used in meaningful and appropriate ways and locations. Here we refer to the fact that little, if any, technologies are left behind for users with ASD to engage with after studies have been completed. There is a real and urgent need ensure that technology is in the hands of those who stand to benefit the most (in this case ASD populations). We, as a research field, need to find better mechanisms to enable this to happen, in addition to using these opportunities to evaluate the impact and measure the changes using technology may have on their users. However, with the development of affordable and easily accessible equipment we are perhaps entering a phase of being able to provide greater applied interventions; in situ contexts and places where technology might have the greatest impact on ASD groups.

Therefore, while there is some evidence for the need to pursue this area of research, it is equally important to recognize and acknowledge this is an emerging field; one that is in the early stages of research outputs and projects. Virtual reality technology used with and by people with autism is an exciting one, with a positive future, but an area that warrants further research to yield a larger evidence-base.

11.4 Virtual Reality Head-Mounted Displays: A Case Study with Autism Users

As highlighted above, virtual reality technology (VRT) offers a unique advantage in rehabilitation interventions by providing a safe, repeatable, and diversifiable environment for the ASD population to maximize their learning [10, 14, 21, 24]. For instance, a simulated scene in which the user could repeatedly practice a particular skill multiple times can be used to train an individual's social interaction with others and thus improve these soft skills. In addition, virtual reality (VR) is also valuable in its capacity to be culturally sensitive and free of restriction.

Specifically, VR simulated scenes without reading requirement can be used by users from any backgrounds. The use of virtual reality and immersive experience is particular helpful for the ASD population when compared to more traditional didactic learning. The reasons for this include: (a) VR does not require language proficiency, and (b) VR offers real life experience and multiple sensory stimuli to maximize learning experiences. The latter is especially influential when considering the style of learning is different for different and diverse individuals. However, sensory inputs and experiences can be two-fold (and quite different). On one hand, VR immersive experience could take full advantage of maximizing sensory input; on the other hand, it could also cause potential negative effects on users such as dizziness, confusion, and anxiety. As such, similar to any intervention, before rehabilitation practitioners could fully utilize VR as a means to boost learning for ASD populations ethical and safety concerns should be a top priority. In addition, we suggest this is especially the case for head-mounted displays and wearable technology such as the Oculus Rift TM. As the first step that leads to our goal of using the VR technology as an intervention to enhance learning for the ASD population, we conducted a study aiming at understanding the acceptance, immersive experience and psychological reaction of the ASD users to the VR head-mounted display and its application.

The acceptance of VR technology among users with ASD has been studied over the last decade. Specifically, Peli [28] suggested that with short-term use, there is little to no negative effect on ASD users. Similarly, Strickland et al. [39] also reported that children with ASD generally do not have any trouble negotiating simple VR scenes. Although these studies have established a safety net for the use of VR technology among the ASD population, the technology itself has evolved and advanced in many ways. For instance, the VR technology and the head-mount display used to be only an extension of the traditional monitors, which merely increase the viewing size and the viewable angle. The users of these VR devices acted as a passive role and did not have control over what was presented to them. Much has changed in the VR arena. The interaction components have since been introduced and integrated in the VR applications. For example, users can now use a joystick to browse and manoeuvre in the virtual space. In addition, more current head-mounted display can also capture and respond to users' head movement to simulate a real world 3D experience. With these advancements for the VR technology, it is necessary to revisit safety and ethically-informed questions again.

VR technology has become popular (in a commercial and research perspective) again since a short-lived resurgence in the 1990s and has been used extensively in the entertainment business in addition to military and architectural contexts. Currently, there is a plethora of head-mounted displays available for the general public to purchase and use in the daily life. Depending on the purpose and the affordable budget, one can select inexpensive VR goggle such as the Google Cardboard, which costs approximately twenty dollars a pair. The design of Google Cardboard is to insert a smartphone into a pair of goggles made of cardboard with two lenses to simulate the 3D immersive effect. Since the low cost design, there is no audio or interactive component built-in for the device.

Users merely use it to simulate visual 3D effect. The more sophisticated options are also available such as Samsung Gear VR, HTC Vive, and Oculus Rift. These VR devices capture the user's head movement and can be used to play video games.

11.5   A Case Study Using Oculus Rift (Developers Kit 2)

To illustrate some of the potential use and ethical concerns, a case study will be discussed next. The main purpose of this case study was to learn the perception of individuals with ASD of a range of VR experiences using the head-mounted display. Specifically, two main goals were sought in the research: a) participants' acceptance of wearing a VR goggle, and b) their opinion on 3D immersive experience. The intention for the first aim was to explore whether there is any adverse effects (e.g., dizziness and anxiety) from the use of the VR HMD. The second aim was to see whether a VR HMD could be a potential intervention for the ASD population in the future.

11.5.1 Equipment

As described earlier, currently there are numerous options for the VR head-mounted display available for the public to purchase and use. In the case study, we chose to use Oculus Rift for three main reasons: (a) its light weight and portability, (b) a laptop can be used to control the simulation in comparison to other VR goggle such as Samsung Gear VR which requires a specific smartphone to use it, and (c) the software is readily available. A list of the equipment used in the study included: (1) an Oculus Rift head-mounted display; (2) stereo headphones; (3) a Microsoft Xbox 36 input controller; and (4) a laptop computer. Figure 11.1 illustrates the technology used in the study. Oculus Rift was to simulate the 3D visual experience, stereo headphones were to simulate the 3D surrounding sounds and helped to ensure the participants felt immersed in the simulated environment, the Microsoft Xbox 360 controller was to allow participants to navigate within the

Fig. 11.1 Equipment (PC, HMD, input device and headphones), as used in the study

simulated environment, and the laptop computer was to control the scene presented and volume.

The population of the study included individuals with ASD who were interested in securing an employment opportunity. It is important to emphasize their desire for receiving an employment opportunity since the intention of the study was to explore the suitability of using the VR technology as an intervention for vocational rehabilitation. A total of 29 participants with ASD were recruited from a private non- profit community rehabilitation organization (CRO) in a Midwestern state in the United State of America. The mission of the given CRO is to provide vocational training for individuals with disabilities to obtain employment. The participants were identified by the case managers from the CRO. The study involved two phases. In the first phase, the participants were invited to experience three VR simulated scenes. Figure 11.2 provides an illustration of visual experiences during Phase one. Figure 11.3 shows one of the participants experiencing the VR simulation.

These three scenes, approximately ten minutes each, were relative low intensity in relation to lengths and sensory stimuli. Specifically, three scenes included:

1. A virtual cinema, in which the participants were in the front row of a cinema. The participants could choose a variety of films to watch. Only head movement was tracked for simulating visual 360° rotation.
2. A virtual café, in which participants sat in front of a virtual character. The virtual character would maintain eye contact with the participants if they looked at the character. Only head movement was tracked for simulating visual 360° rotation.
3. A virtual safari, in which the participants would be driving a Jeep around the Africa Savanah. The participants were prompted to use a controller to manoeuvre the car. The participants could look for animals as part of their

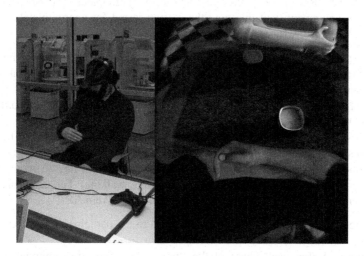

Fig. 11.2 Example of the 3D café environment and how the participant responded to seeing a coffee cup in the VE

Fig. 11.3 Example of a participant with ASD using the virtual reality head-mounted display in situ

'journey'. Head movement was tracked for simulating 360° visual rotation possible. 3D viewing angle would also change according to participants' sitting or standing.

After the participants completed the first phase of the study, they were invited to participate in phase two of the study, which took place few days after phase one. Eleven participants were selected for phase two. During this phase, two longer and more intense VR scenarios were presented to the participants who wore the same set of VR equipment, and this time the session lasted for approximately 25 min. Specifically, these two scenes included:

1. An Apollo 11 mission, in which the participants were taken through a historical tour related to Apollo 11 and experienced boarding the spacecraft, taking off, and entering zero-gravity. Head movement was tracked for simulating 360° visual rotation possible. 3D viewing angle would also change according to participants' sitting or standing.
2. A Tuscan house, in which the participants were taken to a Tuscan house in extensive, grounds where they could see views across a lake and towards hills beyond. The participants were prompted to use a controller to navigate and explore the environment (as though they were walking); full body and head movements were tracked for simulating real life movements.

Figure 11.4 highlight the interface presented to the participants involved in this phase of the study.

Before and after each phase of the study participants were interviewed by a researcher to discuss their expectations, concerns, and reflections about the

Fig. 11.4 Example of the interface/view of the participants. *Top-left*; café, *top-right*; safari, *bottom-left* and *right*; Tuscan villa inside and out

experiment. A selection of assessment instruments were also used to help better understand their overall enjoyment and psychological status from the use of the head-mounted display. Specific instruments included: (a) demographic questionnaire, (b) the Independent Television Commission-Sense of Presence Inventory (ITC-SoPI; Lessiter et al. [17]), and (c) the State-Trait Anxiety Inventory (STAI; Spielberger et al. [37]). ITC-SoPI was used to measure the subjective effects experienced by an individual within the virtual environment (i.e., presence of 'felt'), and STAI was used to assess participants' level of "state anxiety" (i.e., anxiety about an event; in this case the HMD VR environment) and "trait anxiety" (i.e., anxiety as a personal characteristic in general).

Additionally, a researcher was setup to observe and take note on each participant's behavioral reaction during the VR experience. These behavioral observations were especially important since the participants might not be able to truly express their feeling through standardized instruments. Below, Table 11.1 offers an overview of the participant characteristics in this study.

Table 11.1 Demographic characteristic of the participants who took part in Phase I and II of the study

	ASD group in Phase I (N = 29)	ASD group in Phase II (N = 11)
Demographics		
Mean age (SD)	32.02 (9.88)	29.77 (8.66)
Gender (%male)	76 % (n = 22)	91 % (n = 10)
Vocational history		
Full-time employed	34 % (n = 10)	27 % (n = 3)
Part-time employed	66 % (n = 19)	73 % (n = 8)
Formal ASD diagnosis		
Autistic disorder (%yes)	55 % (n = 16)	64 % (n = 7)
Asperger's (% yes)	34 % (n = 10)	18 % (n = 2)
PDD-NOS (% yes)	10 % (n = 3)	18 % (n = 2)
Intellectual ability		
IQ score mean (SD)	83.58 (23.69)	86.63 (30.70)

11.5.2 Results

Twenty-nine participants all agreed to wear the VR head-mounted display during phase one (100 %). Specifically, 86 % (n = 25) of the participants completed all three different virtual scenarios. Four participants requested to discontinue the experience after the virtual café scenario. Upon the completion of phase one, participants were asked to return for the phase two. Among 29 participants, 23 (79 %) agreed to come back for the phase two. All Eleven participants who were selected for the phase two agreed to participate in the study (100 %), and all of them completed two scenarios from the phase two. In relation to overall acceptance and enjoyment of the VR HMD, all participants reported a score of 3.0 or above (1 = not enjoy at all to 5 = enjoy the most), with a mean score of 4.32 (SD = 0.69). As for the likelihood of using VR-HMD again, the majority of the participants (with the exception of two participants) reported a score of 3.0 or above (1 = not likely to 5 = most likely), with a mean score of 3.92 (SD = 1.98).

In relation to immersive experience, the participants reported that the spatial presence was above average (M = 3.8; SD = 0.62) with both engagement and ecological validity reporting high scores (M = 4.1; SD = 0.57 and M = 4.0; SD = 0.33, respectively). In addition, participants reported low negative effects from the use of VR HMD (M = 2.0; SD = 0.34). Specific to psychological influence from the use of the VR HMD, we did not find any anxiety-provoking situation from the use of the VR HMD.

Base on the comments provided by the participants, we obtained positive feedback such as "feels good… it was okay… I can get used to it", "it was nice… feels amazing… the experience is awesome… that was amazing… you're immersed to what you're doing… enjoying it", and "amazing… it is so cool… I love it".

There were, however, negative or somewhat not positive feedback such as "if you look too fast [referring to moving their head], I can see why some people don't like that"; "the headset bugged me" (participant #5), and "a little blurry".

11.6   Implications

This study provided several important insights to the two questions that we sought to address: (a) is it safe for the individuals with ASD to use HMD VR interfaces (wearable-technology), and (b) would individuals with ASD accept and enjoy the experience in the virtual environment? Answering the first question, the participants of the given study expressed a general acceptance of wearing VR HMD during the experiment. In addition, most of them stated that they would be interested in trying more in the future. However, although the majority of the participants gave positive feedback during and after the VR experience, there were still some negative comments. These comments mainly focused on the visual effect of the VR experience, in which some stated that the visual effect can make people dizzy at time, and others claimed that the graphic was not smooth enough. There were still others commenting on the VR HMD which should be made to fit better and more comfortable. While these concerns were important to hear, these issues can be resolved with the advancement and adjustment of the equipment. For instance, users can feel dizzy when using a HMD due to its low or fluctuating frame-per-second rate. However, using a more sophisticated graphic card with a highly specified computer can often solve this problem. In addition, Oculus Rift, which was used in the study, was still in the beta phase (SDK 2) and so was not fully tested and commercially available. Feedback like this can be useful for the modification of HMD VR experiences to help design experiences to better suit users' comfort levels.

In addition, we did not find a significant change in anxiety level among the participants pre and post the use of the HMD. This is encouraging when considering individuals with ASD tend to resist new experiences. For most of the participants in the study, this was their first time wearing a VR HMD. While they did not necessarily feel comfortable wearing a VR HMD, they did not feel it intimidating either. As a conclusion, ASD users appeared to not experience sensory issues from the use of VR HMD.

Lastly, participants reported high spatial presence, engagement and ecological validity within the VR environment. In other words, the experiences viewed through the HMD were seen as 'real' and could happen in real life. This is an important indicator if we are to continue investigating the use of VRTs for populations with ASD. After all, generalization of learning and integration into real life is the ultimate purpose for the intervention.

11.7 Considerations for Future Research

Although VR has a variety of applications in rehabilitation, as well as social cognitive training for individuals with autism, some ethical concerns surround the use of fully immersive VRT, such as the use of head-mounted displays (HMDs). Researchers and clinicians should take special considerations while using VRT for clinical or research purposes.

11.7.1 Beware of Possible Discomfort Experienced by Users and Ensure Safety

Oversensitivity to sensory inputs could be common to individuals with autism who may experience 'cybersickness' in the form of nausea, headaches and dizziness [6, 25]. Prior to undertaking a study using VR-HMD, be conscious of its possible effects (see [33] for an overview) and develop a safe and ethically appropriate framework. One way to overcome this is to engage and communicate with support workers who work closely with the individuals with autism, to gain better understanding beforehand about individuals' tolerance, health condition, previous experience, and/or related behaviors in using VRT/HMD.

11.7.2 Inclusion of Users in Study Design

Users are not usually included in the development of an application [15]. This may lead to applications that are not optimal for the target group. User-centered design methodologies and/or community-based participatory approach (e.g., [46]) should be considered in study design because representative users are included in product design, participate in research development, and provide users' perspectives and input from the start.

11.7.3 Introducing to VRT to Users with Incremental Steps

While various researchers have noted the possible benefits of virtual environments for people with autism (e.g., [5]), Sharples et al. [33] reported that HMDs have been shown to cause greater nausea and disorientation effects than desktop displays. Therefore, researchers should be aware of and cautious about the potential side effects of using HMDs. One way to approach this could be during the recruitment process by providing an information leaflet about the study in addition to a short video showing the technology and explaining the process. This step might help to fully communicate the details of the study to potential participants, to inform what to expect, to prepare them to use the technology properly, and in some cases to answer concerns some participants might have. Also, it might be helpful to allow research participants to try on the HMD in stages (with shorter time to begin and then slowly increase the time period), then safely guided them into the immersive and intensive interactive VR experiences.

11.7.4   Use of VRT as a Tool and in Collaboration with Other People

There is concern that providing a safe, nonsocial environment on the computer somehow 'colludes' with the social disability of autism. Also, VR may become too safe and too attractive so that the user can become a computer addict and be reluctant to re-enter the real world. In fact, Parsons and Mitchell [25] emphasized that the aim of using VE is not to circumvent real-world social interaction altogether, but to provide a teaching aid that would allow practice and demonstration alongside input from a rehabilitation professional, teacher, caregiver or support worker. Thus, real-world social interaction can be incorporated by the presence of teachers and caregivers sitting alongside the person with autism, as in the use of different desktop virtual environments [19, 20].

11.7.5   Promotion of Cognitive Flexibility Within VE to Increase Generalization

There are still debates whether successful transfer can be achieved in the realm of social skills training for people with autism. Generally there seem to be good reasons for thinking that VEs might be particularly useful for people with autism in the context of social skills training. Perhaps most importantly, the promotion of both intra- and intercontext flexibility within a VE training package may allow researchers to tackle a very specific impairment in cognitive functioning common to many people with autism—cognitive flexibility [25].

11.8   Conclusion

VR holds promise for social improvement in autism by offering a platform to safely practice and integrate social cues that may improve social skills, social cognition, and social functioning. Future research in VR platforms and social skills and cognition training should utilize technology that includes facial tracking and or movement on the avatar. This would allow naturalistic and sometimes subtle, real-time facial affect to be projected onto an avatar in the VR thus decreasing the loss of social information from real-life to VR. In addition, studying the feasibility of a remote application of the VR- SCT, such as at home or in a school setting will increase accessibility of interventions and answer if results are consistent across venues. Future research into measures of social cognition will also be useful in understanding the impact of treatments on social abilities and functioning, particularly for adult populations with autism [12, 14]. For example, there is a lack of standardized and published measures of rating individual's personal emotion expression for facial expression and speech/prosody (e.g., timing of what is said and how it is said). Additionally, other measures such as behavioral observations, depression, and quality of life inventories could prove meaningful to the outcomes of treatment studies with this population.

11.9 Summary

We have presented an overview of the state-of-the-art and in doing so have suggested there is a urgent and relevant need to investigate the potential of virtual reality, in this case wearable technology (i.e. head-mounted display), used by autistic populations. The research field in this area, although at the early stages of an evidence-base, has provided indications that virtual reality technologies can have the potential to help aid people with autism in replicating real-world scenarios, but without any undue real-world consequences. In addition, and by using VRTs appropriately, they have the potential to become a training space to help increase confidence and practice of otherwise difficult daily experiences. In terms of moving this field forward, with the re-introduction of mainstream VRT head-mounted display's, and as they have changed significantly since the work of Strickland et al. [38], the need to re-address some key and fundamental questions are apparent. Therefore, this chapter sought to address questions around the acceptance of a HMD along with measuring the presence, immersion and any negative effects. We also looked at anxiety and whether using the HMD had any adverse effects. Findings from this study tend to indicate that the participants we worked with appeared to both enjoy using the HMD and VR experience in addition to reporting high levels of immersion, presence and ecological validity, while negative effects, such as anxiety were self-reported as not heightened post-VRT exposure.

While on the one hand the results tend to reveal that HMDs and VRT might be a good fit for users with an ASD, on the other hand they should be interpreted with some caution. Here we refer to the small sample included in our study, the subjective and exploratory nature of our work, dosage (or exposure) to the VRTs, and the selection criteria for Phase II of the study. We would also suggest future work match/compare results with typically developing users to help to better contextualize findings.

Notwithstanding the limitations of this work, the findings provide some insights to the manner in which people with autism experience HMD VRTs. As such, and due to the possible wide-spread uptake of these evolving technologies, we suggest HMD VRTs could stand to have a large and meaningful impact on how people with an ASD learn and test various skills in situ settings and could be developed as a way to overcome some challenges faced by these populations.

References

1. Brown DJ, Neale H, Cobb SV, Reynolds H (1999) The development and evaluation of the virtual city. Int J Virtual Reality 4(1):28–41
2. Burdea GC, Coiffet P (2003) Virtual Reality Technol, 2nd edn. Wiley, New York
3. Charitos D, Karadanos G, Sereti E, Triantafillou S, Koukouvinou, S, Martakos D (2000) Employing virtual reality for aiding the organisation of autistic children behaviour in everyday tasks. In: Sharkey P, Cesarani A, Pugnetti L, Rizzo A (eds) International conference on disability, virtual reality and associated technologies, Sardinia, Italy, pp 147–52
4. Cheng Y, Ye J (2010) Exploring the social competence of students with autism spectrum conditions in a collaborative virtual learning environment—the pilot study. Comput Educ 54 (4):1068–1077. doi:10.1016/j.compedu.2009.10.011
5. Cheryl GT (1999) Virtual environments for the investigation and rehabilitation of cognitive and perceptual impairments. NeuroRehabilitation 12(1):63–72
6. Cobb S, Beardon L, Eastgate R, Glover T, Kerr S, Neale H, Wilson J (2002) Applied virtual environments to support learning of social interaction skills in users with Asperger's syndrome. Digital Creativity 13(1), 11–22. doi:10.1076/digc.13.1.11.3208
7. Colby KM (1973) The rational for computer-based treatment of language difficulties in nonspeaking autistic children. J Autism Childhood Schizophr 3(3):254–260
8. Cromby JJ, Standen PJ, Newman J, Tasker H (1996) Successful transfer to the real world of skills practised in a virtual environment by students with severe learning difficulties. Proceedings of the 1st International Conference on Disability, Virtual Reality and Associated Technologies (IDCVRAT), Reading, UK
9. Fabri M, Moore D, Hobbs D (2004) Mediating the expression of emotion in educational collaborative virtual environments: an experimental study. Virtual Reality 7(2):66–81
10. Georgescu AL, Kuzmanovic B, Roth D, Bente G, Vogeley K (2014) The use of virtual characters to assess and train non-verbal communication in high-functioning autism. Front Hum Neurosci 8(807):1–17
11. Heimann M, Nelson KE, Tjus T, Gillberg C (1995) Increased reading and communication skills in children with autism through an interactive multimedia computer program. J Autism Dev Disord 25(5):459–480
12. Hillier A, Fish T, Cloppert P, Beversdorf DQ (2007) Outcomes of a social and vocational skills support group for adolescents and young adults on the autism spectrum. Focus Autism Dev Disabil 22(2):107–115
13. Josman N, Ben-Chaim HM, Friedrich S, Weiss PL (2008) Effectiveness of virtual reality for teaching street-crossing skills to children and adolescents with autism. Int J Disabil Human Dev 7(1):49–56
14. Kandalaft MR, Didehbani N, Krawczyk DC, Allen TT, Chapman SB (2013) Virtual reality social cognition training for young adults with high-functioning Autism. J Autism Dev Disord 43(1):34–44. doi:10.1007/s10803-012-1544-6
15. Korpela R (1998) Virtual reality: opening the way. Disabil Rehabil 20(3):106–107. doi:10.3109/09638289809166066

16. Lahiri U, Warren Z, Sarkar N (2011) Design of a gaze-sensitive virtual social interactive system for children with autism. IEEE Trans Neural Syst Rehabil Eng 19(4):443–452. doi:10. 1109/TNSRE.2011.2153874
17. Lessiter J, Freeman J, Keogh E, Davidoff J (2001) A cross-media presence questionnaire: the ITC-sense of presence inventory. Presence: Teleoperators Virtual Environ 10(3):282–297
18. Mitchell P, Parsons S, Leonard A (2007) Using virtual environments for teaching social understanding to 6 adolescents with autistic spectrum disorders. J Autism Dev Disord 37 (3):589–600
19. Moore DC, PaulPowell Norman J (2005) Collaborative virtual environment technology for people with autism. Focus Autism Dev Disabil 20(4):231–243
20. Murray DKC (1997) Autism and information technology: therapy with computers. In: Powell S, Jordan R (eds) Autism and learning. A guide to good practice. David Fulton Publishers, London, pp 100–117
21. Newbutt N (2013). Exploring communication and representation of the self in a virtual world by young people with autism. PhD thesis, University College Dublin, Ireland
22. Newbutt N (2014) The development of virtual reality technologies for people on the autism spectrum. In: Silton N (ed) Innovative technologies to benefit children on the autism spectrum. United States of America: IGI Global, pp 230–252
23. Parsons S, Leonard A, Mitchell P (2006) Virtual environments for social skills training: comments from two adolescents with autistic spectrum disorder. Comput Educ 47(2):186–206. doi:10.1016/j.compedu.2004.10.003
24. Parsons S, Mitchell P, Leonard A (2004) The use and understanding of virtual environments by adolescents with autistic spectrum disorders. J Autism Dev Disord 34(4):449–466
25. Parsons S, Mitchell P (2002) The potential of virtual reality in social skills training for people with autistic spectrum disorders. J Intellect Disabil Res 46(5):430–443. doi:10.1046/j.1365-2788.2002.00425.x
26. Parsons S, Mitchell P, Leonard A (2005) Do adolescents with autistic spectrum disorders adhere to social conventions in virtual environments?. Autism 9(1):95–117
27. Parsons S, Cobb S (2011) State-of-the-art of virtual reality technologies for children on the autism spectrum. Eur J Spec Needs Educ 26(3):355–366
28. Peli E (1998) The visual effects of head-mounted displays (HMD) are not distinguishable from those of desk-top computer display. Vision Res 38(13):2053–2066
29. Riva G, Botella C, Baños R, Mantovani F, García-Palacios A, Quero S et al (2015) Presence-Inducing media for mental health applications. In: Lombard Matthew, Biocca Frank, Freeman Jonathan, Ijsselsteijn Wijnand, Schaevitz Rachel J (eds) Immersed in media. Springer International Publishing, New York, pp 283–332
30. Rose FD, Attree EA, Brooks BM, Parslow DM, Penn PR (2000) Training in virtual environments: transfer to real world tasks and equivalence to real task training. Ergonomics 43(4):494–511. doi:10.1080/001401300184378
31. Rutten A, Cobb S, Neale H, Kerr S, Leonard A, Parsons S (2003) The AS interactive project: single user and collaborative virtual environments for people with high-functioning autistic spectrum disorders. J Vis Comput Anim 14:1–8
32. Self T, Scudder RR, Weheba G, Crumrine D (2007) A virtual approach to teaching safety skills to children with autism spectrum disorder. Topics Lang Disord 27(3):242–253. doi:10. 1097/01.TLD.0000285358.33545.79
33. Sharples S, Cobb S, Moody A, Wilson JR (2008) Virtual reality induced symptoms and effects (VRISE): comparison of head mounted display (HMD), desktop and projection display systems. Displays 29(2):58–69. doi:10.1016/j.displa.2007.09.005
34. Silver M, Oakes P (2001) Evaluation of a new computer intervention to teach people with autism or asperger syndrome to recognize and predict emotions in others'. Autism 5:299–316
35. Smith MJ, Fleming MF, Wright MA, Losh M, Humm LB, Olsen D, Bell MD (2015) Brief report: vocational outcomes for young adults with autism spectrum disorders at six months after virtual reality job interview training. J Autism Dev Disord 45(10):3364–3369. doi:10. 1007/s10803-015-2470-1

36. Smith MJ, Ginger EJ, Wright K, Wright MA, Taylor JL, Humm LB, Fleming MF (2014) Virtual reality job interview training in adults with autism spectrum disorder. J Autism Dev Disord 44(10):2450–2463
37. Spielberger CD, Gorssuch RL, Lushene PR, Vagg PR, Jacobs GA (1983) Manual for the state-trait anxiety inventory. Consulting Psychologists Press, Palo Alto, CA
38. Strickland D, Marcus L, Mesibov G, Hogan K (1996) Brief report: two case studies using virtual reality as a learning tool for autistic children. J Autism Dev Disord 26(6):651–659
39. Strickland DC, McAllister D, Coles CD, Osborne S (2007) An evolution of virtual reality training designs for children with autism and fetal alcohol spectrum disorders. Topics Lang Disord 27(3):226–241
40. Swettenham J (1996) Can children with autism be taught to understand false belief using computers? J Child Psychol Psychiatry 37(2):157–165. doi:10.1111/j.1469-7610.1996.tb01387.x
41. Wallace S, Parsons S, Westbury A, White K, White K, Bailey A (2010) Sense of presence and atypical social judgments in immersive virtual environments responses of adolescents with autism spectrum disorders. Autism 14(3):199–213. doi:10.1177/1362361310363283
42. Wallace S, Parsons S, Bailey A (2016) Self-reported sense of presence and responses to social stimuli by adolescents with ASD in a collaborative virtual reality environment. J Intellect Dev Disabil. 1–11
43. Wilson PN, Foreman N, Stanton D (1997) Virtual reality, disability and rehabilitation. Disabil Rehabil 19(6):213–220
44. Wilson PN, Foreman N, Stanton D (1998) A rejoinder. Disabil Rehabil 20(3):113–115
45. Wilson CJ, Soranzo A (2015) The use of virtual reality in psychology: a case study in visual perception. Comput Math Methods Med. http://www.hindawi.com/journals/cmmm/aa/151702/. Accessed 10 Dec 2015
46. Wright CA, Wright SD, Diener ML, Eaton J (2014) Autism spectrum disorder and the applied collaborative approach: a review of community based participatory research and participatory action research. J Autism 1(1). doi:10.7243/2054-992X-1-1

Part IV
Special Needs

Chapter 12
Nursing Home Residents Versus Researcher: Establishing Their Needs While Finding Your Way

Jon Ram Bruun-Pedersen

Abstract Residents at nursing homes need to exercise to retain self-efficacy. But all the while, many do not seem to want to prioritize exercise routines over leisure activities. The first part of this chapter analyzes the potential reasons for this lack of exercise commitment at a nursing home in Copenhagen, Denmark, and show a solution to overcome such obstacles, by augmenting the exercise routine with the accompagnement of recreational virtual environments. The second part of the chapter shares insights from the experiences from spending 3 years with the unique challenges and complex conditions that researchers face, when operating and navigating the specific field of nursing homes, due to the inherent characteristics of its context and users.

Keywords Older adults · Rehabilitation · Exercise · Nursing home · Virtual environments · Intrinsic motivation · Trust

12.1 Introduction

This chapter bases itself on a couple of studies performed in relation to a PhD project, which has been running over the course of three years. The project has investigated the background behind the problem of inactivity with regards to nursing home residents routinely exercise at a Danish nursing home, Akaciegården, in the Copenhagen area. From the research, it has been possible to map various aspects to nursing home life, which seem to oppose a desire to maintain a regular exercise routine for many residents. It has also been possible to implement and test a solution, which has positively affected some of these aspects, by using virtual environments to augment the exercise into a different type of experience. Last but not least, it has been possible to experience the (possibly) deceptive complexity of

J.R. Bruun-Pedersen (✉)
Department of Architecture, Design and Media Technology, Aalborg University
Copenhagen, Copenhagen, Denmark
e-mail: jpe@create.aau.dk

© Springer International Publishing AG 2017
A.L. Brooks et al. (eds.), *Recent Advances in Technologies
for Inclusive Well-Being*, Intelligent Systems Reference Library 119,
DOI 10.1007/978-3-319-49879-9_12

working as a contemporary media technologist researcher, within the very specific context of older adult nursing home residents as the users. Primarily concentrated around the context of two individual studies performed for the project, this chapter provides an elaborated insight into the process of problem identification, solution design and results throughout the two studies, and afterwards discusses the implications of performing such research.

The initial incentive for engaging this specific area of research was a desire to combine an exercise routine for elderly users with an augmentation design using virtual audiovisual technology. The applications of virtual reality technology have been investigated for some years now, but as this project has come to realize, the field seems to yet lack the maturity of covering requirements and best practice frameworks for user demographics such as older adults. From a societal perspective, the elderly demographic will be growing rapidly over the next decades [1], and there is a need for methods to help maximize the longitude degree of independence for older adults. Exercise is one such method, but while research with the elderly segment and commercial entertainment technology has been given some attention for a period, such as with the Nintendo's Wii console or Konami's Dance Revolution game [2, 3, 4], the body of research on tailored technology and content, designed specifically for exercise purposes for nursing home residents is both sparse and necessary [5]. It is imperative that the sense of meaningful purpose of a newly applied technology such age group demographic is considered [6], and what motivates physical activity behavior between age groups is very different [7].

Throughout its duration, the project has been fortunate to gain close access to the daily-living nursing home residents and their routines, which has facilitated iterations of observation sessions, interviews, private visits and social interactions with residents and staff. The resulting outcome has a variety of insights into what constitutes the nursing home life, as well as an understanding of residents' ability, approach, opinions and practices related to daily physical exercise. This has enabled the project to positively touch upon the somewhat untouched potential of virtual technology and media, specifically for nursing home exercise activities. And it has given an important understanding of what constitutes constructive procedures and necessary precautions to consider, when working scientifically with this type of technological development, with this type of user group.

The first part of the chapter (Sects. 12.1−12.4) takes a closer look at the aspects that form the exercise reluctance problem in the context of the nursing home. Thereby the reasoning behind the solution design is also detailed, leading to a description of both the solution design itself and the results from two studies using the solution; one study which initially validates the solution design [8], and another study which evaluates the motivational implications on its implementation at the nursing home over an extended period of time [9]. The second part of the chapter (Sect. 12.5) addresses a seemingly overlooked aspect to such research, which is the inherent challenges of working as a researcher in the context of a nursing home, with nursing home residents as the target user group. While these considerations are not explicitly detailed in the first part of the chapter, they have been a necessary part of the research approach throughout the project. The second part of the chapter thereby contributes insights to research practices, which is believed by this author to essential for any future endeavor within this field of research.

12.2   Physical Therapy and the Manuped

When moving from their original home to a nursing home, many facets of everyday life is made easier for new-coming residents compared to their previous living. This is partly due to the now closely available assistance from personnel, the optimized demographics of the nursing home increases independence, relating to daily activities such as toilet visits or overall indoor mobility, which increases overall life quality [10].

Rehabilitation is another area where nursing homes often have much to offer residents, in terms of attendance flexibility, availability and variance of activities. In Denmark, many nursing homes encourage physical therapy as part of the everyday activities at the home. The level of independent living and health status is an important factor of the quality of life for any person, and especially for these elderly segments [11]. Physical activity on a regular basis is found to improve sarcopenia, physical function, cognitive performance and mood in elderly adults [12], in addition to being able to maintain independence and retain self-efficacy [13]. Exercise makes a difference in everyday life for residents, as it fundamentally enables other activities. Examples of everyday situations where lacking self-efficacy can influence quality of life, are getting dressed, walking, or even controlling one's own drinking glass and eating utensils.

At Akaciegården, exercise and physical independence is a central part of the nursing home philosophy, but personnel are facing difficulties inspiring a portion of residents to maintain regular exercise routines [8]. Factors such as *independence* and *self-efficacy* could be regarded as rational, reasonable, and qualify as motivational reasons why exercise should be an attractive activity. Meanwhile, studies suggest that it does not reflect the engagement of many nursing home residents [8, 9].

Many residents who do not exercise regularly do not seem overly concerned with their lack of commitment, despite being well aware on the positive effects it could bring to various physical deficiencies. When asked into the reason for not exercising, many residents choose to deprioritize exercise because (a) they don't like it for one or several reasons, (b) they had been exercising regularly at one point, but couldn't find the motivation to return, (c) they feel that exercising too often interferes with other types of leisure-based activities overlapping physical therapy sessions, or d) they are simply too lazy [9].

The physical therapy center at the Akaciegården nursing home is open in two separate 2.5 h sessions a day, four times a week, with two physical therapists present in each session. Average resident participation is 10−15 during morning session, and 6−12 in the afternoon depending on the weekday due to weekly social

Fig. 12.1 The Manuped exercise device

activities on certain days. Therapists are very attentive and social towards the present residents in general, but certain exercises (any walking or standing related exercises) need the complete and attentive presence of either one or both therapists. Physical therapy sessions are therefore a race on resources, in terms of therapists' capability of placing appropriate personal attention to all residents who need it. It also means that exercise activities that residents can perform unassistedly have high value for the daily exercise routine. The most essential exercise activity for unassisted is a machine called a manuped; a chair-based bike device for arms and legs (see Fig. 12.1). In the context of the nursing home, the manuped is an all-round exercise device.

A manuped activates most parts of the body to varying degrees, and affords exercise with a high variance of intensity. Residents can use it unassistedly, as it requires no active balancing, or strength comparable to standing up independently. Nor does it demand sudden reactions or swift coordination changes (anything demanding dynamic muscle activity or quick coordination and balance). This combination of traits makes it safe for the physical therapist to leave a resident alone while using the device. Devices such as the manuped become essential, due to the resource limitations for the two therapists, as they can have more residents exercising at the same time, without having to actively attend all of them.

The manuped exercise is a single-person activity. Practicing residents are facing a wall while exercising, mainly because of the logistics placing them by the device, from e.g. their wheelchair. It does, however mean that once they are in place and start using it, residents are practically left alone for the duration of the exercise (10 min for the weakest, and approx. one hour for the few strongest). From both early interviews and casual conversations with residents, they generally really like the physical therapists. Much of what was appreciated about the physical therapy center had to do with the two therapists. Another very clear opinion among residents was that although using the manuped is good for the body, such activities are extremely boring, and difficult to want to do.

12.2.1   Why Residents Dislike Exercise

Many residents struggle with obstacles such as balance issues, pronounced lack of muscle strength (a combination of lack of exercise and sarcopenia), coordination difficulties, arthritis, or other chronic conditions [14]. Besides these, regular disease generally hit harder and last longer at this age. The effects of this were partly the reason why many residents expressed reluctance to make the effort to go exercise. When thinking about the physical therapy, they were thinking in parallel about the pain and discomfort affiliated with the activities. Of course, the impact of this varied between individuals, partly due to differences in physical performance/limitation, partly due to personal distaste for the associated pain, and partly due to their individual desire to overcome it. The last point is important, as almost all residents experience pain or physical issues to some degree, while exercising. So while the perspective of pain during exercise is a fact for most, it depends on the individual if it holds them back, is not important, or perhaps serves as a reminder to go exercise, as the sensation will only worsen if they do not. All scenarios were met during conversations with residents during visits.

12.2.2   Returning from Illness

Periods of illness are a big part of everyday life at a nursing home, which is why periodical absence from an exercise routine does not equal a lack of motivation. However, the two are potentially linked, as, physical conditioning from exercising decays much faster than at younger ages, during periods of illness. When the illness has worn off, residents are often left with the sensation of having to start all over again. For some nursing home residents, this can mean reestablishing the ability to walk. If a resident has not been able to maintain a regular exercise routine, physical therapy activities can be very physically demanding. Exercises are painful, and the residents cannot help but observe, compare and acknowledge their own inadequacies, when they for example with a physical therapist under each arm to aid the ascent, as they try to simply practice standing up from a chair. In addition, performing (relatively speaking) intensive exercise, affects their body substantially the rest of the day, perhaps even more. Especially if they have not been maintaining a regular routine, many residents are fatigued after a physical therapy session, to the extent that it sometimes means that they cannot engage in much else for the remainder of the day. Residents therefore go into a cycle of training, where they hit a barrier, and then struggle for an extended period of time to get back into a shape where they are actually able to perform simple tasks (such as standing up).

Exercising for the pure sake of long-term physical improvements can thus be regarded a substantial, and unstable investment. Observing a resident return to exercising after a longer break, give several impressions. One is clear happiness about being able to return to exercising and to the nice environment of the physical therapy center. Another is sadness and irritation concerning the lost long-time effort from before the break. It is a cycle that repeats itself for many residents, which has been evident from many conversations along the project. And it has shown to be a factor that makes it difficult for residents to prioritize physical therapy. Some residents do return to their exercise routine after longer periods of illness, however. This has especially been pronounced for residents who from previous occasions have gained personal experience with the physical difference between returning to exercise and staying away from exercise.

12.2.3   The Alternative

Most residents at Akaciegården are physically or mentally unable to leave the nursing home at their own initiative. If they do, they require personnel (or family) assistance, which is often not possible within the in-house resources, on any general scale. This means that social and/or entertaining leisure activities are hugely important for the residents, as most residents are almost completely limited to the experiences that are offered inside the walls of the nursing home. The nursing home arranges a large number of such in-house activities per week, where residents gather with a group of personnel, who deliver or conduct the event. Events are typically quite casual, for example rhythmic and singing gatherings, a movie, coffee and cake in the yard, card or board games, etc. As it is an important priority for the events that any resident should be able to participate, the events have a minimum of physical activity required (such as crossing squares on a paper, eating biscuits or drinking coffee while sitting on a chair or wheelchair). The leisure activities at Akaciegården are enjoyed by residents, and for many reasons. They increase personal bonds between and within personnel and residents, fill the residents' days with enjoyable content and experiences, and activate (or reactivate) residents socially. Meanwhile, with a fixed perspective on increasing the desire to exercise, leisure events can be seen as a diversion, which leads residents away from regular exercise routines. And while these authors would never encourage less leisure activities at Akaciegården, they do represent a challenge for the physical therapy personnel, in terms of keeping residents with their programs.

12.2.4 Laziness?

As mentioned earlier, a fair amount of residents who did not exercise a lot, claimed to be simply too lazy to exercise regularly. When asked into the reason behind their laziness, many residents described that it came from a combination of the areas just described; from pain, physical difficulty and boredom of sitting alone facing a wall while exercising, to the lack of any leisure entertainment during exercise sessions, the knowledge of the potentially lost effort if hitting a barrier, to the calm, relaxing and entertaining activities offered other places at the nursing home. From this outset, it seems that while some residents find themselves to be too lazy to exercise, they have a background of understandable reasons to simply not feel very positive about the activity itself.

12.3 The Conventional Manuped Exercise Experience

From observations of the daily routines in the physical therapy, the manuped exercise was by far the most shared and commonly used. It was the most resource relieving exercise form for the physical therapists, but evidentally also the most repetitive and socially isolating exercise experience in the room. Over the years visiting the nursing home, numerous conversations were had with residents, concerning the manuped experience. Two iterations of interview rounds were done as well with residents in relation to the studies, where residents were asked to describe their opinions on the conventional manuped exercise [8, 9]. Interview asked into the positive and negative aspects of the (conventional) exercise, as well as opinions about possible improvements [9].

Positive aspects predominantly related to the function and usefulness of the manuped, as a facilitator of physical conditioning. Positive responses were, for example *getting the legs working, doing an effort to stay functional, to be able to do things,* or *physical improvement.* Negative sentiments referred more to the actual experience of the exercise, in terms of how it was *trivial, uninteresting, boring, repetitive, demanding,* and *hurtful,* describing the opinion about the activity, as e.g. *"you have to pull yourself there"* or *"you are caught once you enter that place"* [9].

An interesting situation happened when subjects were asked about possible improvements to the experience. Most subjects could not imagine any possible alternative. This was eventually found this to be characteristic for the user group. Nursing home personnel explained that many do not want to seem ungrateful of the offerings placed at their disposal, and have a tendency to humbly accept things as they are at the nursing home without thinking about questioning it.

Some responses did however encourage changes; by e.g. simply needing a *routine shift* or a *diversion*, while others again suggested the same, but coining it *entertainment*. A few had the imagination to associate to *"a real bike ride outside, alone, on your own, peace, and the smell of summer"* or *"to bike to Brighton"* (where she was originally from). The pattern is that positive responses about the manuped exercise were health oriented and rational, while negative responses related to the experience of the actual exercise activity, Potential-improvement responses related to the experience by removing the boredom and repetitiveness from the exercise form.

12.3.1 VE Augmentation: An Extra Layer

Based on the responses, it was decided to augment the exercise experience, to bring a new layer to the manuped, so the exercise activity would be accompanied by an audiovisual experience. A virtual environment (VE) was set up in connection to the manuped, to react to the residents' press on the manuped pedals. When they pressed the pedals, the VE moved forward and gave a sensation of a bike moving forward in space inside the VE. As such, a manuped exercise would (to a certain extent) reflect a bike ride through a (virtual) landscape.

This implementation was based on the following parameters. A fundamental part of the setup was to not change the form and function of the exercise. On the perspective in interaction design, and possible gameplay oriented elements, it was decided to the keep the interaction with the system based on the normal input scheme of the conventional exercise. This would make the augmentation simple and easy to transfer to the users. The background for this decision came from an intermediate focus group session performed with a Nintendo Wii (using Wii Sports), which attempted to establish and evaluate the residents interaction abilities and preferences, in terms of interaction possibilities to perform while exercising.

The focus group experience led to two conclusions:

(a) The residents in the focus group were practically unable to interact with the simple Nintendo Wii gameplay, due to cognitive and motor-based deficiencies,

(b) Aiming, from a developer's perspective, for a gradual improvement of interaction skills along a hypothetical timeline seemed dangerous. It was clear that the residents (all besides one) who were unable to perform the required actions within the Wii Sports game experienced a personal defeat and embarrassment, not a personal encouragement to learn a new skill. The rationale was that if even remotely more demanding interaction requirements were included, a substantial portion of the users might be alienated before they even started.

Even without changing the interaction significantly, it was expected that such augmentation could redeem some of the cardinal complaints of the conventional manuped experience. On a very basic level, the augmentation would change repetitive experience of looking into a blank wall during every single exercise, looking into a dynamic, constantly changing landscape view in motion. Instead of simply performing trivial pedaling, an augmentation would give the effort of pushing the pedals an immediate feedback reaction and possibly an immediately valuable payoff. Another possible result could be that the attention to pain and physical inability, could be changed to the augmented exercise which could provide a diversion of attention, distracting the resident's focus from the activity related pain, as also suggested by de Bruin et al. [15].

An aspect to the VE augmentation that could be have a strong impact is the ability to transform the indoor manuped bike ride to an "outdoor" bike ride. In addition, supposing that the link between actions and system response would feel natural, and that the VEs would be able to provide a convincing experience, residents might be given an option, which they might have thought to be lost - the ability to unassistedly travel to other places outside the nursing home. By that hypothesis, instead of being a lonely, uneventful and trivial single-person experience performed as rationally forced chore, the manuped exercise could suddenly be facilitating a unique, enjoyable and perhaps even personally valuable experience.

With the conventional manuped exercise, the only reward of the exercise actions is the long-term goal of better fitness, leading to gradually higher independence and self-efficacy. This long-term goal would now be supported by a short-term value from the daily ventures inside the virtual landscape.

As such, a hypothesis about the augmentation was that the hard work might transform into something more resembling leisure time—something that was previously suggested to be of high contrast to the manuped exercise. All in all, the goal for the manuped exercise would be a place where residents would want to be, instead of a place "you cannot wait to get away from".

12.4   Solution Design

As explained in more detail in [8], the first study of the concept was done through a small-scale, summer countryside VE, with 10 participants trying out a single exercise run. Results suggested definite promise, with user responses relating to the previously mentioned expectations for the augmentation. The system implementation can be seen in Fig. 12.2. To be able to track the user input actions, magnets and Hall Effect magnet sensors was attached to respectively the manuped pedal arm and manuped frame exterior, one for each side of the manuped to track each pedal push. The sensor was wired to an Arduino UNO microcontroller, which forwarded the sensor output to a desktop PC (Intel Core i5, 8 GB DDR3 RAM, ASUS GTX760 graphics card). The PC then was running a Unity 4.6 based build of the VE, which was visually displayed (running an average of 45 fps in 1080p resolution) on a Samsung 46' LCD TV. In the first study, a pair of Sennheiser HD650

Unity3d Software,
MAX/Msp 6,   46' Flat screen Display (right),
Desktop PC   Logitech Z623 2.1 Speakers

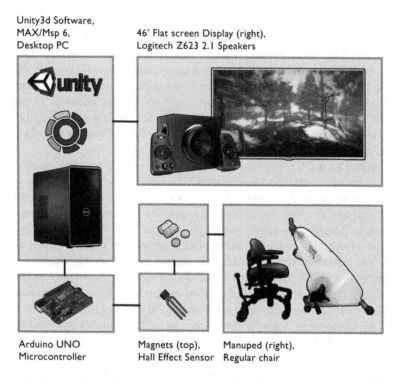

Arduino UNO   Magnets (top),   Manuped (right),
Microcontroller   Hall Effect Sensor Regular chair

Fig. 12.2 The augmentation system

headphones provided the VE soundscape, but as headphones were generally found to be intrusive, the later study used a Logitech Z623 Black 2.1 stereo speaker set.

The dynamic, moving visuals (examples from the first study VE in Fig. 12.3) provided a welcome change of "scenery" from the original static view of the wall. Residents did in fact connect the manuped pedal actions with the forward movement through the VE, and the introduction of a (virtual) landscape to explore were very positively received by most subjects. They accepted the experience of traveling inside the (VE) landscape, and many expressed something close to a thrill, concerning being able to experience the natural beauty of the landscape.

Fig. 12.3 The virtual environment used in the pilot study

Fig. 12.4 The four virtual environments used in the longitudinal study (with Danish titles)

The feedback of the system was perceived to place the (short term) purpose to the exercise actions, and one subject even mentioned that she now felt the ability to move herself forward outside the nursing home, which was a feeling she had not experienced for a long time (due to her weight issues in a wheelchair). In this sense, the pilot gave a real sensation of the exercise being transformed into something that could qualify as an enjoyable experience, which in turn placed it much closer to being classified as a leisure activity. However, the earliness of the implementation, as well as the fact that this experience was a first test, did not elude some of the subjects, who expressed the need for a higher diversity in VE content and more VEs to choose from, if the augmentation was to retain leverage over linger periods of time [8].

For the next study [9], a larger collection of (four) more content- and size-wise complex VEs was designed and implemented, using a design framework developed for the exact purpose [16] (as can be seen fin Fig. 12.4). The purpose was to test the concept over a longer period of time and to evaluate the residents' intrinsic motivation to exercise, comparing the experience of the conventional manuped exercise with the augmented one. During the development of the new implementation, over the course of approximately a year, the nursing home residents were only subject to the conventional manuped exercise. Prior to installing the new system, a select group of subjects were responded to a set of questions relating to their positive, negative experiences with the exercise, their suggestions to improvements, and their level and orientation of intrinsic motivation (as defined by Ryan and Deci [17] in relation to Self-Determination Theory) in relation to the manuped exercise.

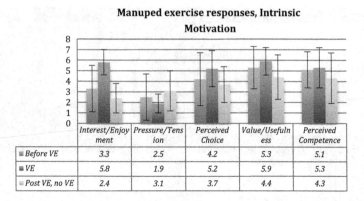

Manuped exercise responses, Intrinsic Motivation

	Interest/Enjoy ment	Pressure/Tens ion	Perceived Choice	Value/Usefuln ess	Perceived Competence
▪ Before VE	3.3	2.5	4.2	5.3	5.1
▪ VE	5.8	1.9	5.2	5.9	5.3
▪ Post VE, no VE	2.4	3.1	3.7	4.4	4.3

Fig. 12.5 The levels and factors of intrinsic motivation from the longitudinal study [9]

The level was measured on a 7 point scale, whereas the orientation was measured based on items relating to different factors of intrinsic motivation, taken from the Self-Determination Theory related Intrinsic Motivation Inventory[1] (IMI). The orientations (factors) chosen for the study were Interest/Enjoyment, Pressure/Tension, Perceived Choice, Value/Usefulness, and Perceived Competence [9]. Subsequently, the new augmentation was installed and ran at the nursing home for more than four months, after which the residents were asked to respond to the same questionnaire, only this time concerning their experience with the augmented exercise experience. In addition, they were also asked to respond accordingly to how they would feel about going back to exercising without the augmentation [9]. Results concerning the IMI orientations were very interesting, and can be seen in Fig. 12.5. Subjects responded as expected before the augmentation had been introduced ("Before VE"), if not slightly more positive than expected. As previously stated, it was the impression that they did not want to seem ungrateful, and that they were generally trying to be content with the opportunities they were offered. Results concerning the augmentation ("VE") were consistently more positive towards the augmentation, and most said it had become an integrate part of their reason to want to exercise on the manuped. And results from residents concerning returning to the conventional exercise ("Post VE, no VE") showed a clear negative curve, compared to both the augmentation and especially the original responses to the conventional manuped exercise. Four of the five motivation orientations had similar patterns between the three "conditions", with a not too high increase from the conventional towards the augmented exercise, and then a larger decrease in motivation in relation to returning to the conventional exercise. One motivation orientation, Interest/Enjoyment followed the same trend, but stood out by having much larger jumps between values.

---

[1]http://www.selfdeterminationtheory.org/intrinsic-motivation-inventory/.

The subject responses to the positive, negative and possible improvement aspects of conventional manuped exercise, all suggested that which was previously explained in this chapter; a useful activity for the functionality of the body but boring and repetitive. The augmented was praised highly in these categories, and the thought of returning to the conventional exercise form was poorly received. Meanwhile, by far the most clearly significant difference between "conditions" in Fig. 12.2 is that of Interest/Enjoyment, whereas the remaining orientations are much more similar. What the original publication suggested as the reason, was how many aspects of the other orientations were already quite ok for the conventional exercise form [18]. The results suggested improvement by the augmentation over the conventional form, but none even remotely close to the difference of Interest/Enjoyment. Combined with the user responses on positive/negative/ improvements, which very clearly suggested the augmentation experience to be a substantially superior exercise experience, it looks logical to suggest that the leisure and entertainment aspect of the activity itself has proven to be a strong factor in improving the desirability of such type activity for nursing home residents. From that perspective, the inclusion of the augmentation, as a transport to another place, as well as being an entertaining layer on top of a meaningful exercise, could be argued to make a lot of sense.

Both studies showed also suggested how content variety, details and consistency was important for the users experience of riding through the VE, as well as liveliness inside the VE (animals, water, wind, etc.) [8, 9]. These aspects to the experience give believability to the experience, and add to the sensation of a recreational ride through nature, as also highlighted more in a paper by Bruun-Pedersen, Serafin and Kofoed, discussing various key considerations to design challenges when creating custom VEs for this purpose [18].

12.5 Research Challenges: Nursing Home Residents

Essential to the research just described was a constructive collaboration relationship with the nursing home residents. It demanded to meet them in their environment, during their everyday life routines at the physical therapy, and even required disturbing their privacy, unannounced, on occasion in their private homes.

Nursing home residents have unique characteristics as a user group, and it is necessary to consider those contextual properties in the planning and execution of studies. Many of these characteristics were not considered when beginning the project. Neither was the importance and necessary degree of empathetic understanding of the individuality and context of the residents, all of which were gradually discovered to be cardinal parts of conducting the research with residents.

12.5.1   Planning of the Study

On the practical level, it became necessary to recognize the daily rhythm, capabilities, and limitations of the residents as subjects. In this sense, planning is extremely important, but so is a high level of flexibility. Sudden changes cannot be expected from residents, but at the same time it is not possible for a researcher to rely upon agreements or schedules bring upheld by residents. This is not due to conscious neglect, but most often to the constant risk of illness, otherwise immediate loss of strength to participate or simply dementia. In this perspective, studying nursing home residents requires patience and studies require a lot more time than what would seem rational on paper.

Numerous unforeseen occurrences can happen from one day to the next. In the first study, 15 residents signed up, 10 participated, and the study took twice the time to perform the single trial in the pilot VE than expected. In the 4-month study, 17 agreed to participate and 8 completed the study (new subjects were added to the pool along the study, however), due illness, dementia, amputations, death, or hospitalization. It has also occurred a few times that a subject suddenly declined participation with no explicit explanation or reason, but overall, this type of situation has been rare. Over the course of a longitudinal study, a pre-defined group of subjects is therefore bound to change, sometimes dramatically, and it is a part of this type of research to follow its course.

12.5.2   Keeping the Residents Reminded

Especially when performing longitudinal studies, it can be very challenging to keep a position in the minds of residents unless the routine is soundly established in their schedule and mind. As many suffer from dementia and have poor recollection of people or faces, a resident would forget the face of the researcher, or the purpose of an arrangement made prior, on multiple occasions during the studies. Being part of their everyday life routines to the extent of being considered a part of the everyday actions is very constructive for many aspects of this type of research. It relates to residents being confident that social interactions will be pleasant and respectful has shown noticeable differences in the type of working relationship possible within the user group.

Another crucial way to keep focus is through a clear and constructive relationship to personnel. For longitudinal studies, it simply heightens the probability for support of the agreements made with residents for various study procedures, as well as performing practical tasks such as noting measures, etc., under circumstances where the research team not present at the nursing home. Perhaps even more importantly, the all-important relationship between researcher and residents (which will be addressed later) goes partly through the personnel as well.

Personnel essentially work as a gateway to be accepted by the residents, both directly and indirectly. The direct part is to introduce the affiliation of the researcher and the nursing home towards the residents. Personnel often have the trust of the residents, who will accept foreign people if they are clearly trusted and introduced by the personnel. Indirectly, the residents also observe many things, for instance the relationship between the personnel and the research team. Residents have described using this as part of the 'measuring stick', to evaluate whether a researcher was initially someone they want to interact with.

12.5.3   Establishing and Maintaining a Routine

Nursing homes like Akaciegården are much about rhythm and habit. Every day has a standard operation schedule (morning/evening assistance for individual residents, meals, etc.), and every week has a repeating activity schedule for specific week-days, such as bingo, singing class, physical therapy, rhythmic gymnastics, the in-house hairdresser, a beautician, etc. Nursing home living thereby follows a recognizable pattern, which residents should be able to comfortably learn and rely upon. In addition, singleton events such as concerts, movie viewings, or various celebratory/traditional events occur on a regular basis throughout the year. Such general format for week schedules is necessary, as residents with decreased memory/cognitive capabilities are challenged even when navigating such recurring schedule. In addition, singleton events are not in small quantity per year. And while residents enjoy them, some need personnel to help keeping track of the activities interesting to them.

When performing especially longitudinal studies with residents, it is crucial that resident participation is structured, so that it becomes part of the fixed weekly rhythm. Otherwise, there is a high risk that they will forget or not enable themselves to go, because they might need help to get to the physical therapy and either won't ask (out of politeness) or simply cannot get help from the resources of personnel at a given time. A routine needs to be established around the group of residents who have agreed to participate as subjects. It thus becomes necessary to involve all personnel affiliated with each individual resident participating, and make detailed arrangements and agreements concerning each individual resident's study proce-dure activities. Residents regularly need personnel to remind them of their daily activities, and the study procedure needs to become part of this routine.

It even became necessary to schedule when individual subjects would be able to access the manuped, simply because of an overweight of subjects initially (and randomly) wanted to use the single available manuped device at the same time each session. To ensure the procedure to run smoothly, access to the manuped for residents was then scheduled into specific time slots, so certain divisions of resi-dents had access to the manuped, at least on certain days, at least in certain time frames. Of course, none were unwelcome at any time during physical therapy

Fig. 12.6 A3 poster placed at all common areas to help residents remember their engagement

hours, but typically residents would follow a routine given to them, and not deviate too much.

Setting up such type of longitudinal study at nursing homes involves a surprising amount of people, besides the administration for permission and the physical therapy team working at the clinic. The nursing home facilities used in the studies are split into 8 separate departments. The group of residents used for the studies were spread on 6 of these departments. Each department has individual teams of personnel, and each team has changing shifts, substitutes and volunteers, whom all need to be informed to ensure that the routine is running every week. Each department needs to know which of their residents need to go and when, and make sure to fit this into their schedule concerning preparing and aiding the resident with clothing, baths, etc. outside the research schedule.

Besides having personnel assisting residents to uphold the scheduled research activities, residents' general connection to their participation agreement seemed to benefit as well from exposure to various types of paper media. A3 posters with images of the recreational VEs and a member of the research team with the physical therapists (as seen in Fig. 12.6) were placed in all central traffic areas inside the nursing home, serving as a constantly occurring reminder for subjects on their daily tours round the building.

In the monthly (in-house) nursing home newspaper, a page was dedicated to the purpose of the project, using short text and large pictures (see Fig. 12.7). This was confirmed by many residents to add to their awareness of the exercise initiative.

Fig. 12.7 A3 page from the in-house news paper

In conjunction, it was important for the research team to be present and visible at the nursing home as much as possible, despite not always having an active role in the data gathering.

12.5.4 A Trusting Relationship

No matter the methodology, studies that require inter-personal interaction of any kind with nursing home residents will include a variable degree of qualitative aspects. Whereas this inter-personal interaction might not be directly linked from the empirical aspects of the research, it is very difficult to separate the person and the subject participation, when working with residents.

According to Truglio-Gallagher et al. [19], a personal relationship between researcher and subject is fundamental for qualitative inquiry with older adults, and most advantageously obtained through the establishment of trusting relationship between older adults and researcher. As qualitative approaches such as questioning, conversation or interviews invites for replies to inquiry, the quality of subjects' responses, meaning the quality of insight into the older adults' experience, obviously depends solely on the willingness of respondents to converse and share their experiences [19]. But whether or not the method is in fact qualitative, a trusting relationship and faith in the researcher is vital for the sheer possibility of retrieving information, as showed from the previous example, where the collection of quantitative data had to be conducted through a (for residents) demanding, pseudo-qualitative interview approach. Their hardship and effort in completing the cumbersome quantitative responses, clearly supports the notion from Truglio-Gallagher et al., on how trust and faith between the parties are essential for any meaningful communication, partnership and quality of data with older individuals [19].

12.5.5 Personal Boundaries

During the described studies, the trusting relationship to residents sometimes meant the difference between subject participation or not. Residents with whom it was not possible to build a personal connection or a type of social comfort, were difficult to recruit as subjects, and if recruited, showed low engagement into the routines. They also displayed very limited willingness to share their experiences when asked. Residents with whom it was possible to build a personal connection, showed the opposite traits.

According to Haal et al., trust can generally be regarded as the "the optimistic acceptance of a vulnerable situation in which the trustor believes the trustee will care for the trustor's interests" [20]. For older adults such as nursing home residents, vulnerability is a central phenomenon due to their gradually increasing physical and mental limitations, and thereby overall frailty [21], and a relationship depends on the inherent knowledge that they will not be harmed, by the impression of the good intention of the other person [19], and researchers must place careful attention to how they are to 'connect' to residents [22]. Connecting is not trivial, as individual residents have a varying degree of curiosity or acceptance to new elements in their everyday routine, which ranges from very open minded, to very cautious and alienated. Being mindful of the personal boundaries of a resident can prevent unfortunate situations, where a personal space or boundary is overstepped. It much depends on the personal situation of the individual resident (private life, illness, death of a friend, etc.), but to a degree, which was not initially considered sufficiently during the studies.

Moreno-John et al [23] point out in their literature review on trust relating to older adult participation in clinical research; how many groups of older adults have a general mistrust to the healthcare system as well as researchers, in addition to some older adults showing reluctance to sign consent forms [19]. The research team representatives was not initially aware, and thus did not consider this in one situation, where a female resident was asked to fill out a letter of consent. A signature would simply allow us the usage of video footage of her verbal and behavioral responses, exclusively for transcription purposes. When presented with the document for signature, she looked at the researcher with fear, took distance and expressed that she would not sign or further participate—and that she would now like to be left alone. After the situation had played out, staff gave background into how this particular resident had just placed a series of signatures for power of attorney, for her family home to be sold and most of her belongings to be taken to storage, only a few days before. This had left her very sensitive to formal documents, and people who wanted to intrude her personal life with demands and restrictions of freedom over her property. In reference to the above quote from Hall et al [20], is clear that this particular situation had no optimistic acceptance with the so-called trustor of the interests of the trustee, leaving the resident feeling exclusively vulnerable. The example illustrated the importance in creating a comfortable and trustful social space between researcher and residents, before initiating active processes or making direct requests. Introducing the written consent too early in the process resulted in mistrust and a loss of a subject for further studies. It will never be known whether the just described situation could have been avoided, but the experience served as an onwards reminder to remember, to patiently develop a personal connection to residents before proceeding to personal requests or commitments.

12.5.6 Establishing a Connection

Trojan and Yonge recommend investigators to think carefully about how to connect to potential participants [22], and in the case of the nursing home, establishing a personal relationship was eventually found not to be very complicated, but require time and personal investment.

In the beginning, it was found to be very difficult to approach residents and make simple, meaningful conversation. In hindsight, this was primarily due to generation-cultural differences (such as choice of language and terminology, as well as conversation topics and rhythm). Becoming accepted to a degree where individual residents would open up to a researcher, took very different amounts of time of obtain. Depending on the individual, this could literally take between minutes and months. And due to certain "less accessible" residents being obvious choices as subjects, some of the long periods spent to create a personal connection were deemed necessary.

Gilson points out how inter-personal trust evolves over time [24]. A substantial amount of effort to be present and visible at the nursing home was therefore placed into activities such as participating at the physical therapy center, in planned social events, visiting residents at their private apartments for a casual chat, or bringing family (kids) to the nursing home, in the attempt to become familiar face. The many hours spent allowed a lot of social interaction and personal connection. There is no doubt that ever since, making conversation, asking for participation, or conducting studies felt natural, and became significantly easier in relation to acquiring subjects, requesting their time, making appointments and retrieving information.

Spending time at the physical therapy was necessary and insightful in many ways for the purposes of understanding the rhythms, capabilities and limitations of the residents in an exercise-oriented context. However, for building personal and trustful relationships, the most constructive place to do so was at residents' private apartments, clearly being the location inside the nursing home, where most residents tend to feel relaxed and comfortable. In addition, apartments had a useful application when performing the first couple of visits, as some residents would not contribute much in the beginning of sessions, not yet knowing the researcher or procedure well. Most private apartments had many objects of meaning to the resident, such as images of grandchildren, or other possessions of personal significance. If the initiating conversation (for instance before the start of an interview or questionnaire) did not fuel itself naturally, such objects were extremely well suited conversation starters/topics, sometimes resulting in long conversations. In these cases, such topics gave residents a sensation of good intentions and focused, personally directed interest elicited by the researcher. The result was most often a strong foundation for further conversation, and leeway into research related topics. This conversation environment was presumably reminiscent to what Truglio et al. calls the "cornerstone in all qualitative studies" [19], being the point where trust is completely established and present, for proper dialogue to take place. With most residents, this was typically marked by a certain "critical" point, from which they will start talking and telling stories almost endlessly. Not only (and sometimes almost not at all) in relation to responses on asked topics or items, but past life stories, thoughts, inquiries to the researcher, etc. It would not be uncommon in this situation, to almost have to struggle for speaking time for the researcher.

In short, some residents have a barrier that needs to be 'broken down' gradually, respectfully and patiently by personal engagement from the researcher, into whatever captures their interest of conversation. After a certain time spent together, it has been the experience that most residents happily accept the relation, and thereby have become ready to partake in most interactions or challenges later proposed by from the researcher. To the experience of this author, the personal relation is *the* central aspect to achieving leeway for the research collaboration.

To fully appreciate the rationale behind this, it is important to remember the context of the nursing home, in terms of why the impression of an honest, personal engagement and interest with the residents has value to them. The nursing home, for all its merits, can be an extremely lonely place for many residents. While a certain group of the residents have (the luxury of) actively visiting family and friends, many do not. Friendships do establish themselves between residents, but are subject to a combination of high risk of illness, high level of cognitively limited residents, and the frequent exchange in residents at the nursing home. When approached, many cognitively capable residents really like a meaningful dialogue whenever they can get one; meaningful in this context is (as described above) personally directed, attentive, interested and respectful. In the context of their everyday life, it is the impression that this is one of the things to which many residents do not have access to the degree they used to have. And for the cognitively well functioning residents, the experiences from conversations leaves this seeming like a substantial loss of everyday value and meaning.

Once personal relations are properly established, there are very few things this user group will not do for e.g. a research group. The experience has been that if personally engaged residents are in any way capable of helping the researcher towards his or her goal, making the effort to form relations a desirable investment to the resident. The positivity of this engagement is elaborated by Truglio-Gallagher et al., in how participation in research endeavors is able to provide some older adults a personal learning exercise, as well as a sense of importance from the personal role they play, providing a service and contribution which benefits others [19]. This has been recognizable throughout the project, by a sense of personal pride with certain resident individuals, from their contributions to the studies. A typical scenario would be a resident greeting the researcher, eagerly describing how many times or how long (etc.) the individual resident had been exercising with the augmentation since last time the researcher and resident had last seen each other. Or how the resident could actually feel a physical improvement since beginning the exercise routine anew, from following the established participation schedule.

12.5.7 The Difficult Conversations

There are some aspects to the interaction with residents that a researcher (in the role suggested in this chapter) has to be aware of and prepared to handle. For many reasons, the context of the nursing home also somewhat signifies conclusion for many. It is the last place they will most probably live, if not for a hospital in the very end. It is a place they *have* to live, because they or their relatives have shown incapable of properly maintaining an independent lifestyle. And for many, it is a place they live because they have outlived their contemporaries, amongst those partners or even children.

The nursing home lifestyle serves as constant a reminder of this, not only due to the change in environment from their life as it previously was, but due to the nursing home environment. The sheer amount of people partly or completely sharing this lifestyle is substantial, and could possibly serve as an overwhelming reminder of a specific individual's situation. As such, many residents find them selves looking backwards more than forwards. At this advanced age, the combination of social isolation and loneliness, as well as comprised functional ability, leave many residents very vulnerable at times [19].

In the process achieving or maintaining a personal relationship through conversation, some residents pose the unique challenge of intense conversation topics concerning loss, death, and loneliness. Insights provided by senior nursing home personnel, such conversations are very important to the residents. Thus it is necessary for anyone who has ambitions to be a meaningful person for a resident, to fully submerge and engage into these conversations. Its importance was further highlighted by the notion that such conversations are not possible that often for the residents, as many social and health (Sosu) assistant personnel deviate from these topics when confronted. Despite most likely wanting to help residents, many Sosu assistants are simply not personally equipped to handle such comprehensive conversation topics.

Engaging in such conversations in the role of a researcher can make a real difference, for the purposes of establishing or maintaining personal relations. They *will* arise with most residents, and it is important to make the choice early on, to not refrain from the subject matter when it arises.

12.5.8   Advancing with VE Technology

One of the challenges with initial studies was for residents to participate with unknown and exotic technology, being unsure of whether they would have the necessary skills or perform as expected. Explaining the setup and making agreements with a majority of the residents was therefore not trivial.

When initiating the studies, many residents needed much time to get familiar with the technology and VEs. Similar to establishing personal relations with patience, introducing technological advancements would need to happen slowly. This was the reason the studies presented in this paper were performed using an LED TV, instead of using a Head Mounted Display (HMD). The HMD would almost certainly provide a more immersive and convincing experience of being inside the VE. But it was feared that pushing technology too fast would overload residents' upper threshold for novel technology experiences per time.

Over the course of the 4-month study, many residents developed a relationship to their favorite VEs, as well as a good relationship to the VE augmentation system. Increasingly positive attitudes towards the augmentation were seen from from residents. In periods where it has been shortly unavailable, residents have asked if they could get it running again soon. This has aspired confidence in relation to commencing with HMD studies, now that residents are experienced with the VEs, and confortable with the LED TV based exercise augmentation.

12.5.9   Gathering Data

The items from the Intrinsic Motivation Inventory (see Footnote 1) (IMI) were in one of the studies used to measure the level and orientation of the motivation effect of the VE augmentation of the manuped exercise. However, using quantitative measures with nursing home residents was found to be a substantial challenge, and will not be replicate in future studies, if avoidable. First and foremost, written generally conflicted with often occurring instances of compromised vision for many residents, where small letters would be practically invisible, not to mention that some residents was actually lacking the ability to hold and place a pen to set their cross. What showed to be a central challenge however, was how most residents were conflicted by the arbitrary nature of placing their rating.

In the IMI questionnaire, items are statements to which subjects need to state their level of agreement to, corresponding to a 7-point Likert scale, (1 corresponds to "not true at all", 4 is "somewhat true" and 7 corresponds to "very true"). While this method could seem straightforward to most, many nursing home residents, with the variance of cognitive limitations between them, seemed to severely limit their ability to fully comprehend this, perhaps slightly abstract concept. Only very few residents were able to fill out the form by themselves. Almost all residents required assistance from the present researcher, who was forced to adapt and read the items aloud, and subsequently try to conceptualize and step-wise isolate a correct answer. An example would be, having to read an item aloud and asking, "*would you say that you agree or disagree?*" Depending on the answer, the procedure would lead to a follow-up question such as "*so would you say that you completely agree, or only somewhat agree?*" A last follow-up question would be "*so, would you say that you lean towards better or worse than (answer)?*" In practice, the 17 IMI items would most often transform from a written form into a structured interview guide for the researcher, conducted through verbal conversation. In this adapted form, the response sessions were highly uncomfortable for both researcher and resident, due to how a verbal depiction of something quantitative is a challenging exercise. Based on this experience, the advice concerning this user group and data gathering would be to keep to purely qualitative methods, at least in anything related to user responses. In addition, qualitative measures, which encompass any sort of interview, speak to the residents' desire to partake in conversations and have another moment to share their life story to someone who genuinely wants to listen.

12.6   Conclusions

This chapter has looked into various conditions relating to the lives of nursing home residents, predominantly through the search for an explanation for why many residents lack the desire to perform regular exercise, despite the obvious benefits. It has described the circumstances for the choice to either exercise or not, and presented one suggestion to a solution, in the form of the VE augmented manuped exercise. The new manuped form sought to satisfy a need for short-term entertainment, to complement the long-term benefits of the exercise. It also provided an immediate feedback to the exercise actions, which gave a different sense of purpose to the exertion of the exercise, among other things. Besides that, the chapter suggests that the augmented manuped experience offered something not found in other regular nursing home activity offerings, which was the ability to travel to another place, and see and experience recreational landscape environments otherwise impossible for the residents to experience. The second part of the chapter looked into more circumstances and characteristics of the nursing home environment, seen from a researcher's perspective, in term of some concrete challenges and experience-based suggestions to constructive approaches to constructively improve the chances of success, when performing research with nursing home residents. There are many obstacles, but also many constructive solutions, which should be able to ease the working conditions of the researcher, as well as making the collaboration a better experience for the nursing home resident.

References

1. Lutz W, Sanderson W, Scherbov S (2008) The coming acceleration of global population ageing. Nature 451(7179):716–719
2. Bainbridge E, Bevans S, Keeley B, Oriel K (2011) The effects of the Nintendo Wii Fit on community-dwelling older adults with perceived balance deficits: a pilot study. Phys Occup Ther Geriatr 29(2)
3. Sugarman H, Burstin A (2009) Use of the Wii Fit system for the treatment of balance problems in the elderly: a feasibility study. In: Virtual rehabilitation international conference. IEEE, 111–116
4. de Bruin PDE, Schoene D, Pichierri G, Smith ST (2010) Use of virtual reality technique for the training of motor control in the elderly. Z Gerontol Geriatr 43(4):229–234
5. Laver K, Ratcliffe J, George S, Burgess L, Crotty M (2011) Is the Nintendo Wii Fit really acceptable to older people?: a discrete choice experiment. BMC Geriatr 11(64)
6. Ijsselsteijn W, Nap HH, de Kort Y, Poels K (2007) Digital game design for elderly users. In: Conference on future play (Future Play '07). ACM, NY, pp 17–22
7. Brunet J, Sabiston CM (2011) Exploring motivation for physical activity across the adult lifespan. Psychol Sport Exerc 12(2):99–105
8. Bruun-Pedersen JR, Pedersen KS, Serafin S, Kofoed LB (2014) Augmented exercise biking with virtual environments for elderly users—A preliminary study for retirement home physical therapy. In: Proceedings of VR 2014 - Workshop on Virtual and Augmented Assistive Technology (VAAT 2014) (Minneapolis 2014)
9. Bruun-Pedersen JR, Serafin S, Kofoed LB (2016) Motivating elderly to exercise—recreational virtual environment for indoor biking. In: 4th international conference on serious games and applications for health (Orlando 2016), IEEE Xplorer
10. Kofod J (2008) Becoming a nursing home resident: an anthropological analysis of Danish elderly people in transition. Doctoral dissertation, Department of Anthropology, University of Copenhagen, Copenhagen
11. Spirduso WW, Cronin DL (2001) Exercise dose-response effects on quality of life and independent living in older adults. Med Sci Sports Exerc 33(6):598–608

12. Landi F, Abbatecola AM, Provinciali M et al (2010) Moving against frailty: does physical activity matter? Biogerontology 11(5):537–545
13. Lieberman DA (2013) Effects on knowledge, self-efficacy, social support. health promotion and interactive technology: theoretical applications and future directions. In: Street RL, et al (eds.) Health Promotion and interactive technology: Theoretical applications and future direction, 1st edn. Routledge Communication Series
14. Silveira P, van Het Reve E, Daniel F, Casati F, de Bruin ED (2013) Motivating and assisting physical exercise in independently living older adults: a pilot study. Int J Med Inform 82 (5):325–334
15. Bruin PDE, Schoene D, Pichierri G, Smith ST (2010) Use of virtual reality technique for the training of motor control in the elderly. Z Gerontol Geriatr 43(4):229–234
16. Ryan RM, Deci EL (2000) Intrinsic and extrinsic motivations: Classic definitions and new directions. Contemp Educ Psychol 25(1):54–67
17. Bruun-Pedersen JR, Serafin S, Kofoed LB (2015) Simulating nature for elderly users—a design approach for recreational virtual environments. In: Proceedings of the 2015 IEEE international conference on computer and information technology; ubiquitous computing and communications; dependable, autonomic and secure computing; pervasive intelligence and computing. IEEE Computer Society Press, pp 1566–1571
18. Truglio-Gallagher M, Gallagher LP, Sosanya K, Hendrickson-Slack M (2006) Building trust between the older adults and researchers in qualitative inquiry. Nurse Res 13(3):50–61
19. Hall MA, Dugan E, Zheng B, Mishra AK (2001) Trust in physicians and medical institutions: what is it, can it be measured, and does it matter? Milbank Q 79(4):613–639
20. Fried LP (2003) Establishing benchmarks for quality care for an aging population: caring for vulnerable older adults. Ann Intern Med 139(9):784–786
21. Trojan L, Yonge O (1993) Developing trusting, caring relationships: home care nurses and elderly clients. J Adv Nurs 18(12):1903–1910
22. Moreno-John G, Gachie A, Fleming MC, Napoles-Springer A, Mutran E, Manson SM, PÉrez-Stable EJ (2004) Ethnic minority older adults participating in clinical research developing trust. J Aging Health, 16(5), 93S–123S
23. Gilson L (2006) Trust in health care: theoretical perspectives and research needs. J Health Organ Manage 20(5):359–375
24. Ryan RM, Deci EL (2000) Self-determination theory and the facilitation of intrinsic motivation, social development, and well-being. Am Psychol 55(1)

Part V
Ethics and Accessibility

Chapter 13
DigitalEthics: 'Ethical' Considerations of Post-research ICT Impact

Anthony Brooks

Abstract This contribution posits a supposition on the topic of 'ethical' consideration when advanced, engaging, and playful ICT, i.e. beyond that available to the public, is researched. In focus are participants without communicative competence —especially young children diagnosed with Profound and Multiple Learning Disabilities (PMLD)—who are unable to express their desire for continued access/use/play. The position results from a mature body of empirical work that began in 1985 that, through specific case studies, stresses the need to contemplate affect on participants, thus promoting funding applications to include post-research access to content for participants as well to support staff intervention training to optimize content use.

Keywords Ethics · ICT · Healthcare · Games · Virtual reality

13.1 Introduction and Grounding—A Retrospective of the Field (Selected)

Advances in digital content and means to interact with such content are ever-present and similarly advancing to enable wider accessibility and thus inclusion for all. Contemporary natural user interfaces (NUIs) advanced the keyboard, mouse and joystick paradigm and control of content can these days be via held, worn, or gesture-based devices with varying profile. Aligned to this are the various 'maker' communities that use microcontroller platforms to develop affordable sensor-based control devices/interfaces, e.g. the Arduino, Raspberry Pi, etc.

At the start of the 21st century commercial games by Nintendo, Sony, and Microsoft opened the doors for healthcare and others to explore in research investigations on potentials for motion-focused sessions in treatment programmes. Such pervasive use of sensor-control were preceded by smaller entities e.g. in

A. Brooks (✉)
Aalborg University, Esbjerg, Denmark
e-mail: tb@create.aau.dk

273

United Kingdom (Soundbeam[1] and their ultrasound linear profile sensors mapped to generate MIDI to sound synthesizers). Similarly, in UK from mid 1980s and subsequently in Denmark from 1992, the author realized HandiMIDI—later known as SoundScapes from the mid-nineties [1]. Besides biofeedback worn sensors, bespoke infrared volumetric sensors that enabled gesture in invisible space to generate MIDI were created that could be mapped to control multimedia e.g. sounds, images, animations, and robotic devices [2]. The research realized an early (circa 1999) games-based company titled Personics.

In Canada, Axel Mulder began his Infusion[2] sensor hardware focused company to create and market a plethora of sensor devices to satisfy early demand in the 1990s and ongoing...such 'maker' outlets are now commonplace for home hobbyists, students, and inventors.

The human sensing field advanced with the advent of improved camera-based interfaces alongside more powerful computers and dedicated vision processing cards resulted in commercial product—albeit expensive initially. Pioneering the field in Canada were Vincent John Vincent and Francis MacDougall who claim to have invented video gesture control with green screen background subtraction in 1986 under their start-up incorporated company Very Vivid, Inc. Using an Amiga computer with monochrome cameras and a simple video capture board an interactive musical instrument was created where a performer (usually Vincent John Vincent) could manipulate his/her hands to trigger predefined regions of the surrounding air space resulting in the generation of musical sounds on MIDI driven synthesizers. The system was performed live on stage at Siggraph and they received their base patent in 1996 for the GestPoint video gesture control software based on computer vision techniques. This led to the company GestureTek[3]—who licensed their patented technology in the field to Sony for EyeToy on PlayStation, to Microsoft for XBOX Live and Kinect, and to Hasbro for Ion Education Gaming System. These systems utilized camera capture with background subtraction gameplay so that affordable unencumbered gesture control (i.e. using free air space) was available in commercial devices. However, certain of these systems restricted hacker access and game mappings were closed. A step towards the desired developer access to the sensing devices was the Kinect where a SDK was available (similarly the Nintendo Wii was explored by developers via third-party software).[4]

Related work that likely influenced Vincent and MacDougall was likely that by Myron Krueger, who it is claimed—

---

[1]http://www.soundbeam.co.uk.

[2]https://infusionsystems.com.

[3]http://www.gesturetekhealth.com/about/history.

[4]The Nintendo Wii was able to be accessed via the OSC protocol— http://www.osculator.net.

… envisioned *the art of interactivity*, as opposed to *art that happens to be interactive*. That is, the idea that exploring the space of interactions between humans and computers was interesting. The focus was on *the possibilities of interaction itself*, rather than on an *art project*, which happens to have some response to the user.[5]

—see Krueger's books Artificial Reality (1983) [3] and Artificial Reality 2 (1991) [4] for further. The work of Krueger, as exemplified in this statement, is an influence on this author.

More recently, such legacies continue e.g. in software such as in Italy Complementing was numerous software that enabled visual programming of digital signals so that various routings, scaling and other manipulations to the signal could be easily changed. Examples include The Eyesweb Project[6] by the InfoMus Lab, University of Genoa, Italy, which began in 1997; in Germany Palindrome and Frieder Weiss' camera-based EyeCon[7] software has been distributed since 1995; in San Francisco, USA in the form of software via Opcode—now Cycling74—in the form of their Max/MSP/Jitter[8] product, which was adopted early by academia and artists. Max/MSP/Jitter was the author's main tool to map and adjust the sourced gesture data to control responsive audio and visuals throughout the 1990s and first decade of the 21st century. The software grew from the original Max software to include MSP for audio control and Jitter for visual control within the same visual programming environment. Also, external routing was possible. Subsequently Ross Bencina's AudioMulch[9] (auditory) and Mark Coniglio's Isadora[10] (visuals) are also notables explored amongst the growing field of easily accessible visual programming environments. Through these offerings developer adopters could programme and/or design interactions, stimuli, and mixed 'n' matched systems according to context and use across applications. The created environments, being data-based, also offer opportunities for archiving and analysis to evaluate affect through use. The author's experience of these cited apparatus and softwares provide grounding upon which comparative selections were made towards optimized system for each specific intervention and targeted outcome.

Reflecting these advances in sensing, as cited, bodies of research emerged that included examining impact in special needs, health and wellbeing fields. These are elaborated elsewhere in the authors publications. It is with this focus and specific to a case study involving four PMLD young children who had no (or very limited) means of communication, that this contribution posits a supposition questioning the need for increased consideration of post-research impact.

---

[5]Cf https://en.wikipedia.org/wiki/Myron_W._Krueger.

[6]http://www.infomus.org/eyesweb_ita.php.

[7]http://www.frieder-weiss.de/eyecon/.

[8]https://cycling74.com.

[9]http://www.audiomulch.com.

[10]http://troikatronix.com/isadora/about/.

Case Study

The case study referred to took place at the turn of the century in Denmark's second city Aarhus. The Centre for Advanced Visualization and Interaction at Aarhus University CAVI[11] hosted the investigation. Three CAVI technical rooms were used. The four children, all male, were of an average age of approximately five years. The rooms consisted an active Virtual Reality (VR) environment, a Holo-bench, and an Experimentarium, the latter being a large empty room with white walls. The focus of this case is activity in the VR room. Design was for a non-formal intervention so that the children could experience motion control of responsive stimuli via their residual abilities. It is important to state that the children were so profoundly disabled that it was extremely difficult to determine reactions without the aid of their nearest carer who interpreted the represented signals from each child. Non-verbal semiotic representations, utterances and individual nuances of gesture are typically used.

The children controlled robotic moving head lighting devices in the 'Experimentarium' via the author's bespoke infrared volumetric sensing system using a MIDI to DMX translation to communicate with the robot lights. All parameters are available such that options on mapping are available with two sensors set up for navigation channels—the most dynamic change parameter as it entailed a Human-Physical-to-Robotic-Physical coupling—with a third sensor selectively determining color, effect, gobo according to gesticulations. Audio feedback was also set up using a MIDI synthesizer. The auditory content was in the form of sounds, musical phrases, loops and effects as discussed with care-provider in attendance regards each child's preferences. This strengthened associated feedback stimuli i.e. motion to visual feedback, motion to physical robotic feedback, and motion to auditory feedback. No on-body sensing was used to capture gestures and this was astute as typically to prepare and establish optimal positioning and minimize noise is a challenge. A challenge that in the past has resulted in subjects falling asleep!

In the 'Holobench room' the author's game (a simple flying plane) navigation was set up with the author's bespoke infrared sensing system. This was not an optimal set up as the children had difficulty in viewing the monitor.

For the active Virtual Reality game content a 3D Star Wars fighter jet was set up to be navigation-controlled by head gesture within a double infrared volumetric sensor profile assigned to appropriate axis on the model—thus two sensors instead of the typical three.

Programming of the environments was conducted to reflect individual competencies with additional fine-tuning of parameters during sessions when and if needed.

Across the three rooms, many positive experiences were observed of all children. These observations have been reported elsewhere in the author's body of work and are not the main focus of this contribution. The catalyst of this contribution

---

[11]http://cavi.au.dk/about-cavi/.

developed following a post-research period contact to the institute where the children were volunteered from and especially to the parents of one of the children who we shall denote as M.

To conduct the initial research all necessary ethics and approvals were in place for showing the archived videos in research presentations. However, such was the impact in the research presentations, additional permissions were sought to further show the outcome videos in the public domain so that families, friends and professionals in the field could be made aware of potentials. Thus, contact was made to request the additional permissions. The parents gave their full permit for such dissemination; however, they also informed the researcher that M had deceased in the 6-month period since the research. This eventually led to the author's post-research study reflections and the speculated position stated herein.

Post-research Study Reflection

It was a number of years later that reflection was made by the author of the potential impact when access to a "super toy" is given for a day to a child such as M, but it is then taken away. In this case the child could not communicate what the access meant for him. In the research it has been questioned if there is a sense of awakening or self-agency through the empowerment and a resulting realization of potentials. In the author's other publications this is referred to as 'intra-subjective' (an inner comprehension and learning) associated to the 'inter-subjective' interactions from the created environment. It can also be that such entities bring about a realization of impairment. One can metaphorically suggest that the VR game was a carrot that dangled in front of the child, one that could not be fully 'grabbed' because of limited ongoing access. A speculation questioned whether the experience was so powerful for the child that he became depressed when unable to access and play with "his toy" anymore. Could the experience have been so powerfully positive that it ended up being extremely negative because of no further access granted, which the research team had not considered as a possibility of post research impact. Did he dream of controlling the space craft after the research? These are questions that will remain unanswered.

This contribution suggests that when there is such positive reactions to a research entity such as a 3D VR game controlled by motion (something more common a half decade later) a discussion needs to be undertaken to see if there is possibility to extend access to that being researched and to support training—potentially as further research in the same application—so that longevity of applied testing within a defined research window evolves to be something that can continuously be used at the institute where the volunteers originated from to become participants.

13.2 Discussion

The author, although from a utopian perspective as outlined in the previous section, believe it important to point out the possible impact, the potential reflected need not further fulfilled, that emerged within his body of mature research where a case study is shared to illustrate the position. In this case research subjects had no verbal competence to be able to communicate and were severely disabled in that specialist carers close to each child had to interpret meaning from their signs. A publication on such a sensitive issue relating to subjective speculation that has limited ground for argument is not a strong position. However, despite the speculative and weakness in position, it is shared due to the authors' experiencing numerous interventions where influence on the subjects has been profound, which has led to state this case. Granted what is shared considers those most marginalized through their severe impairments. That involved an inability for the child to communicate verbally in order to inform of their desire to continue in playing post-research interventions; to further interact with something experienced as out of this world where access was momentarily gifted by the researcher (author) but then removed. It can be argued that potentially the child(ren) could have communicated to their carers of family who understood their alternative non-verbal means but those people likely informed that they could not have further access as it was a visit, a research intervention. It can also be argued that giving the initial access, which motivated such active engagement and impactful experience as illustrated on archived video and evaluated by care-providers in attendance, was better than the child(ren) not having had access at all. However, it is not the purpose of this short chapter to argue these points. It is rather to point out a concern that as researchers we may ignore the post-research impacts of studies on subjects.

In situations where this speculative position has been presented orally (e.g. author keynotes at numerous international conferences) fruitful discussions have resulted with peers acknowledging 'dues paid' to be able to make such a claim. These peers informed their understanding of the grounding and reasoning behind the self-critique and more holistic reflections. However, it is not always the case as exemplified at an initial EU wide network meeting towards an Inclusive Music, Disability, and Well-being network on10th June 2016 at Queen's Graduate School, Queen's University Belfast, Ireland. At this meeting the position was ridiculed as overly 'dramatic' following a brief single introduction to the concept, commenting that what is hypothesized is not valid. However, given the differences in systems used i.e. solely audio interactions by the critiques (i.e. music inclusion specialists/academics), versus the more contemporary and sophisticated multimodal Virtual Reality that was core of the case study shared. An argument can be made regards the level of engagement difference, impact potentials, and their position as being limited in multimodal intervention. Their argument was limited and poten-tially beyond their understanding of impact potential of VR in this field. It can be argued that we are seeing such potentials as serious games are becoming more ubiquitous and adopted in practices where previously solely auditory feedback was used. This is exemplified in the case of a third-party randomized intervention study investigating effects of traditional physical training versus visual computer feedback training, using the commercial interactive game-based product that the author's research was responsible for realizing, in frail elderly patients. Results were that the computer feedback-training group (using unencumbered gesture-controlled games) showed a marked improvement that was up to 400 % in the training specific performance and Clinical Rehabilitation Impact was concluded [5].

Logically, making a Virtual Reality system available for participants based upon the hypothesis at the time of the study presented was not a strong argument due to the prohibitive costs involved and need of a designated room/space. However, contemporary advancements in e.g. game platforms (including, power, sensor-based controllers, display resolution and size); 'Smart' home theatre systems incorporating 4 K Ultra High Definition (UHD) large screen monitors with optional 3D and Dolby Atmos/DTS:X5.2 or 7.2 surround speaker systems; computer processing power; and the age of commercially available affordable Head Mounted Displays…and related technologies…could provide a solution for an institute-level post-research accessible system that could be utilized in such cases as presented herein. It is thus posited via this chapter that such solutions should be afforded within research project budgets now that prices have lowered significantly. Additionally, training of use should be budgeted for both end-users and families (i.e. for home use) and at an institutional level for care-providers and therapists so that optimized intervention can take place. Supplementing would be an authoring tool that is easy to access via a GUI such that changes can be made to fine-tune the individual systems to each end-user.

13.3 Holistic and Contemporary Reflections

This chapter began by introducing how advances at the turn of the century were explored for impact in offering potential benefits for those diagnosed with profound impairments. At the time researched systems were only affordable by university research labs with significant budgets. The chapter then illustrates speculatively how alongside potential benefits there exists possible negatives to consider whereby such metaphoric 'hi-fidelity toys' are experienced by participants in a study with such engagement and enjoyment yet they are unable to access post study. When such participants have no verbal competence to express their desire to access and continue it is likened in the text to a child whose 'favorite toy' is taken away and the consequences when the resulting depression has to be internalized and bottled up because of the profound and multiple disabilities. However, changes in affordability and availability are evident as such technologies are increasingly becoming ubiquitous such that many homes have games, music and entertainment set ups. Thus, changes are already evident in signal processing protocols used in the field with advances from MIDI to realize wider creative and expressive boundaries as witnessed through advances such as Multidimensional Polyphonic Expression (MPE). Wireless MIDI running under Bluetooth is now more reliable, low-latency, and easy-to-configure setups with pervasive mobile devices. RTP-MIDI (also known as AppleMIDI) is a protocol to transport MIDI messages within RTP (Real-time Protocol) packets over Ethernet and WiFi networks compatible both with LAN and WAN application fields. Also evolving is Audiobus enabling inter-app routing of audio. Aligned are how hardware manufacturers are bringing out new expressive controllers that take advantage of these changes such as Eigenharp, GECO for Leap Motion, Haken Continuum, LinnStrument, Seaboard by Roli, Soundplane, and the Moog Theremin. The additional nuances offered by advancements such as MPE suggest a next-generation control set-up to research where fine-tuning exhibits higher resolution opportunities than what is available at present.

Whilst many see such technological advances positive there is once again a human component that needs to be questioned as presented. Additionally, to reflect and critique it can also be considered on whether the selected media investigated in such studies are too narrow. In the case presented the young participants were exposed to unencumbered gesture control of digital music, digital painting, robotic device control, and Virtual Reality gameplay. Qualitative evaluations (by those present to witness) pointed to the selected gameplay and robotic light control as being the most engaging for the children. However, such evaluation was subjective at best and it could be questioned what media stimulates 'the best'…in other words it is human researchers that select the media for the human subjects to experience, and from the authors' perspective there are numerous examples where a selected entity just does not stimulate as optimally as others. Often there is response bias involved where a verbal competent participant does not wish to upset a researcher (e.g. in the author's brain injury rehabilitation research studies this has been experienced). Thus, subject profile details beyond diagnosis and age, ethnicity etc., need to include preferences, desires, and influences and remain an open document for ongoing development.

Contemporary physiological real-time analysis technology in the form of heart rate, brain activity, GSR, and especially pupil dilation matched to an eye tracker offers opportunities for improved assessment. The author's prior (and ongoing) research in multimodal biofeedback systems is reported elsewhere. Thus, when using physiological measures, whatever is experienced offers physiological readable signals that are quantifiable against the more qualitative signals that represent self-agency of interaction, empowerment, alongside fun, enjoyment and play. Topics oft considered indeterminate from intervention by those from the 'hard science' field i.e. expecting a basis of perceived methodological rigor, exactitude, and objectivity.

13.4   Conclusions

This terse contribution reflects on state-of-the-art advanced ICT in the form of playful entertainment and creative expression from around the turn of the 21st century. A short historic timeline is shared with some luminaries in the field mentioned. Following, a case study is presented that led to a speculative argument stated to awaken considerations. Contemporary advances in apparatus and software was introduced and opportunity in this field suggested—including further nuances of expression that can be utilized to give additional opportunities in future work. This is expounded in this document with an example in the form of MIDI generators with MPE (Multidimensional Polyphonic Expression) capability and the digital communication protocols elaborated herein. It will be up to creative individuals with insight to alternative use and application to adopt such advances and develop the next generation systems in line with others who have paid their dues and pioneered.

Recently, in the related world of display apparatus, Facebook-owned Oculus Rift, Sony Playstation console, and HTC Vive all launched Virtual Reality hardware for the first time. Samsung has produced the Gear VR and Google launched the cardboard HMD, both enabling a smart phone to be used as the stereoscopic monitor. Reflecting trends, Denmark, since the 3rd of February 2016 has been home to the World's first Virtual Reality store in Copenhagen under the name of Khora.[12] As Virtual Worlds and Games for Serious Applications become increasingly pervasive it is predicted that in the near future it will not be unusual that a contemporary home has a HMD to access such worlds and play such games, optimally immersed in the gameplay. Immersed in a Virtual World subjects can interact within environments, manipulate objects, experience multimodal stimuli; new levels of experience are attainable. Given the argument herein regards consideration of post-research impact and the specific case, such home-based HMDs and home theatre systems could alleviate the situation. In other words, home adoption of apparatus means that researchers can plan towards post-research continued compliance with minimal investment from their research budgets. Such compliance also means return to study opportunities to question e.g. transfer to Activities of Daily Living (ADLs) longevity.

This contribution anticipates how research participants will desire continued access of such engaging ICT experiences, especially if they feel benefit, joy, and positive outcomes from use. Accordingly, to optimize uses and experiences, facilitators, therapists, carers and other helpers will be required who are proficient in training fully all the innate potentials of the hardware and software. The argument is thus posited that discussions are necessary to consider impact of participants if such experiences are delimited to a research study period such that post-study the systems are no longer available. As suggested in the text—it is possible that home-based set-ups, suitable GUIs and authoring tools for fine tuning to each user can alleviate this situation in future.

Acknowledgements All volunteer participants, associated staff, and supporters.

References

1. Brooks AL (2011) Intelligent decision-support in virtual reality healthcare and rehabilitation. Stud Comput Intell 326:143–169

---

[12]http://www.khora-vr.com.

2. Brooks AL (2004) Robotic synchronized to human gesture as a virtual coach in (re) habilitation therapy. In: 3rd International workshop on virtual rehabilitation (IWVR2004)
3. Krueger M (1983) Artificial reality. Addison-Wesley Professional
4. Krueger M (1991) Artificial reality 2. Addison-Wesley Professional
5. Hagedorn DK, Holm E (2010) Effects of traditional physical training and visual computer feedback training in frail elderly patients. A randomized intervention study. Eur J Phys Rehabil Med 46(2):159–68 Epub 2010 Apr 13. http://www.ncbi.nlm.nih.gov/pubmed/20485221

Chapter 14
Accessibility: Definition, Labeling, and CVAA Impact

Anthony Brooks

Abstract This contribution is timely as it addresses accessibility in regards system hardware and software aligned with introduction of the Twenty-First Century Communications and Video Accessibility Act (CVAA) and adjoined game industry waiver that comes into force January 2017. This is an act created by the USA Federal Communications Commission (FCC) to increase the access of persons with disabilities to modern communications, and for other purposes. The act impacts advanced communications services and products including text messaging; e-mail; instant messaging; video communications; browsers; game platforms; and games software. However, the CVAA has no legal status in the EU. This text succinctly introduces and questions implications, impact, and wider adoption. By presenting the full CVAA and game industry waiver the text targets to motivate discussions and further publications on the subject that could significantly impact industries targeted by this volume (In citing the CVAA the author does not posit himself as a specific expert in the act or field—reference is made to awaken discussions to the important issues surrounding accessibility and inclusion.).

Keywords Accessibility · Advanced communications services (ACS), games software · Game platforms · Entertainment Software Association (ESA) · Consumer and Governmental Affairs Bureau (CGB or Bureau) · Federal Communications Commission (FCC or Commission) · Twenty-First Century Communications and Video Accessibility Act (CVAA) · PEGI

A. Brooks (✉)
Aalborg University, Esbjerg, Denmark
e-mail: tb@create.aau.dk

© Springer International Publishing AG 2017
A.L. Brooks et al. (eds.), *Recent Advances in Technologies for Inclusive Well-Being*, Intelligent Systems Reference Library 119,
DOI 10.1007/978-3-319-49879-9_14

14.1 Introduction

The author, amongst his other credits, is the Danish representative for three working groups under the International Federation for Information Processing[1] (IFIP) within the Technical Committee (TC) 14: Entertainment Computing, namely WG (working group) 14.7—Art and Entertainment; 14.8—Serious Games; and 14.9—Game Accessibility.[2] IFIP is a global organization for researchers and professionals working in the field of Information and Communication Technologies (ICT). This is introduced as this text presents reflections on the Twenty-First Century Communications and Video Accessibility Act (CVAA)[3] with a delimited focus of impact on the game industry. Selected activists are included in the text such as special interest groups, associated charities, and others. Their contribution to this text enables clarity of meaning that is not deciphered by the author—or that is the intention through direct text usage. Such delimitation is made in line with the wide spread adoption of games in segments of society populated by persons diagnosed with impairment (also referred to as 'disabled' or 'handicapped'—depending on global location, culture, and nationally enforced political correctness). Professional adopters associated to this societal segment (e.g. healthcare therapists, care-providers, etc.) state their positive responses to the proposed increased accessibility in game platforms and games software, which reflects innate potentials being recognized. This is in line with the title of this volume, which reflects such adoption and impact of games as alternative reality content (i.e. Virtual Reality [VR], Augmented Reality [AR], and Mixed Reality [MR]) and related Information and Communications Technologies (ICT) and related digital media However, research informs that the CVAA is not legally binding outside of USA, thus should not directly impact in Europe. This impact is questioned herein, which is considered timely as it addresses accessibility in regards system hardware and software in the games industry: an industry that can affect wellbeing and quality of life issues that this volume focuses upon. The author's body of mature work is presented as an open vehicle for the discussion address as it has a long history of originating and utilizing compendiums of sensor-based hardware mapped to selected multimedia software content. Mapping of this content to respond to human input in a meaningful and entertaining manner is a design art necessitating knowledge of the human potentials of function, motivation, and engagement. Mechanics of medium (e.g. game mechanics, art software programming) are essential to master to achieve successful output. Alongside are such aspects of play, fun, and agency for the end user player. These and other elements contribute to the field that now has to address

---

[1]http://www.ifip.org.

[2]http://www.org.id.tue.nl/IFIP-TC14/WGs.htm.

[3]https://www.gpo.gov/fdsys/pkg/BILLS-111s3304enr/pdf/BILLS-111s3304enr.pdf.

the implementation of the CVAA, which was signed by President Barack Obama on October 8, 2010 with compliance extended to January 2017 for software. This is an act created *to increase the access of persons with disabilities to modern communications, and for other purposes.*

The act seemingly focuses solely on Americans with disabilities. This text predicts wider adoption outside of USA to support differently abled persons' access to games produced outside of North America. A hypothesis is that if wider global enforcement is not forthcoming such that developers of games outside of CVAA impact zone are not legally bound then it is likely a work-around for industry to establish production outlets outside of USA to not be impacted. Thus, the independent smaller developers could be impacted to closure and the larger mega-corporations could even further dominate what is available in the market. However, certain of these corporations are initiating accessibility initiatives as reported herein through second reference.

In positing the position taken herein, the author states his non-expert status in these matters and thus the chapter text additionally offers as appendixes the full CVAA and games software waiver that comes into force (unless extended) in January 2017. This is to enable readers who prefer to access the act and waiver in non-digital form—potentially offered as some may prefer to fully digest in paper format in order to optimize their opportunities to critique and reflections so as to offer future discussion input. The purpose of this contribution is to inform readers that may be of a different network than would typically browse the accessed sources herein. The text uses materials directly sourced to ensure meaning is maintained in the passing of information and the author does not claim authorship of the directly sourced text extracts. This fact is pointed out throughout the chapter, as the contribution is somewhat different than typical material found in such an academic volume. The author naturally thus has trepidation with presenting the text with so much unaltered contribution from others. However, the significance of the passing of information so that others may be advised of activities in this field is believed imperative to share.

The text begins with offering the author's mature body of research as a vehicle upon which a reference can be made for what follows. In the following the state-of-the-art is retrieved from a number of sources. The chapter closes with discussions and conclusions that illustrate proactive communities that are concerned about game and other accessibility in order to support those with impairment to a richer wellbeing and quality of life through enjoyment and assessable communication within the genres herein approached.

14.1.1 Background

The author's work is considered one of many possible vehicles from where to position an argument for wider adoption of the CVAA. As an early researcher and developer of sensor-based unencumbered motion control interactive games to use for societal benefit experiences point to the need for non-corporate control of the market so that individual needs can be addressed in created content systems. Recent advancements in authoring tools, in the author's opinion, are a step in the right direction to such tailoring of interactive systems. In the author's work, complexities of mapping requirements (input to output—stated simply) have been a major system consideration according to needs of intervention/use. Aligned with this is the added requirement to measure the interactions to be able to assess/evaluate the accessibility of the system and the effectiveness and efficiency of the design and intervention method implemented. Whilst CVAA impacts game platforms (hardware) and content (software) there are many tools available for independent game developers (programmers and non-programmers) such that other sensor systems that do not require such hardware consoles (e.g. cameras, ultrasound, infrared, laser…). Providing the developer company is under thirty employees the company does not come under the CVAA. This offers another opportunity for work-around in the author's opinion such that corporates will limit size of company segments and list them under a corporate umbrella entity such that CVAA is non-enforceable. Noted this is a speculation. However, again it is pointed out that this text has a target of stimulating further discussions rather than being a definitive publication. The position is argued from the author's research having been at the core of a commercial start-up game company from the turn-of-the-century. The research resulted in hardware and media selection tailored to offer manipulation options for optimizing to an end-user's needs and profile—as well as to the designated medical professionals desires for patient/user's progression (usually motion based physiotherapy). In this work, creative expression (music making, digital painting, robotic device control—'playfully applied') was used interchangeably and complementary as a variation option to game genre environments, where both supplemented traditional therapeutic rehabilitation in treatment programmes. End-user accessibility issues were found as innately individual with differing levels of complexity according to the participant's impairment; thus the compendium of interfaces optimized potentials providing the facilitator was knowledgeable across offerings. A finding was to the importance of the facilitator who led intervention sessions and their ability to select accessible content (hardware and software under a mix'n'match strategy), then to be able to 'improvise' and initiate parameter (and/or 'goal') change to motivate engagement, enjoyment and outcome. Assessment in such treatment programmes can be overviewed as progression nuance toward therapist goal for the intervention. Observations by those closest to the participant, dedicated therapists, and other medical professionals informed results. An outcome of the research is an evolving model titled Zone of Optimized Motivation (ZOOM) that directly relates to accessibility for multimedia responsive content. ZOOM has previously been introduced and published, thus this contribution builds upon the ongoing work to widen discussion with delimited focus on the CVAA. Readers who wish wider knowledge of the author's research may access via his publications site.[4]

---

[4]http://vbn.aau.dk/en/persons/pp_add659c1-98b1-4ddc-b5be-28eb4b5f1eb7/publications.html.

14.1.2 Interactive Accessibility Community Reaction

In an industry news posting on the Interactive Accessibility site titled "New Requirements for Video Games this Fall", Miller [1] pointed out how:

> Video game consoles, services and software are covered by the 20-First Century Communications and Video Accessibility Act of 2010 (CVAA.) The CVAA requires video game equipment software and related services to be accessible and to be usable by individuals with disabilities. However, the Federal Communications Commission (FCC) is allowed to grant waivers of the CVAA'S accessibility requirements in cases where the equipment, services or software is designed primarily for purposes other than communication. The FCC previously granted a waiver until October 8, 2015 for the following categories of video gaming services and equipment.
>
> Class I—Game consoles, both home and handheld, and their peripherals and integrated online networks, which are designed for multiple entertainment purposes but with a primary purpose of playing games.
>
> Class II—Game distribution and online game play services designed for the primary purpose of distributing online game software or enabling online game play across a network.
>
> Class III—Software designed for the primary purpose of game play. Game software means playable games on any hardware or online platform, including, but not limited to, dedicated game consoles, PCs, mobile devices, and the Internet (i.e. browser based games).

The Entertainment Software Association (ESA) applied to the FCC for a 15-month waiver only for class III video game software, which resulted in an extension being granted to the ESA by FCC. As of writing, this is to January 2017 (see note of further waiver application in Appendix 2 opening.). The CVAA relates to the WCAG 2.0 Guidelines,[5] which assists and advises developers in creating web experiences that are more usable and accessible by people with impairments. Web accessibility guidelines and standards define how to make web content and applications accessible to people with disabilities so that, ideally, web content is available to all individuals, on all devices. Aligned to both the CVAA and the WCAG 2.0 is the American's with Disabilities Act (ADA)[6] is a Federal law requiring equal treatment of people with disabilities in public accommodations online as well as offline.

14.1.3 Accessible Games = Accessible Virtual Reality?

Virtual reality (VR) has as a main commitment to produce an immersed experience/appearance of the real world such that the user's illusion is of being present in a virtual environment. VR first appeared more than 20 years ago in universities and industry whereby originators such as Nintendo explored associated

---

[5]http://www.w3.org/TR/WCAG20/.
[6]https://www.ada.gov.

technology with the Power Glove and Virtual Boy. Subsequently, other originators in the industry such as Sony and Microsoft produced more advanced devices capable of increasing interactivity and immersion between the user/participant and the videogame virtual environment, namely: Sony Eye Toy, Sony MOVE, or Microsoft Kinect. More recently other companies have entered the market producing similar systems, e.g. ASUS and their Xtion Motion Sensor peripheral; Leap Motion and Orion; Intel's RealSense™systems… and more. However, the lack of content, knowledge about virtual reality techniques and, above all, the computational performance in the technology existing at that time, delayed virtual reality impacting as predicted until the turn of the twenty-first century whereby the game industry aligned themselves with graphical card manufacturers to reciprocally and ongoing push each other in creating the games industry as a major player. Alongside Personal Computer and game console platform advances arrived new delivery apparatus in the form of Head Mounted Devices such as Oculus Rift, Samsung Gear, HTC with its VIVE project or the Morpheus Project for PS4. These manufacturers realized low-cost virtual reality head-mounted devices designed for videogames and more general entertainment. Such devices are becoming adopted into schooling, travel/tourism, archaeology and healthcare. In the authors research experiences have shown that an optimal strategy toward motivated engagement is to mix and match system to content to prevent boredom and mundane experiences for the user/participant.

14.1.4 Mixing and Matching for Accessibility

The focus of the author's research has been on observing use by volunteers with impairment from birth and acquired. The mix'n'match strategy is wider than at a systemic level where an interface is matched to responsive content as it targets matching the person using the system towards optimal experience in training.

Exemplifying this is that the author's system was first introduced to patients with acquired brain injury at dedicated hospital and clinic rehabilitation training departments in Aarhus and Copenhagen, Denmark, following being featured at the "Year of the Brain" in Euro Brain '97, the joint European conference on brain diseases, hosted at the 'Aalborg Kongres & Kultur Center', Denmark. The author led intervention sessions alongside traditional staff members e.g. neuropsychologists, physiotherapists, etc. In the sessions it was clear how each patient had individual needs, preferences and desires that could be formulated into an alternative strategy to supplement the traditional training programmes. The author's systems comprising various interfaces and digital media was set up so that individual differences could be implemented as recallable presets so that each session cessation corresponded to next session start. Data logging of progress and session video analysis informed incremental steps to each 'next-session' expected presets so that the facilitator could initiate around five steps of varying challenge via preset change. Patient engagement and motivation was clear and post-session discussions informed how the alternative to the traditional intervention was positively received—an oft-mentioned phrase was 'through the fun that was had'. The training still targeted patients' limb movements and balance in a similar way as the traditional methods (as witnessed/observed by the author) did. The patients reported how instead of focusing on the specific movements as instructed by the staff, through the digital media and author method of intervention; focus was on manipulating the media that acted as feedback stimulus. They explained that even knowing it was their movements that generated interface-sensed data resulting in the media manipulations and they did have a level of focus on that primary action, once the feedback had started and the causality cycle had initiated they stated how it felt as if the digital feedback was assisting them. In this regard the author has published previously on the afferent-efferent neural feedback loop closure involved to invite uptake and collaborations using the method and apparatus. However, despite the positive outcomes with patients and staff, dissemination of results presented at international medical-focused conferences during the time of residency, such as the World Congress for Physical Therapy in 1999 (Yokohama, Japan), Integrative Medicine & Expressive Therapies, hosted at the Omega Institute. Rhinebeck, New York, USA, also in 1999, led to small responses. Technical conferences also led to limited uptake and adoption, e.g. dissemination at the User Interface Technology in the 21st Century—APCHI 2002: 5th Asia Pacific Conference on Computer Human Interaction—User Interface Technology in the 21st Century, which was presented by the author in Beijing and published by Science Press China in 2002.[7] Because of the limited reactions by the academic and scientific international communities the work was opened up to include associated academics, who were related to the research (though not actively involved), as co-authors, and the work then presented by the author at ICDVRAT 2002: The Fourth International Conference on Disability, Virtual Reality and Associated Technologies, at Veszprém, Hungary. This event presented digital media systems as used in healthcare whereby accepted research assessments and evaluations were not solely quantitative, thus, more aligned to the author's approach. Since these early sharing of the research, method and apparatus, the work has evolved and is considered an on-going work in progress towards meaningful impact. However, the author questions whether the research and application approach, and thus potentials, are fully understood in the field.

This is important for the author as The European Brain Council (EBC) inform that around 165 million Europeans are living with a brain disorder, causing a global cost (direct and indirect) exceeding 800 billion euros for the National Health budgets.[8] The contribution as a supplementary training may not realize full impact until a greater understanding and intervention of reopening neural networks of those suffering physical injury is achieved. In other words, the training concept presented may require realization of advanced drug treatments that result in brain plasticity

---

[7]http://www.worldcat.org/title/proceedings-of-the-apchi-2002-5th-asia-pacific-conference-on-human-computer-interaction-user-interaction-technology-in-the-21st-century-november-1-4-beijing/oclc/843436791.

[8]http://www.braincouncil.eu.

opportunities so that the complementary training herein referenced is ready to support in intervention within a treatment programme.

In line with this, contemporary research on acquired physical injury such as stroke is taking strides toward such advances. For example, Takao Hensch, a neuroscientist at Boston Children's Hospital and Harvard University and an expert in brain plasticity stated how "For those with physical injury, re-opening neural networks might one day improve the chance of full recovery. Patients who lose function after a stroke, paralysis or other forms of neurological injury can often relearn tasks with intensive rehabilitation, in part because the brain is trying to repair itself through a period of heightened plasticity—a glimmer of what the brain did when you were very young. 'The brakes go down. For a short period of time, it seems like there is an effort there to increase plasticity,' Hensch says. 'What we are aiming to do is extend that window, and judiciously pair it with training, so that maybe there's an opportunity to restore function.'"[9]

The training in question as experienced when the author introduced his concept of digital media manipulation via gesture at the brain injury rehabilitation training centers between 1997 and 2002 seemed too traditional and too focused on solely direct training of a redundant limb (besides other training e.g. speech, etc.). Impact of the author's work on staff was clear and resulted in one Cognitive Psychologist, who was assigned as co-worker in the project at the Centre for Rehabilitation of Brain Injury (CRBI), writing his Diploma of IT specializing in software development and user interfaces, at the IT University, Copenhagen 2001–2003 on digital media using the author's concept and original system from the period.[10]

In closing this presentation of the linkage to the author's research it becomes clear how it acts as a vehicle for discussion alongside other bodies of work. To bring focus back to the CVAA relative to the need for global accessibility to digital content, the next section details specific segments from the act that is fully presented in the appendixes, including waiver.

14.1.5 Twenty-First Century Communications and Video Accessibility Act (CVAA)[11]

Twenty-First Century Communications and Video Accessibility Act of 2010—Pub. L. 111-260—[Highlights below[12] with full original report in Appendix 1 – granted waiver extension specific for video games software in shared in Appendix 2].

---

[9]https://aeon.co/essays/can-neuroscience-give-my-brain-the-plasticity-of-a-child-s.

[10]Dipl. IT med speciale i udvikling af software og brugergrænseflader, IT Universitetet [E.B. Lyon].

[11]https://www.gpo.gov/fdsys/pkg/BILLS-111s3304enr/pdf/BILLS-111s3304enr.pdf.

[12]https://www.fcc.gov/consumers/guides/21st-century-communications-and-video-accessibility-act-cvaa.

On October 8, 2010, President Obama signed the Twenty-First Century Communications and Video Accessibility Act (CVAA) into law. The CVAA updates federal communications law to increase the access of persons with disabilities to modern communications. The CVAA makes sure that accessibility laws enacted in the 1980s and 1990s are brought up to date with 21st century technologies, including new digital, broadband, and mobile innovations. The following are highlights of the new law.

Title I—Communications Access

- Requires advanced communications services and products to be accessible by people with disabilities. Advanced communications services are defined as (1) interconnected voice over Internet protocol (VoIP) service; (2) non-interconnected VoIP service; (3) electronic messaging service; and (4) interoperable video conferencing service. This includes, for example, text messaging, e-mail, instant messaging, and video communications.
- Requires access to web browsers on mobile devices by people who are blind or visually impaired (a "ramp" to the Internet on mobile devices).
- Creates industry recordkeeping obligations; requires changes to complaint and enforcement procedures; tightens deadlines for the FCC to respond to consumer complaints; requires biennial reporting by the FCC to Congress; and directs the Comptroller General to issue a five-year report on the FCC's implementation.
- Requires an FCC clearinghouse on accessible communications services and equipment.
- Applies the hearing aid compatibility mandates to telephone-like equipment used with advanced communications services.
- Updates the definition of telecommunications relay services (TRS) to include people who are deaf-blind and to allow communication between and among different types of relay users.
- Requires interconnected and non-interconnected VoIP service providers to contribute to the Interstate TRS Fund.
- Directs the allocation of up to $10 million per year from the Interstate TRS Fund for the distribution of specialized equipment to low-income people who are deaf-blind, to enable these individuals to access telecommunications service, Internet access service, and advanced communications.
- Authorizes FCC action to ensure reliable and interoperable access to next generation 9-1-1 services by people with disabilities.

14.1.5.1 Title II—Video Programming

- Restores video description rules promulgated by the FCC in 2000 and authorizes some expansion of those obligations over the next 10+ years.

- Requires video programming that is closed captioned on TV to be closed captioned when distributed on the Internet (does not cover programs shown only on the Internet).
- Establishes deadlines for the FCC to respond to requests for exemption from the closed captioning rules.
- Requires video programming distributors, providers, and owners to convey emergency information in a manner that is accessible to people who are blind or visually impaired.
- Expands the requirement for video programming equipment (equipment that shows TV programs) to be capable of displaying closed captions, to devices with screens smaller than 13 inches (e.g., portable TVs, laptops, smart phones), and requires these devices to be able to pass through video descriptions and emergency information that is accessible to people who are blind or visually impaired, if technically feasible and achievable.
- Requires devices designed to record TV programs to pass through closed captions, video description, and emergency information so viewers are able to turn on/off the closed captions and video description when the TV program is played back, if achievable.
- Requires interconnection mechanisms (cables) to carry (from the source device to the consumer equipment—e.g., TV set) the information necessary to permit the display of closed captions and make video description and emergency information audible.
- Requires user controls for TVs and other video programming devices to be accessible to people who are blind or visually impaired, and requires TVs and other video programming devices to have a button, key, icon, or comparable mechanism designated for activating closed captioning and video description.
- Requires on-screen text menus and program guides displayed on TV by set-top boxes to be accessible to people who are blind or visually impaired and requires set-top boxes to have a button, key, icon, or comparable mechanism designated for activating closed captioning (when built-into the set-top box).

14.1.6   Related Literature from the Community—Abstract (cf) from Powers et al. [2]

Video game accessibility may not seem of significance to some, and it may sound trivial to anyone who does not play video games. This assumption is false. With the digitalization of our culture, video games are an ever increasing part of our life. They contribute to peer to peer interactions, education, music and the arts. A video game can be created by hundreds of musicians and artists, and they can have production budgets that exceed modern blockbuster films. Inaccessible video games are analogous to movie theaters without closed captioning or accessible facilities. The movement to have accessible video games is small, unorganized and misdirected. Just like the other battles to make society accessible were accomplished through legislation and law, the battle for video game accessibility must be focused toward the law and not the market.

As stated at the start of this document, the author is the Danish representative for IFIP TC14 Working Group 14.9—Game Accessibility. This is a relatively recent initiative established in 2015 under the TC 14 (Entertainment Computing) and the International Federation of Information Processing (IFIP).

The aim of the IFIP TC14 Working Group 14.9—Game Accessibility is:

- *To promote software and hardware research on game accessibility*
- *To provide, encourage and facilitate the use of methods and tools for inclusion in the game industry*
- *To establish a shared understanding of current and emerging requirements for users with special needs in the field of video games*
- *To connect interdisciplinary approaches/groups and encourage cooperation and collaboration in research and development projects*
- *To develop methodologies and guidelines for game designers and game developers for accessible games*
- *To build a common knowledge base, making it available for practitioners and for academics/lecturers, for teaching purposes.*

The scope of the working group encompasses:

- *Games, video games and entertainment objects, in their broadest sense*
- *Simulations and virtual/augmented or mixed reality*
- *Transmedia and crossmedia*
- *Mobile phones, tablets and emerging platforms and devices*
- *Development tools, game engines*
- *Social, cultural and ethical impact and considerations*

(see http://www.ifip.org/bulletin/bulltcs/tc14_aim.htm#wg149)

Game accessibility research has been ongoing since the start of the game industry for inclusion of people with impairment and having special needs to access. On the whole it can be said that mainstream games have been inaccessible for many in this segment of society. Since October 2015 the Communications and Video Accessibility Act (CVAA) in the USA requires game consoles and distribution platforms to be accessible. Following a petition a waiver injunction was granted so that game software is excluded until January 2017 but there after accessibility issues need to be addressed. The CVAA is an act that enforces legislation in the USA for companies with more than 30 employees and as one of the largest game markets—see next section briefly presenting an extract from the Video Games in the 21st Century: The 2014 Report—impact is being felt. This strong change in the digital entertainment landscape has to be taken into account by researchers in order to aid the game industry to meet this new requirement.

(cf) Video Games in the 21st Century: The 2014 Report was an economic impact study conducted by Economists Incorporated and released by the Entertainment

Software Association in 2014. The report quantifies the U.S. video game industry's contributions to the American economy, including how:

- *From 2009 to 2012, the U.S. video game industry increased in size by more than 9 %—four times the growth rate of the U.S. economy during the same period.*
- *In 2012, the entertainment software industry added over $6.2 billion to U.S. Gross Domestic Product.*
- *The computer and video game industry directly and indirectly employs more than 146,000 people.*
- *The average salary for direct employees is $94,747, resulting in total national compensation of $4 billion.*
- *Direct employment for the industry grew at an annual rate more than 13 times the growth of the overall U.S. labor market (9 % vs. 0.72 %) between 2009 and 2012.*

(see http://www.theesa.com/article/u-s-video-game-industrys-economic-impact)

The CVAA will predictably have an impact on these attributes of the game industry in USA. If adopted globally the act would have significant roll-over effect.

Laws and regulations in the game industry are numerous and wide reaching. Most address issues on piracy, censorship, and addiction. One such act dealing with copyright is the Digital Millennium Copyright Act,[13] which forbids unlawful copying and distribution of electronic media, including video games (2014)—(see https://kb.iu.edu/d/alik). Others consider pornography, violence and suitable content by age with a defined labeling strategy—see PEGI in this document.

The CVAA refers to "the communications functions offered via gaming" (Richert 2012—see https://www.afb.org/afbpress/pub.asp?DocID=aw131205). Gaming industry representatives requested that they be exempt from the CVAA's requirements; but the FCC denied the request after noting, "how important gaming has become for social interaction, education, and even the fostering of professional and intimate relationships" (Richert, 2012). Research highlights how the FCC's determination concerns the communication elements of video games and not the design elements.

The act impacts game platform giants in regard hardware platforms need to comply as well as software (the latter following appeal from January 2017). Such all-encompassing accessibility is a complex issue and it is likely that should there be global adoption (as predicted in this text) there will need to be a resulting accessibility rating scale in line with The Pan-European Game Information (PEGI) age rating system format.[14] However, this has innate complexities and challenges.

---

[13] https://www.congress.gov/bill/105th-congress/house-bill/2281/text/enr.

[14] http://www.pegi.info/en/index/id/33/.

14.1.7   The Pan-European Game Information (PEGI) Age Rating System

PEGI is a games packaging labeling strategy whereby labels on front and back indicate one of the following age levels: 3, 7, 12, 16 and 18. The labels provide a reliable indication of the suitability of the game content in terms of protection of minors. The age rating does not take into account the difficulty level or skills required to play a game. Descriptors further advise on reasoning behind the rating and an extended consumer advice text explains why a game received its classification. The PEGI system is created and owned by the Interactive Software Federation of Europe (ISFE), which was established in 1998 to represent the interests of the interactive software sector with regard to the European Union and international institutions.[15] The UK-based Video Standards Council (VSC) and the Netherlands Institute for the Classification of Audiovisual Media (NICAM) are the two independent bodies that administrate the PEGI system. NICAM reviews 3 and 7 games whilst the VSC reviews 12, 16 and 18 rated games. Consumers who disagree with a rating can contact the administrator and file a complaint.

The PEGI process involves that prior to a game going to market the publishers complete an on-line content assessment and declaration form, whose first part questions legal provisions according to targeted markets by country. This is followed by a self-assessment on possible presence of violence, sex and other sensitive visual or audio content, which leads to PEGI allocating a provisional age classification and rating (Fig. 14.1) along with content descriptors and extended consumer advice as appropriate. VSC or NICAM, according to age rating, then examine aligned with the PEGI criteria in order to confirm or adjust the rating/classification and supporting text. Once all is approved a license is authorized with the necessary label and descriptor(s) to include.[16]

In detail (cf PEGI with full permission including images—abridged US English):

PEGI labels appear on the front and the back of games packaging indicating one of the following age levels: 3, 7, 12, 16 and 18. They provide a reliable indication of the suitability of the game content in terms of protection of minors. The age rating does not take into account the difficulty level or skills required to play a game. Text descriptors below for each label:

- PEGI 3: The content of games given this rating is considered suitable for all age groups. Some violence in a comical context (typically Bugs Bunny or Tom & Jerry cartoon-like forms of violence) is acceptable. The child should not be able to associate the character on the screen with real life characters; they should be totally fantasy. The game should not contain any sounds or pictures that are likely to scare or frighten young children. No bad language should be heard.

---

[15]http://www.pegi.info/en/index/id/1183/.

[16]http://www.pegi.info/en/index/id/1184/.

Fig. 14.1 PEGI classification rating labels

- PEGI 7: Any game that would normally be rated at 3 but contains some possibly frightening scenes or sounds may be considered suitable in this category.
- PEGI 12: Videogames that show violence of a slightly more graphic nature towards fantasy character and/or non-graphic violence towards human-looking characters or recognizable animals, as well as videogames that show nudity of a slightly more graphic nature would fall in this age category. Any bad language in this category must be mild and fall short of sexual expletives.
- PEGI 16: This rating is applied once the depiction of violence (or sexual activity) reaches a stage that looks the same as would be expected in real life. More extreme bad language, the concept of the use of tobacco and drugs and the depiction of criminal activities can be content of games that are rated 16.
- PEGI 18: The adult classification is applied when the level of violence reaches a stage where it becomes a depiction of gross violence and/or includes elements of specific types of violence. Gross violence is the most difficult to define since it can be very subjective in many cases, but in general terms it can be classed as the depictions of violence that would make the viewer feel a sense of revulsion.

Descriptors shown on the back of the packaging (Figs. 14.3 and 14.4) indicate the main reasons why a game has received a particular age rating. There are eight such descriptors: violence, bad language, fear, drugs, sexual, discrimination, gambling, and online gameplay with other people. The latter labeling, as depicted in the next image and detailed further on next page, being directly related to the new USA CVAA (Fig. 14.2).

The 'Online Gameplay' classification typically in multiplayer modes would have integrated communication between players and thus fall directly into the CVAA.

Interesting is that PEGI since 2015 has declined to use this online descriptor (see rightmost image in Fig. 14.3 and lowest image in Fig. 14.4), which may or may not be linked to the CVAA and potential of global adoption as speculated by the author.

There are complexities innate to classifying such games. PEGI uses a combination of content declaration and game review to determine the appropriate PEGI rating for each game. Initially, the publisher of a game will complete an online declaration form that is sent to the administrator of the system. The completed form is then reviewed and used as a basis for checking the content of the game. The content declaration aspect of PEGI is a significant strength as only the developer/publisher of a game has a complete overview of the content in the game.

Fig. 14.2 PEGI descriptors

Bad Language
Game contains bad language

Discrimination
Game contains depictions of, or material which may encourage, discrimination

Drugs
Game refers to or depicts the use of drugs

Fear
Game may be frightening or scary for young children

Gambling
Games that encourage or teach gambling

Sex
Game depicts nudity and/or sexual behaviour or sexual references

Violence
Game contains depictions of violence

Online gameplay
Game can be played online

Fig. 14.3 PEGI descriptors—definitions

The overview allows the administrator to focus on sections or aspects of the game that will most likely to affect the rating. This is more efficient and reliable than the administrators attempting a full play through (assuming it was possible) as the only basis for the rating. Should a customer disagree with the classification they can

Fig. 14.4 Author in Porto workshop programming MIDI-to-DMX gesture control of robotic lighting units (seen on the right in the picture with GOBO of the commercial games company 'Personics' that resulted from the research)

complain to PEGI who call an expert committee to evaluate the complaint. However, more recently PEGI works with the International Age Rating Coalition (IARC)—established in 2013—in improving the categorization of games by simplifying the process by which developers obtain age ratings from different regions around the world by reducing it to a single set of questions about their product. The questionnaire is programmed with unique algorithms that generate ratings reflecting each participating rating authority's distinct standards, along with a generic rating for the rest of the world. IARC rating assignments also include content descriptors and interactive elements identifying apps that collect and share location or personal information, enable user interaction, share user-generated content, and/or offer in-app digital purchases. The IARC system currently includes rating authorities, which collectively represent regions serving approximately 1.5 billion people, with more expected to participate in the future. IARC is administered by many of the world's game rating authorities in providing a globally streamlined age classification process for digital games and mobile apps, helping to ensure that today's digital consumers have consistent access to established and trusted age ratings across game devices. (cf.[17])

---

[17]https://www.globalratings.com/about.aspx.

This contribution questions if such a rating system is developed for accessibility how it will be implemented due to the individual traits associated to human condition—beyond the generic. As a point of departure the author's mature body of research and associated work is introduced as a vehicle for discussion wherein experiences have suggested that it is not solely the 'technical' solution that should be questioned but also human aspects of intervention strategies. From this point of departure, which relates to the author's unsuccessful approach to major game publishers in the start of the 21st Century to increase accessibility for more inclusive games, an attempt is made to stand back from one's own work and experiences toward stimulating emergent discussions on the issue and how it may be addressed so as to best serve those the CVAA targets to aid.

Interestingly is that in a recent international conference on the 8th International Conference on Virtual Worlds and Games for Serious Applications (VSGames2016[18]) attended by delegates across industries likely affected by CVAA —none knew of its existence when the author questioned the audience in his keynote address.

14.2 Informing Actions Using the Author's Work as Vehicle

The vehicle referenced to in this contribution is the author's body of research and work that transcends and bridges disciplines toward cross-informing actions and direction in human condition and performance whereby digital technology can be utilized as a beneficial tool in rehabilitation intervention.

Actively doing 'something' that involves social interaction, creativity, and enjoyment such that a fun and rewarding activity is experienced evokes the human condition. Associated terms can be regarded e.g. empowerment, play, artistic expression, making and controlling, agency, … such that when the activity/action is observed the linked human performance to condition can be identified. However, in such work, human condition and ability to perform differ. In this case performance is widely interpreted beyond that as narrowly defined in certain higher education establishments e.g. in the author's experiences in the United Kingdom. Findings made clear that to enable as wide access, and thus opening inclusion to as many as possible, a mix'n'match strategy was deemed a best fit. This is elaborated in the text.

The author's history includes being born into a domestic situation with severely impaired family members. Tacit knowledge gained from interactions over many years has contributed to originating and developing the research presented herein. As a teen, experiments with alternative means and methods to enable the impaired individuals in the family to creatively express and play were many and varied. The most successful from those times being a wheelchair-mounted apparatus where the

---

[18]http://vsgames2016.com.

method to operate was via a users' weight transfer via residual upper torso movement that enabled an elbow to depress and raise a variable toggle 5-volt guitar foot pedal that was positioned in the signal chain of a music system. A simple causality resulted in that a user's torso motions that controlled the volume of the music. Experiments continued as an engineering degree was attained; interactions with an expanding array of persons with impairments were ongoing to gain insight in the field; and activity in creative 'art' endeavors increased to develop understanding, experience and learning that acted as precursors to the research conceptual framework.

There was deemed a need to have 'hands-on' experiences beyond simply self-researching and self-testing and then introducing to participants with distinct medical diagnosis where upon a situation that involves potential stress for the user. Stress e.g. through additional social contact—potentially environmental 'set-up' distractions—and being observed and needing to 'perform'. The author thus extended the concept such that following self-researching and self-testing the concept became central in his performance art, both on stage in his own performances and as core building blocks to his Museums for Modern Art (MoMA) exhibitions that were commissioned as interactive installations. Both situations were for public audiences. In the former case (stage performances), the author would use various interfaces mapped to various multimedia in real-time improvisations based upon themes narratives. At the time the use of the bodies' electrical signals were studied.

The biosignals used for mapping in rehabilitation sessions were the same as used in stage performances. On stage live signal sources were used as a real-time traced element of the projected performance visuals as corresponding 'picture-in-picture' (PiP). Thus, acting as correlational components for the audiences to have opportunity to associate stimuli changes in what they saw or heard (i.e. mapped data to media). For example, a purposeful muscle spasm could be controlled by the trained performer with a goal of generating a specific feedback as an audience perceived element of a performance. Thus, when 'wired up' in a biosignals-monitoring set-up (piezoelectric transducers that are attached directly to the performers skin to sense the stimulated muscle), the muscle contraction results in an acceleromyographical measurement of the muscle activity. A voltage is created within the muscle when it accelerates and that acceleration is proportion to force of contraction and this electrical force data is sensed by the transducer(s) and routed to the biosignals interface. Such an effort would typically result in a noticeable change to corresponding biosignals, for example, the performer's heart muscle's electrophysiological pattern of depolarization and repolarization; spectral content fluctuations of the performer's brain wave activity (neural oscillations); and/or electrodermal activity (EDA)—aka galvanic skin response (GSR). In addition to the worn biosignals system sensors, the author's performance set-up evolved over the years and included multiple active sensor spaces that could be triggered and used to manipulate via linear, planar or 3 dimensional (volumetric) performance gesture. These sensor spaces could be mixed and cross-hatched to offer additional performance attributes and mapping opportunities in performance situations. The use of sensor-spaces gave rise to the MoMA concepts as outlined in the following.

In such performances the performer experiences increased social contact and stress through observation and need to perform. At the time—beginning of the 1990s—such performances with biosignals was rare and the author experienced that audiences found it problematic to associate the human actions, matched to the biosignals trace changes (PiPs). Post performance interviews indicated that they believed it as 'playback' i.e. pre-recorded media, versus the reality of what it was as an abstract avant-garde improvised performance where the performer activated media in real-time via gesture and biofeedback control. This was even the case when such performances were commissioned at the start of the 21st Century for intellectuals, business, and academic luminaries. An example is the author's installation and performance for the 'i3 village' Orbit/Comdex Europe Basel Sept. 25–28 2001 where he designed the focal point for the central plaza installation and further where he performed for the annual dinner of the European Network for Intelligent Information Interfaces (http://www.i3net.org). The 420 square meter installation at Orbit/Comdex Europe was an environment showing state-of-the-art research results organized theme-wise. Philips Design supervised the overall design of the village and the author designed the central meeting area titled the 'EyeCube' piazza. Comdex is America's foremost information technology and communication show, now also organized in Europe under the auspices of Messe Basel. Approximately 95.000 professionals visited the event—see www.messebasel.ch/orbitcomdex. The author's performance for the annual dinner had an audience of luminary professors, industry specialists, and related. However, despite this, it was expressed afterwards that it was a belief that playback was used and this, in the authors eyes, reflected the limited (and some would say square) knowledge and insight that the audience had at the time of such avant-garde performance utilizing state of the art sensor-based technologies. The author's comment in 2001 following the performance was that it was unfortunate that those present did not comprehend that "In future, other aspects of our sensorium could be uncovered that further alter and expand our methods of perception and change our approach to the creative process. I am convinced that the human visual system and psyche bear far more potential than we presently realize."[19] Aligned to the Basel installation and performance was a workshop for the i3 Spring Days 2001 in Porto, Portugal.

The author's workshop at Porto featured live performance examples of gesture control of robotic devices as well as gesture nuance from the bespoke sensors programmed/mapped to sound, image and tactile/haptic feedback manipulation—see Figs. 14.4, 14.5, 14.6, and 14.7. Featured was the games company that the author's research was responsible for creating titled Personics. Personics ApS was established as a company to develop personal interactive communication systems. The company offered clinical and personal multimedia systems that were used for entertainment, rehabilitation, disabilities training, art installations, and group

---

[19]http://www.nislab.dk/Publications/i3mag11.pdf (p. 39) © author.

Fig. 14.5 The author (back to camera, bald with dynamic shirt) demonstrating live performances of unencumbered gesture control at i3 Spring Days 2001. Additional performances include worn biofeedback sensors (Images © author unless denoted)

Fig. 14.6 Auditory, visual, and tactile/haptic feedback stimuli controlled by the human performance attributes of the child lying on a Vibroacoustic apparatus whose residual functional abilities are positioned within volumetric infrared invisible sensing spaces—some limbs having retroreflective tape for performance enhancement. Parent and care-providers attend

Fig. 14.7 A typical MoMA set up is illustrated where numerous cross-hatched interactive zones are created as a Virtual Interactive Space awaiting human participation to trigger responsive multimedia. Interestingly is that objective observation analysis showed how '*some get* it *more than others, some not at all*'

| Mobility | Hearing | Vision | Cognitive |

Fig. 14.8 Example of a biofeedback/biosignals state diagram where mapping of sourced signals is to multimedia and human interaction data measurement

interaction. Personics also organized workshops and provided training services. It was founded in 2000 and was based in Arhus, Denmark. However, due to mismanagement and conflict the author departed the company thereafter resulting in liquidation and closure.

Figure 14.6 illustrates how Vibroacoustic Therapy was investigated to supplement audiovisuals (and robotic light physical control—not shown). Knowing that the human body is over 70 % water and that since sound travels 5 times more efficiently through water than through air, sound frequency stimulation directly into the body is a highly efficient means for total body stimulation at a deep cellular level.[20] The term Vibroacoustic Therapy (VAT) was coined in the 1980s by the

---

[20]http://www.vibroacoustictherapy.com.

Norwegian therapist and educator, Olav Skille.[21] Research in Europe and throughout North America has shown that audio waves in the range of 30–120 Hz transmitted through the vibroacoustic device have a very positive effect on various systems of the body (cf.[22]).

Core of the installations and performances over the years of the author's research is that all should be accessible, or be able to be tailored to be accessible for all. Such installations were featured at leading Museums for Modern Art as commissioned exhibitions for eighteen months as well as at major international events such as the Olympics and Paralympics where cultural events support the sports occasion overall (in 1996 Atlanta, USA and in 2000 Sydney, Australia). A typical MoMA set up is illustrated where numerous cross-hatched interactive zones are created as Virtual Interactive Space awaiting human participation to trigger responsive multimedia (Fig. 14.7). The technic of using infrared means a no-noise environment and one that can be increasingly designed for higher sensing resolution of interactions via the areas assigned for cross-hatching. This method aligns with that used by the author in installation performances where control and non-control limits and delimitations were investigated (by the author as performer/designer/producer/director).

14.2.1   The Basis of Mixing'n' Matching to Human Input to Optimize Access: Tailoring

In order to best fit a human condition and related ability to perform to give maximum access and inclusion it was necessary to explore both human—from those most marginalized by impairment—and technology. In this case technology refers to both interfaces, that source data from the human, means of routing and as required manipulation e.g. scaling, etc., of that data (if not built into the interface), and multi media (content) that responds to the incoming data and acts as stimuli for the human. In addition, various forms of presenting means to best represent the stimuli so that the human can be optimally stimulated through targeted immersion in the experience. Furthermore, to be able to evaluate the concept and especially the mixing and matching decisions towards a bespoke tailored and developed 'system' learning for future iterations, a means to collect and analyze the data, use of the components, and human responses of the designed interventions was needed. This core design was the initial author-conceived framework behind how his sole research realized the 'Communication Method and Apparatus' patent—US Patent 6,893,407—see next but one section following.

Upon reading the actual patent text a knowledgeable practitioner would see how the core principles were adjusted out of proportion (by the second listed 'inventor'

---

[21]http://www.olavat.com/research–casesudies.

[22]http://www.vibroacoustictherapy.com.

who was a non-researching businessman) in an attempt to be all-encompassing beyond the real world. This businessman was 'the other party' in the company start-up (in fact it was started without the author as partner as agreed—hence my resignation and departure, and subsequent claim of patent ownership, that led to a court case, liquidation and closure). The fuller story is elaborated in a publication titled "The Rise and Fall of a Serious Games Company: How not to conduct business, when its an artist's work" ('in press').

14.2.2 Accessibility Exemplified: The First Cultural Paralympiad

This contribution, authored in 2016, marks two decades since the author's research and body of mature work was presented as a performance art production under the umbrella of the first cultural event that supplemented the Paralympics—The Cultural Paralympiad—when the Olympics were hosted in Atlanta, Georgia, USA. The Cultural Paralympiad 'showcased and highlighted the talents of artists with disabilities during the Paralympic Games in Atlanta. It was the first time a cultural program of this sort had been presented in conjunction with the Paralympics and it was a rousing success as music, theatre, dance, art exhibits, films, and performance art by professional artists with disabilities in mainstream settings were all presented' (see Lewis [3]). The event was realized by coordinator/organizer Deborah Lewis, the Community Events Coordinator, Administrative Director, and Executive Director of Special Audiences in Atlanta, Georgia, who received the ABBY Award for the effort. Special Audiences was an arts organization whose mission was to make the arts accessible to all Georgians. Founded in 1974, their initial funding was made possible by government sources both the Department of Human Resources and the Georgia Council for the Arts. In a supporting role was the local branch of 'Very Special Arts' (VSA), an international organization on arts, education, and disability that was founded more than 35 years ago by Ambassador Jean Kennedy Smith to provide arts and education opportunities for people with disabilities to increase access to the arts for all. VSA provides arts and education programming for youth and adults with disabilities around the world.[23]

The author's SoundScapes event for the first Paralympiad was hosted at the Rialto Center for the Arts, an 833-seat performing-arts venue in downtown Atlanta, owned and operated by Georgia State University. The author produced, directed and performed in the two and a half hour multimodal performance that featured performers (mixed-ability dancers, musicians etc.,) from around the world. Performances were cored on the SoundScapes concept of sensor-based augmented

---

[23]https://en.wikipedia.org/wiki/VSA_(Kennedy_Center).

performance where digital content could be selectively gesture and biosignal manipulated. Direct control of audio-visual stimuli was thus available for performers to control in order to compliment their own performance for the audience experience—as well as to enhance their own experience via interaction cycles. Videos suggest at this impact of gesture-control of multimedia in a human performance situation (e.g. https://www.youtube.com/watch?v=KrJhBpbwN1Q).

Deborah Lewis, the Community Events Coordinator, Administrative Director, and Executive Director of Special Audiences in Atlanta, Georgia, as the coordinator and main organizer who was responsible for realizing the Cultural Paralympiad, received the ABBY Award for the effort. Special Audiences was an arts organization whose mission was to make the arts accessible to all Georgians. Founded in 1974, their initial funding was made possible by government sources both the Department of Human Resources and the Georgia Council for the Arts. Very Special Arts (VSA), Atlanta, supported the event. VSA is an international organization on arts, education, and disability that was founded more than 35 years ago by Ambassador Jean Kennedy Smith to provide arts and education opportunities for people with disabilities to increase access to the arts for all. Based at The John F. Kennedy Center for the Performing Arts, with international affiliates and a network of nationwide affiliates, VSA provides arts and education programming for youth and adults with disabilities around the world. IBM Corporate Social Responsibility (CSR) Scandinavia, who had already sponsored the author's work since 1992, sponsored the author's production in Atlanta.

Installations and performances both typically included audience participation. In the former where the MoMA contract included the author conducting workshops in the installation for impaired groups on the day the museum was 'officially' closed. This meant that those in wheelchairs could explore the interactive spaces unhindered by those who walk. An example of the MoMA installations is available as a YouTube video.[24] In the latter, i.e. performances, wherever possible the interactive environments were left intact on stage and an offer for audience participation to try was announced e.g. https://www.youtube.com/watch?v=KrJhBpbwN1Q.

Experiences and structured analysis made clear that the accessibility to play within such interactive spaces was important and offered potentials in healthcare. The author's research thus resulted in a patent, the details of which he dictated to the legal firm responsible for filing. Assignment unfortunately was to the start-up company as dictated by the businessman, despite that the author's research funded the patent. Fittingly, as the author was more inclined to open source the systems created, the patent is now void since the problems with Personics. Below is a brief introduction that includes accessibility aspects.

---

[24]https://www.youtube.com/watch?v=3W4VznlgiU4.

14.2.3 Patent

May 5th 2000 was the date of filing a patent application resulting from the author's research as detailed to a patent lawyer. The application and eventual published patent was titled 'Communication method and apparatus'[25] that resulted in a family of patent publications (i.e. US Patent 6,893,407; also published as DE60115876D1, DE60115876T2, EP1279092A1, EP1279092B1, and WO2001086406A1). A second name is listed on the patent as inventor; however, this was granted through business interest. In line with this claim is that the four non-patent citations by the examiner refer solely to the author's solitary work i.e. (1) "Bridging Cultures", Program from The 13th International Congres of WCPT, May 23–28, 1999, Yokohama, Japan; (2) "Pushing the Limits" Congress Program and Abstract Book, 5th Scientific Congress, Sydney 2000 Paralympic Games, Oct. 11–13, 2000, Sydney, Australia; (3) Letter from Clifford Madsen of the American Therapy Association; and (4) Program and Letter from the Integrative Medicine Expressive Therapies International Symposium, May 26–28, 1999. The patent text has been referenced by fifteen patents between 2004 and 2014 with titles/subjects ranging from "System and method for delivering medical examination, treatment and assistance over a network"; "Systems and methods for reproducing body motions via networks"; "Poker training devices and games using the devices"; "Device and method for determining and improving present time emotional state of a person"; "Psychological Testing or Teaching a Subject Using Subconscious Image Exposure"; "Personal Growth System, Methods, and Products"; and "Monitoring system for monitoring a patient and detecting delirium of the patient".

The author's text relates to a non-lingual communication method and apparatus, wherein a physical or physiological signal consciously created by a first subject is detected and converted into a transmitted output signal presented to a second subject in order to communicate information from the first subject to the second subject. The invention further relates to rehabilitation of handicapped or otherwise disabled people.[26] The text further elaborates how "Though the primary object of the invention is a non-lingual communication method and tool between subjects, the invention may also be used for:

- distant therapy, where a clinician is able to instruct and monitor a patient through a communication link, as for example the Internet or wireless communications links, in this case, the invention may be combined with a camera to monitor changes and improvements during therapy,
- body language and personal interaction training, for example of corporate leaders or of schoolchildren in a pedagogic environment,
- spectator interactivity whereas the participants in an event can interact with the performers and/or other participants locally or at a distance,

---

[25]https://www.google.com/patents/US6893407.
[26]https://www.google.com/patents/US6893407.

- theatrical movement, where dance or musical actions are transmitted to another group locally or at a distance, for example through television,
- personal entertainment or recreation, whereas a subject or subjects can, for example through the Internet, interact with others through a user interface or game
- controlling hardware, where the detected signal is used with a suitable transformer to control technical and/or electronic commands for the control of an apparatus."

As is evident in the patent publication the author assigned interest to the Personics Company that was owned solely by the second named 'inventor': The author, whose research had realized national and international projects and was responsible for all the company stood for, was under the impression of shared partnership. This was through naivety, duress and the promised partnership never transpired. The author's sizeable projects, realized on his prior work, funded the company set up, the patent expenses, and day-to-day operation. Once it was found that there was no partnership and that there had been financial imbalance, the author resigned from the company in 2001 and instigated a 2004 legal battle over the patent rights. In the court case the author was offered by the judges a 50 % ownership, however, he refused due to the work being his sole invention. It can be seen in company records how Personics subsequently changed ownership through liquidation and later ceased trading with outstanding debt to the author and cessation of patent. The next section introduces, for the interest of readers, related patents with reflections to accessible related devices that in future may come under CVAA ruling.

14.2.4   Related Patents and Products Exemplified

An interesting patent document details what is titled as 'Free-space human interface for interactive music, full-body musical instrument, and immersive media controller' (US 20070000374 A1). Claims discus multiple sensory stimuli as perceptions experienced at the same time to achieve what is termed 'Kinesthetic Spatial Sync biofeedback entrainment effect', a psychological Gestalt effect wherein the unaided body in continuous motion becomes subjectively perceived as the sole and precision instrument. The claims relate to the author's research in that "the traditional concept of "instrument" (defined as something beyond and separate to the human body) appears to disappear, or at least, becomes greatly reduced in emphasis." An 'Effortless Entrainment' claim relates to the author's defining of inter-subjective and intra-subjective aspects aligned to the human afferent-efferent neural feedback loop closure. The patent text states that by "directing feedback to sustain the players' focus of attention to the immersive media responses, which are perceived as precisely and kinesthetically coupled to the body in empty space, the system evokes spontaneous and effortless entrainment into a continuous Gestalt of:

"My body IS the instrument." Whilst this patent (US 20070000374 A1) relates to "interactive music, full-body musical instrument, and immersive media controller" the claims should be considered from the user-experience perspectives in relation to the associated author's claims of how such digital media used in rehabilitation training offers potentials.

The author's research led to a need to learn about human function and condition in order to be able to design for interaction. Towards such learning the author investigated biofeedback systems, i.e. sensing systems that source human bio signals that can be mapped to multimedia content whilst at the same time providing data to analyze performance. The next section introduces the concept.

14.2.5   Biofeedback

Using biofeedback signals even profoundly disabled can input data to responsive media. The author purchased the Waverider/Wavewave system created by [4] Jonathan Purcell.[27] The system was more affordable than most and provided a fast response to biological signals for effective feedback. Individual sensors act as sources that each exhibit display data for easy comprehension and user-created windows process and output biological signals that can be mapped via MIDI externally to media. Double-clicking on any window opens a dialog box that controls the processing and display characteristics of that window. Musical feedback using an internal sound bank responds with a single note, or a user-defined range of notes, instruments, and scales. All session data could be archived for subsequent analysis—see text below and Fig. 14.8.

The system includes Pre-Programmed Protocols that provide an easy entry pathway for the beginning user and once a user is familiar with the software, Processed Data Modes can be used to design advanced user-specific protocols. Data export to Ascii creates a text file of off-line processed data. The text file can be opened in a spreadsheet and will display the output from each general graph present in a configuration file. Time resolution of the text file is set to the finest time resolution of the collected data streams and the WaveRider DDE creates a live link between raw and processed data outputs, and user-created client applications [4].

The human sourced data at the start of the author's research allowed access for all to express themselves creatively. All had access through the use of biofeedback sensor systems that could source any signal generated by the human as well as environment sensor systems that could source human motion in its owners Kinesphere (see Brooks et al. [5]). These data had two purposes—one, to be mapped to selective content according to a subject's profile, and two, to be archives as thick data for analysis of the interactions. Such analysis informed interested professionals e.g. therapist, medical professionals, facilitators, designer...on how the

---

[27]Purcell J. (2004) WaveRider Operating Manual. USA: MindPeak 2004.

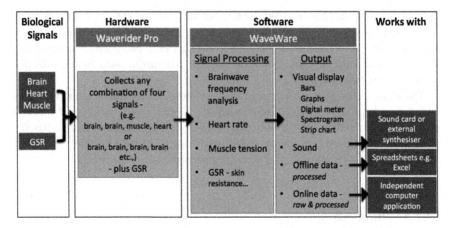

Fig. 14.9 AbleGamers foundation game accessibility guidelines labels

performance was operated, if design was optimal, if in-action intervention was optimal, if the presets innate to design were appropriate. This meant that the author's method and apparatus concept had an integrated level whereby the inter-subjective bespoke environments innately contained intra-subjective impact/learning, whereby the intra- related to self-agency that was indicated via non-verbal bodily representations in the form of human semiotics and nuances that over time became an information channel for the in-action improvised intervention aspect as well as for the on-action analysis.

Near the end of the twentieth century, in the mid nineties, the author explored interactive games as sensory stimuli for participants to complement the creative media options (music making, digital painting, effects, robotic control etc.) Games have innate qualities to motivate engagement and self-challenge in the form of task-based activities and competitions that the author introduced into research studies where there were more than one subject/participant. Each subject was made aware by a paper and pen result chart of 'scores' and 'times' achieved from playing a gesture-based game. The study was conducted at the turn of the century with acquired brain injured patients in a national rehabilitation-training clinic in Copenhagen. Motivation was elevated such that it was clearly driving their engagement to beat the other player. The success of this strategy is reported in the author's 'Humanics' series of papers.

14.2.6 Virtual Worlds and Games for Serious Applications

Game accessibility has been researched since the beginning of the game industry for inclusion of people with special needs. However, mainstream games have generally been inaccessible for many people with disabilities. The author, upon seeing results from using his 'Virtual Interactive Space' to source and input to manipulate created 'Virtual Worlds' of music-making, visuals, and robotics, pursued game-based content controlled from gesture in space. Once he had realized the bespoke system potentials of games that enabled alternative control (i.e. not mouse, keyboard and joystick) in healthcare were clear. However, 'video games present dramatic technical challenges and a unique—and difficult—environment in which to deliver accessibility solutions.[28] Extracts from the CVAA waiver exemplifies the challenges in the following: the full waiver is included as an appendix. For example, it remains technically challenging to develop accessible solutions for Advanced Communications Services (ACS) in an in-game environment given the often fast-paced nature of the experience and the number of gamers simultaneously involved in a possible chat session.[29] In addition, and critically, video games have not typically been a focus of the assistive technology industry, which has expended effort on other areas deemed a higher social priority—such as education and employment—and on applications whose core purpose is communication.[30]

Yet, following the author's research with significant outcomes that consisted of a two-site (Sweden and Denmark) project with Danish Sony distributer sponsorship and SCEE London collaboration in investigating the PS EyeToy in a hospital setting [6], an approach was made following conclusion of this study to Sony asking for support of additional research in such game accessibility—however there was limited positive responses. However, recently Sony Playstation Scandinavian distributer has become a sponsor, notably with the platform VR equipment.

Online gaming offers a person with disability opportunity to play and interact 'socially' from own home, thus, extending social circle and network. The Consumer and Governmental Affairs Bureau concluded that the role of "ACS[31] in online gaming systems" was expanding—at least relative to online gaming systems that historically had not featured ACS at all. In particular, the Bureau cited the "increasing role that ACS [was] beginning to play in online gaming systems and services... both with respect to the ability to compete effectively, and with respect to engaging in communications that are unrelated to game play."[32]

Terminology experts may question interpretations of the statement "Primary Purpose of Game Play" when in context of the research presented herein "Game Play" is the user-targeted experience, however, behind that experience is the use of the game to motivate and enhance in therapeutic rehabilitation intervention.

Advanced Communications Services ("ACS")—the primary purpose of the video game industry's products and services remains game play, and not ACS. Is this communicative aspect debatable and at the discretion of the intervention/user? Will work-around strategies be created to bypass enforcement?

---

[28]https://apps.fcc.gov/edocs_public/attachmatch/DA-15-1034A1.pdf.

[29]https://ecfsapi.fcc.gov/file/60001049004.pdf (Petition of the Entertainment Software Association for partial extension of waiver p. 11).

[30]https://apps.fcc.gov/edocs_public/attachmatch/DA-15-1034A1.pdf.

[31]Advanced Communications Services ("ACS").

[32]https://ecfsapi.fcc.gov/file/60001049004.pdf.

Video games make use of specialized hardware and software environments means that accessibility solutions for games (whether first-party or third party) cannot simply re-use accessibility technologies developed for other environments. Should then companies offer authoring tools that enable alternative interfaces data to be mapped to specific game data—thus reflecting differences in accessibility needs. Could such tools be common to open up a reusable gateway into game software?

ESA anticipates that as platforms develop accessibility solutions for the ACS systems that they may offer, these features may be available for reuse by game publishers, and could help inform other solutions to publishers for in-game ACS.

The online materials on this act points out how ACS rules contemplate that the Commission may waive the ACS obligations for multipurpose services and equipment that, while capable of accessing ACS, are designed for multiple purposes, and primarily for purposes other than using ACS. In evaluating the "primary purpose" of the equipment or services, the Commission plans to consider on a case-by-case basis whether the manufacturer designed the offering primarily to be used for ACS by the general public or for another primary reason, and whether the manufacturer or provider marketed the equipment or service primarily for its ACS functions. In addition, the Commission may also consider the extent to which ACS supports another feature, purpose, or task, as well as the impact that removal of the ACS feature would have on the "primary purpose" of the equipment or service. For class waivers, the Commission will also consider whether the services or equipment at issue "share common defining characteristics." (Appendix 1 and 2)

CVAA's captioning requirements are for broadcast video only: they do not include video in the games. The FCC does not regulate captioning of home videos, DVDs or video games; DVDs have captions most of the time because TV shows, movies, etc. must contain captions in order to be broadcast (to be in compliance).

Any device that offers Advanced Communication Services (ACS) must make those communication services available to people with disabilities unless it is not "achievable" to do so. The CVAA defines ACS as:

- *interconnected VoIP service*
- *non-interconnected VoIP service (does not require connection to the public switch telephone network); for example TRS (Telecommunications Relay Service)*
- *electronic messaging service (including text messaging, instant messaging, email and two-way interactive messaging through a social networking site)*
- *interoperable video conferencing service.*
- *Three categories exist in the CVAA for video games: (I) game consoles, (II) game distribution and game play networks, and (III) game software. Products released before the expiration of the waiver are exempted (PS4, Xbox One, Wii U, etc. are technically exempted). Businesses and organizations with less than 30 employees are exempted.[33]*

---

[33]https://igda-gasig.org.

At the FCC's discretion, requirements can be waived for equipment and services that are capable of accessing Advanced Communication Services, but are designed primarily for purposes other than using ACS. Seemingly this could be a work-around.

Consoles, software and online distribution services that offer broadcast video, such as television shows or movies, must provide users with a set of options so they can manipulate how subtitles are displayed. For example, YouTube, Netflix, iOS, Android, etc. currently offer full control over text size, font, and letterboxing when viewing online videos.

Accessibility features on consoles[34]

When a keynote in 2011 at the Transmedia International Masterclass Marseille (TIMM)[35] the author presented following Boyd Multerer, General Manager, Microsoft Xbox Division, wherein accessibility issues of game platforms was discussed using the author's research as a vehicle for the debate. Interesting at this event were the different interpretations of 'TransMedia'.

The below is a list of what is possible on current consoles:

- Xbox One

  screenreader
  magnifier (also in-game)
  closed caption presentation (with API)
  high contrast
  limited button remapping (at the system level)

- Playstation 4:

  limited screenreader
  magnifier (also in-game)
  closed caption presentation
  high contrast
  bold text, large text, text speed
  limited button remapping (at the system level).

- Nintendo Wii U and 3DS:

  Nothing as of yet—probably won't appear until the next round of consoles are released.

---

[34]https://igda-gasig.org.

[35]http://www.gameblog.fr/blogs/karlito123/p_36765_transmedia-international-masterclass-marseille-timm.

- Steam:

  Nothing as of yet—this potentially puts the company out of compliance with the CVAA, but it may depend on how the exemption for preexisting devices and services is interpreted (for example, what constitutes a major update to the service?)

In line with previous, a sector of ICT that has paved the way for advancing technical solutions is acknowledged as experimental music. The next section presents how new and finite advancement of input, in the form of creative expression data, can have a higher resolution offering a new modality that could be utilized by the games industry. This modality is titled Multidimensional Polyphonic Expression.

14.2.7 Multidimensional Polyphonic Expression—MPE Midi—New Potentials

As a part of this contribution the author wishes to highlight (with a smile) "you read it here first" how Multidimensional Polyphonic Expression (aka MPE midi) offers new potentials in the field of healthcare and (re)habilitation... and potentially games (including alternative reality environments) and creative expression via digital media. MIDI has commonly been used as signal protocol for routing sensing interface data from devices to content such that it responds to a human user's input —e.g. gesture. The responding content can be auditory (i.e. original use of MIDI), or other multimedia e.g. visuals, robotics etc. The new modality offers increased control and is potentially the next-generation controller for systems aligned with the author's invention. The next includes text cf. https://cycling74.com:

An example of devices utilizing the new and enhanced MIDI is Roli, who has created one of the most interesting keyboards of this century. It takes the problem of expressiveness and control to a new level with 5-Dimensions of touch—Velocity, Pressure, Aftertouch, Pitch Bend, Slide. The beauty of this is that these dimensions can be accessed and manipulated on a per note/finger basis, completely independent of each other (using the MPE extensions to the MIDI standard). Evaluation comments are positive—"My first impression on touching the Rise, "Wow, this feels great!!" Then I touched it again, and again, and again, pressing down wiggling my hand. Perhaps it's the rubbery skin-like flesh, but I was definitely drawn to it and enjoyed the reactiveness of the touch and the pushback from the keywaves—remarkably different to a normal keyboard or Piano." - "Muscle memory certainly plays a part with the Rise, the more time I spend with it the better I am able to play it, also increasing my control of subtle data manipulation" – "The data the Roli Seaboard Rise generates can be leveraged in a way that allows the possibility of a live performance (or studio recording) where one can simply play whilst controlling post note manipulation, fx and data without the need for any other controller or even needing to move your hands off of the keyboard, as a real possibility."[36]

---

[36]https://cycling74.com/2016/01/19/max-meets-roli-seaboard-rise/#.V4lFCMeIcgU.

The above text illustrates how *5-Dimensions of touch—Velocity, Pressure, Aftertouch, Pitch Bend, Slide*—can have potentials in healthcare increasing accessibility in creative expression and consequently due to the MIDI signal protocol use in some games. Changes are already evident in signal processing protocols used in the field with advances from MIDI to realize wider creative and expressive boundaries as witnessed through advances such as the example Multidimensional Polyphonic Expression (MPE). Wireless MIDI running under Bluetooth is now more reliable, low-latency, and easy-to-configure setups with pervasive mobile devices. RTP-MIDI (also known as AppleMIDI) is a protocol to transport MIDI messages within RTP (Real-time Protocol) packets over Ethernet and WiFi networks compatible both with LAN and WAN application fields. Also evolving is Audiobus enabling inter-app routing of audio. Aligned are how hardware manufacturers are bringing out new expressive controllers that take advantage of these changes such as Eigenharp, GECO for Leap Motion, Haken Continuum, LinnStrument, Seaboard by Roli, Soundplane, and the Moog Theremin. Such adoption of an extended signal protocol offers opportunities in accessibility within the author's work.

14.2.8 Bridging to AbleGaming: A Reference for Game Accessibility/Developer Design

This contribution reflects on an accessible vehicle in the form of a state-of-the-art advanced ICT system enabling access for many, including those most marginalized, to become a "user" a "player": This via a virtual environment offering interactive experiences in the form of playful entertainment and creative expression from the end of the 20th century. Latest advances, planned to include as herein presented, are further nuances of expression that can be utilized to give additional opportunities in future work, for example in the form of MIDI generators with MPE (Multidimensional Polyphonic Expression) capability and the digital communication protocols elaborated herein. It will be up to creative individuals with insight to alternative use and application to adopt such advances and develop the next generation systems. In utilizing a system with core digital signal communication protocol that lies at the heart of a patent titled 'Communication method and apparatus'—the author questions whether the system if realized today would come under the CVAA ruling that targets communication systems and accessibility.

This contribution anticipates how participants will desire continued access of such engaging ICT experiences, especially if they feel benefit, joy, and positive outcomes from use. Accordingly, to optimize uses and experiences, facilitators, therapists, carers and other helpers will be required who are proficient in training fully all the innate potentials of the hardware and software to enable tailoring to individual needs, desires and requirements for meaningful access.

Everyone has a different way of playing and there is no one right or wrong way of playing for someone with access limitations through impairment. Just as society 'disables' persons with impairment through access situations e.g. steps and no ramp for wheelchair users, the CVAA impacts through attempting to make accessible digital media, ICT and notably games. Disabled gamers (term used such that the game industry disables through not catering for an impairment) just want to be able to play a game and typically that means accessibility options are a must. Online guidelines inform developers in how to design with accessibility in mind offering suggestions that will not negatively affect the design of the game in any way, shape or form through making accessibility options innate and available for those who need them. Those who don't need the options won't even know they are there. Related is how there can be a deeper level that is more challenging for designers such that mapping of input to content to result in a game mechanic outcome can be a design issue for developers. This is subject of a following publication the author intends to write.

Suffice to say that, in addressing the CVAA, developers will more than ever need to make it a priority to include accessibility in order to conform. Whilst many will shy away, some may embrace the challenge such that they will find themselves a new audience that truly appreciates the extra effort. In the world of disability, small changes can make a big difference and relatedly the next section introduces the AbleGamers Foundation[37]—Extracted text in the following (cf/abridged from resources at AbleGamers Foundation—with permission given).

Representing over 33 million gamers with disabilities, The AbleGamers Foundation is a 501(c)(3) nonprofit organization that enables people with disabilities in video games. For nearly a decade, AbleGamers has helped countless gamers with disabilities either return to games after a traumatic injury or play games for the first time despite a disability. In line with this AbleGamers Foundation created an online Game Accessibility Guidelines site[38] (promoting 'includification'), written by developers, and gamers with disabilities as a resource tool to aid developers in adding accessibility to their game. Focus is on four areas as depicted in the Fig. 14.9 and as detailed in a publication[39] which this section extracts from.

The Ablegamers publications are created to assist developers and are exemplified in the following (cf. see footnotes) where a general focus is on four main areas, i.e. Vision, Mobility, Hearing and Cognitive. With each area, information is offered across different levels of classification—aligned to this, the following refers to subcategories as well as existing games that cater for the impairment(s).

Vision (cf/abridged from resources at AbleGamers Foundation—with permission)

---

[37]http://www.ablegamers.com and http://www.ablegamers.com/about-us/.

[38]http://www.includification.com.

[39]http://www.includification.com/AbleGamers_Includification.pdf.

Vision (3 levels): One in seven men have some sort of color deficiency, 75 % of all adults worldwide need some sort of vision correction and over 1 million are legally blind. Developers can take some very easy steps to ensure their developed game is accessible to those with visual impairments, e.g. -

Level One (5 categories)

Changeable Text Colors: The name of this says it all. For those that have difficulty distinguishing between colors, often referred to as color deficiency, the ability to change the color of text associated with a particular function improves overall gameplay. Note that color deficiency differs from color blindness. Color-deficient people can still see a certain color, but they cannot distinguish shades of the color, red vs maroon, or lime green vs dark green.

Often found in MMORPGs, various colors are used to indicate what type of input is being received. Green is often used for guild conversations, white is often used for local communications, light colors of pink or purple are used to indicate private message, whereas red is often used to indicate combat or enemies. This range of color usage can be challenging to players with color deficiency or color blindness.

World of Warcraft does not allow the user to change the color of any text input, but through the use of modifications, gamers can make the colors more distinguishable from one another.

In SWTOR, the individual gamer can define text colors, and this enables those with any color blindness or color deficiency to interpret the game data more easily.

Comment: *It may seem like something that is relatively simple, but for those that need the option, it's important to have. After years of development, many colors have become standardized as meaning certain things (i.e. green is almost always used for the guild chat), which is exactly why allowing colorblind or color deficient users to change colors to what is more comfortable for them is so important. As an industry, games have realized the importance of separating information based on color for easy dissemination. Therefore, it is important for those who need to alter colors to the spectrum they can see is also important. Remember this: 1 out of every 7 men are color deficient.*

Example:

A woman with a strong color deficiency has trouble interpreting chats in her favorite MMORPG because the color of her guild chat is green, her group chat is a very light purple and her instant message (whispers) are pink. If the game incorporated the ability to change the colors of any text to something she could see more easily it would improve the quality of her gaming experience. This a clear example where CVAA would impact to support the case.

Changeable Font Sizes: Since the invention of text and graphical displays, one of the most basic options has been the ability to change the size of text. However, video games do not often provide the option. Gamers with visual difficulties including those that are legally or almost totally blind can still participate in many games, but reading text in the subtitles, directions/instructions and chats can be frustrating if the size of the text is just barely too small to read.

One example of a game that does this well is EverQuest II, which allows users to change text size.

Text size matters to more of an audience than one might think. Up to 50 % of the aging gaming population has or will have vision problems in their lifetime.

Example:

A man loves playing an online adventure game, but his eyesight is slowly fading as he ages and his macular degeneration continues. He squints at the screen with the most powerful glasses money can buy, but the text continues to be harder to read. If the game allowed him to increase the size of the text in his chat window and on quests, he would be able to continue playing the game without interference.

Color-Blind Options: Color-blind options are the most popular and in-demand feature for those with visual issues. Ideally all games should have colorblind options that can be enabled to compensate for a variety of color deficiency issues. The most prevalent of these disorders is the inability to tell the difference between the colors green and red. This is most often avoided by providing a way to change the color green to blue when indicating friendly information such as group mates or health bars.

There are a wide variety of color deficiency disorders in the color-blind spectrum. Few people are monochromatic, and far less common forms of color-blindness besides red-green blindness include the inability to tell the difference between blue and yellow, and the ability to discern everything except blue.

Using color is not a bad thing at all, but to support color-blind players, a secondary indicator needs to be added when color is used to indicate an important status. For example, consider a game with a map that shows players which team has control of territory. The standard implementation of this would be to add a RED and GREEN overlay on the map to indicate status; this would be useless to most color-blind gamers.

A much better way to handle the situation would be to include symbols on the map to indicate who owns each area with the normal color scheme. Perhaps a 'Check Mark' for those things your team has control of and an 'X' on those you don't or fog of war mechanics. Any theme will do, just as long as it is visually distinctive. This can also be an option that can be turned on or off depending on gamer preference.

Popcap[40] is largely considered the front-runner in color-blind accessibility. It is becoming increasingly more commonplace to see colorblind options in casual games, but Popcap was using color-blind friendly options in titles such as Peggles long before it was considered mandatory.

---

[40]http://www.popcap.com.

An example of a problem most can relate to is in Star Wars: The Old Republic. There, "Portals" that you can go in are GREEN, but those you cannot are RED. To the red/green colorblind gamer, these both look the same, and navigation is a mystery.

Example:

A woman purchases the sequel to her favorite casual puzzle game. When she gets the game home, she finds she can't tell the difference between the red and green puzzle pieces defined by the color scheme of the game. If the options were available to change red or green to an alternate color, or set user-defined colors for both, then she would be able to enjoy the game.

High-contrast Target Reticle: For First Person Shooters, it is extremely important that the target reticle uses colors that are easy to distinguish from the other environmental elements in the game. In situations where both the enemy and reticle are red, the user is unable to identify between the target and the sight, leaving no way to tell where the actual target is aiming.

In Max Payne 3,[41] Rockstar enabled an option that allows the end user to choose from coloring the target reticle red, white or blue. The color differential is important to the user with color deficiency. This is particularly true when the point of the game is to test accuracy and reflex actions.

The easiest solution is to change the target reticle to blue or white.

Example:

A man loads a new shooter game that just came out on the market. While the company did a good job by not making the enemies highlighted red and allies as green, the target crosshairs are still green making the environment of an open field a nightmare for him to see the difference between where the target begins and grass ends.

Enemy Marking: The ability to tell friend from foe is one of the most fundamental differentiators in all video games. Some games use naming, visual bars, symbols or color markers. For those who have trouble differentiating between similar colors, small markings can help gamers tell the difference between friend, enemy, or really difficult enemy.

In Guild Wars 2,[42] enemies are marked by red bars with numbers representing mob level and accompanying symbols to declare the level of the faculty in either a gold or silver star.

In Battlefield 3,[43] an option can be enabled to place small blue 'Doritos' over the heads of allies. This allows gamers to quickly recognize that those without the marking are enemies.

---

[41]http://www.rockstargames.com/maxpayne3.

[42]https://www.guildwars2.com/en/.

[43]https://www.battlefield.com/games/battlefield-3.

Example:

A teen has trouble seeing enemies on certain backgrounds of a game with a variety of maps. Without the ability to change the colors of the enemies or mark them in some way, he will be caught unaware when an enemy walks on-screen.

Level Two (3 categories)

Customizable Fonts: Font customization is not a new trend in game development, but the use of very stylized fonts has a negative effect on low-vision gamers. While its understood that most game developers want to run with a theme throughout the project, many of these more stylized fonts are difficult to read by most gamers with good vision; and for the vision impaired gamer, it's an utter roadblock.

The ability to swap out more elaborate fonts with something more simplistic, like Arial or Times New Roman, will allow the low-vision gamer to make text easier to see. Combine this with the font-size changes in level one, and you are on the cutting edge of text support for the low-vision gamer.

Customized HUDS: For those with vision impairments such as color deficiencies, low vision, tunnel vision and difficulty seeing rapidly moving information, the ability to organize the UI to the end users' preference helps to mitigate problems with seeing important information at a glance.

Additionally, the ability to change the colors, size and position of various elements on the UI help those who cannot differentiate between certain colors. While those with other eyesight issues can independently arrange elements to benefit their style of gameplay without needing to petition the developers for a set up that would work for them individually.

Giving multiple alternatives helps alleviate the need to add multiple schemes further down the road in the development cycle.

The most preferred method for dealing with customizing the graphical user interface is to allow the user to individually define how they would like the elements to be arranged.

Understandably, this is an expensive endeavor. However, it may be an additional argument for iterative interfaces that can lead to smoother development processes and better overall design. Currently, consoles rarely allow the user to change the position of UI elements. It is far more common in PC. But in those circumstances, gamers that have a difficult time seeing certain positions on screen benefit greatly from those games that do allow custom interfaces.

Example:

A woman with macular degeneration has difficulty adjusting her sight quickly to various points on the screen. If she is able to redesign the UI so that the elements of most importance are within her visual range, the time spent in that game will be dramatically less frustrating.

Map Recoloring Options/Alternative Views: The next logical progression from the colorblind options laid out in level one, is giving users the option to recolor the 'mini map' or other 'situational awareness' screens and tactical maps away from the traditional red and green to alternative colors such as blue and orange. This allows those with color deficiencies to make better use of the interface.

Demigod[44] from Stardock uses a mini-map that is imperative for gameplay, yet cannot be altered. For users with red/green color deficiency the game is extremely difficult to play without such vital information at a glance.

While not an ideal implementation, World of Warcraft allows mini-map modifications to be done via scripting that enables variations on the shape, color and size of the mini-map.

Allowing the gamer to find a color scheme that best suits their visual needs is the best possible solution in this scenario.

Example:

A man who has monochromatic color deficiency can't see the difference between red and green. His favorite MMO has a mini-map that displays both enemy and allied movement. Unfortunately, the mini map can only be displayed in green and red, so everything looks the same to this man, and the battle lines are a mess of misinformation.

Level Three (2 categories)

Speed Settings: This is a top-tier option for those with mobility, low vision and cognitive disorders, having the ability to slow down the game clock allows those that just need more time to process what is going on, and how they need to react to it.

A disability case, from recent times, refers to a veteran coping with a traumatic brain injury and has difficulty processing information. He has no issue understanding what is going on in the game; he just does not do it as fast as most others. By slowing down the game up to 80 % he can successfully play his favorite sports game that he and his friends played while over in the sands.

My Football Game[45] by VTree LLC allows the game clock to be slowed down to 20 % for those with difficulty moving the mouse easily or reacting quickly. This enables gamers to complete the game at a comfortable rate further particular ability.

Text-to-Speech Input: Another top rated accessibility option would be to include the ability of the game to read the text on the screen and repeat it in audio form. Many programs now do this for users on the web, but it has not yet made its way into the game universe. The successful implementation of this feature would be to offer text-to-speech on user created text, like guild chat, or other chats created by end users.

A gamer with difficulty reading information because of brain injury or eyesight issues has a very hard time interpreting text information at a comfortable rate. The ability to have the game audibly announcing the information displayed on the screen would enable the gamer to participate in the game more effectively.

---

[44]http://www.demigodthegame.com.

[45]http://vtreeentertainment.com.

Mobility: Mobility is the largest category in AbleGamers' accessibility guidelines. Not only are there numerous hardware and software solutions to inaccessibility, but also disabilities that cause physical limitations are the most prevalent. Ranging from Muscular Dystrophy to the loss of a limb in war, mobility impairments make up the largest group of gamers with disabilities. These barriers can often be lifted by simple yet effective accessibility options.

Level One (3 categories)

Alternative Configurations: The fallback position, if programing remappable keys is just not an option in your title, is alternative configuration setups. The idea is simple: develop predefined controller configurations that allow a gamer to select one that best matches their play style, and their disability.

Conventional console developing logic requires including both normal configuration, southpaw, reverse and a few random layouts as a catchall. In order to be truly accessible the configurations must include right-handed, left-handed, button combinations near each other and one-handed.

Developer Exercise: Take a look at the controller, and see if you can play your game with one hand. Place the controller on your knee, your desk, or your thigh and give it a go. If your game is not at least semi-playable with one hand and you are not pleased with the results, alternative configurations can be a reasonable fix if fully remappable keys are not an option.

Example:

An example of multiple configurations being done properly is in Halo Reach[46] by Bungie where a player has six out-of-the-box configurations, such as BOXER, GREEN THUMB, and BUMPER JUMPER. The developers also thought about those gamers who are left-handed by including a SOUTHPAW configuration. While this may seem to be a luxury, when a disability determines what buttons are easy to press and which ones cause fatigue, it is imperative that the disabled gamer be allowed to choose the path of least resistance and maximum enjoyment.

Camera Controls: In first-person shooters and MMORPGs, the ability to control the speed, angle and distance of the character in relation to the field of view is important. In these games, the gamer's ability to move the camera often defines their movement direction, making camera movement one of most important aspects in the game. Camera controls need to allow comfortable movement for both users that are only able to make larger, less precise movements, and also users that can only make small, precise movements.

Many disabilities, such as Cerebral Palsy, limit the ability to manage range of motion, making it very difficult to move the mouse back and forth small distances. This makes precision a massive challenge. A developer can give these gamers the ability to translate large mouse moves into slower camera motion, preventing wild nauseating swings of the camera and allowing total control of the character.

---

[46]http://halo.bungie.net/projects/reach/.

At the other end of the spectrum, some gamers with Muscular Dystrophy have range of motion issues that let them move the mouse only 1/16th of an inch in any direction. For these gamers, precision is easy, but macro movement is impossible. Even with this limited range of motion, a developer still can give these gamers the ability to move their character like everyone else by letting them set the camera further out from the character, and manipulate with extremely sensitive camera movement–one full 360° camera rotation should be able to be accomplished by moving the mouse 1/8th of an inch or less.

Star Wars: The Old Republic and Rift[47] are good examples of games that allow players at both ends of the spectrum to tailor camera movement to their needs. The sensitivity of both games has sliders with a cap that allows for full camera rotation at 1/100th of an inch, or conversely, an area larger than a mouse pad depending on the user's need.

Guild Wars 2 is an example of a game that only allows larger, imprecise movements (very low sensitivity). It does not support extremely small movements. The sensitivity by default is set low and can only be increased slightly. The cap is set near a full 1 inch turning radius.

Providing the option to find a comfortable way to use the camera and allow the gamer to move their character can mean the difference between buying a game after a playing a demo, or walking away from an unplayable experience.

Example:

A gamer with muscular dystrophy uses a mouse that allows 3500 DPI. He logs into a brand-new title fresh off the shelf. The game uses it's own mouse driver emulation code slowing the mouse cursor movement speed and thus making the movements needed to control the direction of the character much bigger. There are no camera or mouse sensitivity settings in the game, which forces the gamer to return the game or consider the purchase a waste of money as the game is unplayable to him.

If the game either used Window's mouse sensitivity or allowed the cap on emulated mouse drivers to be set extremely high, the gamer would be able to adapt the sensitivity to an acceptable level.

<u>Remappable Keys</u>: One of the most widely requested accessibility features across the entire gaming universe, for both PC and console, is the ability to reconfigure keys. This easy-to-implement and cost-effective feature can make a difference for both disabled and non-disabled gamers alike. While adding multiple configurations on the PC and console is a step in the right direction, the ability to truly remap the keys to whatever configuration best suits the play style of the gamer, or to put the most important features in an area that fits the gamer's 'mobility sweetspot.'

Those who have muscular dystrophy, cerebral palsy, multiple sclerosis, war trauma, serious injury and a range of debilitating diseases that affect the ability to use one's hands "normally" benefit greatly from the ability to define their own

---

[47]http://www.swtor.com/info.

personal ways to play because each disability is different in the way it manifests itself.

Examples of games with remappable keys done well include: World of Warcraft from Blizzard and Star Wars: The Old Republic from BioWare on PC. Resistance: Fall of Man from Insomniac, Skyrim and Fallout 3 from Bethesda on the console.

There are numerous examples of games that received serious push back from the community for NOT allowing remappable keys.

However, remappable keys are not a silver bullet that solves inaccessibility and allows anyone to play, as suggested by some in the disabled community. Remappable keys are one of the three must have (along with close captioning and colorblind) options for minimal accessibility, because this simple feature enables play for players with any of a large number of disabilities, it gets the most requests.

Level Two (9 categories)

Third Party Access: AbleGamers state that they understand that developers do not want people to cheat in a created game; developers want the gamer to enjoy the game as intended. However, the common strategy of blocking any non-standard input devices besides the mouse and keyboard will also make it impossible for many gamers with mobility disabilities to play the game. The ability to use third-party devices and assistive technology, like the default on-screen keyboard installed on all Windows PC's, and gaming peripherals is critical.

In situations where technologies, such as Game Guard, block any third-party application or hardware from accessing the game it is designed to protect, it prevents people from being able to use the very technology they have become dependent upon in order to use their computer and play games.

When developers use Game Guard to 'protect' their game, the software prevents the use of on-screen keyboards and head-mice and everything else, virtually locking out anyone who could not use a standard keyboard and mouse. Aion[48] from NCSoft is an example of a game that is problematic for this reason.

There are viable alternatives to protecting a game, such as Battlefield 3's "Punk Buster" and Blizzard's "Sentinel" program, that search out cheating in the game without interfering with assistive technology.

It is important for developers and publishers to be careful when choosing the proper scheme to watch over their games and/or DRM. If every guideline in this paper is followed, but the wrong protection software is included, then a developer's work implementing other parts of this document (see http://www.includification.com) will be for naught.

Example:

A gamer that was in a car accident and now has no use of her limbs uses head-tracking software and an on-screen keyboard with dwell technology–software that allows the mouse cursor to hover over a graphical display of the keyboard, which presses the key the mouse is hovering over after a predetermined amount of

---

[48]http://na.aiononline.com/echoes-of-eternity/.

time. This allows for her to point the mouse with her head onto a software representation of the keyboard. By hesitating the pointer over the corresponding key for long enough, the key is pressed. It is a slow way to type, but from her perspective, it is the only way she has. She is very proficient at this, and this combination of technology means she is able to play certain games to her heart's content.

Movable/Resizable UI: The ability to move and/or resize each individual element on the HUD interface is great for both disabled and able-bodied gamers alike. For those gamers with low stamina, strength, or dexterity, the ability to place the most essential elements where the gamer feels most comfortable is critical. Many disabled gamers position certain key elements such as hot bars in their 'sweet spots' to conserve energy and allow an enjoyable experience as opposed to a workout.

UI customization requires a lot of development work and is often not tackled, but many games support total customization of the UI. An interesting case is World of Warcraft, where Blizzard exposed the XML underpinnings of its interface, to allow modification of the UI with minimal developer support. Perhaps unsurprisingly, there are specific WoW mods developed just for gamers with disabilities such as colorblind and interface altering modifications. These enable critical buttons that require repeated pressing by the player to be centered in an area where the end user often leaves the mouse.

For gamers with disabilities, customization can make it easier to manage situations that need quick reaction time. The less frustrated the gamer is, the more likely they are to continue enjoying the game.

Example:
A woman with strength and stamina deficiency due to Multiple Sclerosis loves to play a strategy game, but its buttons are locked in the lower right-hand corner of the screen. This causes her to move the mouse repeatedly and quickly around the screen, between the UI and other game elements. This action is difficult for her, and often cuts her game time short. If the buttons were placed in a movable element, she could determine the easiest place for her to control the flow of the game. She'd be less tired, able to play longer, and have a more rewarding experience.

Macroability: Games who find it difficult to use the keyboard, perform actions quickly, or press multiple buttons simultaneously will look to macros to help level the playing field against those with better dexterity and speed. The ability to create macros used to be common in many PC games, but the ability to macro anything in some games led to the rise of "bots," which in turn led to many companies scaling back or even eliminating macros, to the detriment of the disabled player.

Why are macros so important? Players with issues like Muscular Dystrophy or the loss of a limb use in-game macros to simultaneously press multiple buttons or button sequences with a single button. This is even more important for those with low strength and stamina; the ability to macro often-used commands lessens the burden of pressing multiple buttons, and therefore extends game playability.

Another use for macros is to provide the ability to store saved text to be used as conversation. For some with slow motor skills, taking in guild chat, looking for a group, or even responding to simple 'tells' can become a painful nightmare. People think they are rude when the only reason they do not engage is the time it takes for the gamer to type out a command. Macros allow that gamer to create a library of text so that he can communicate and enjoy the social aspects of gaming.

As noted above, some companies regard macroing as cheating regardless if it is in game. Star Wars: The Old Republic is an example of a game that disabled macros after the community complained about people being able to use them in player versus player combat. Disabling macros has not dampened the ability of elite PVPers, but it has definitely limited the ability of those who need to press multiple buttons in quick succession in order to compete at a higher level.

The now defunct Star Wars Galaxies allowed for abilities to be queued one after the other in a macro, which reduced 4–6 button presses down to only one. Dark Age of Camelot from Mythic Entertainment has the ability to macro text, which can enable someone who uses an on-screen keyboard to hold entire conversation with single button presses.

Example:

A child with Cerebral Palsy who loves playing her favorite wizard MMO with her brother and father uses an on-screen keyboard to play; which makes using the hotbar buttons a repetitive task. The ability to combine a common key sequence—TAB to target the nearest enemy, and A to start attacking—into one button would decrease the stress in her shoulder, leading to less fatigue. Even the ability to queue abilities one after the other would lead to a better gaming experience for the young lady, more bonding time with her family, and an enhanced feeling of normalcy.

Difficulty Settings and Fail Safes: For years in the videogame industry, difficulty settings were almost standard in game design. Although seen less frequently in current games, difficulty settings are a feature that is enjoyed by disabled and non-disabled gamers alike. Note: The needs described here span both mobility and cognitive disabilities.

As an example of a game that did this well, Mass Effect 3[49] from BioWare allowed the user to become nearly invincible and simply enjoy the story. In this mode you can one-shot most mobs and run through without using anything to regain hit points. This only aided in making Mass Effect 3 a success to a wide range of gamers. Some games also provide 'hardcore' difficulty levels that provide challenges that would alienate all but the most driven of players; in both the casual and hardcore cases, difficulty levels are tools to provide a tailored experience to all potential members of your audience.

In most situations, the goal of publishing and developing a game is to make an emotional connection with the player, and tell a good story. Giving the end user multiple ways to enjoy the experience means all development hard work gets enjoyed. Difficulty settings support the needs of everyone from the most casual

---

[49]http://masseffect.bioware.com/.

gamers, and the most hard-core gamers, letting both enjoy the same title regardless of ability.

Examples:

Imagine a teen with ADHD and learning difficulties—one with cognitive disabilities that do not affect motor function. He is having trouble completing the steps necessary to advance in his favorite action game. In the game's ONE difficulty setting, the player must manage: ducking behind cover, jumping over obstacles in a timed manner, and aiming and shooting successful headshots. This complexity is just too hard, so he gives up and turns to another game.

If the title he was playing had difficulty settings, he would be able to complete these actions in a more forgiving manner and enjoy the game in his own way. An even better approach would have the game recognize when he failed in this task a few times, and display a dialog asking if he would like to skip this; he chooses "yes," there is small cut scene showing his character making the shot, and he keeps progressing.

A woman with Multiple Sclerosis loves playing her favorite RPG, but it's difficult for her to use the mouse for extended periods of time. If the game has multiple difficulty levels, she can enjoy the storyline without fear of being slaughtered because she has times she can't move the mouse another inch.

Save Points: Save points are incredibly important for those who have stamina issues such as Muscular Dystrophy, Multiple Sclerosis, and other neurological disorders. It can be difficult for some gamers to sit down and game for long lengths of time. Save Points allows for the gamer to participate in the game for as long as they are able to without being penalized for not being able to have epic gaming sessions. Many games implement save points at certain intervals throughout the game, but the longer the gaps are between save points, the more difficult (in an un-fun way) the game becomes for these players.

Ideally, each game should be able to be saved at any given moment allowing those who need to stop at a seconds notice the freedom to do so without fear of loss of progress.

Sensitivity Sliders: Simply stated, a critical component of accessibility giving players with strength or dexterity issues the ability to master movement of input devices like the mouse or joystick. For those who have low strength or dexterity, the ability to set the sensitivity level to an incredibly high level lets players transfer small movements of the controller into large movements on the screen. This allows for reduced effort on the part of the gamer and allows for increased time playing. On the other side of this is giving gamers with reduced control of their movements, like those cerebral palsy and other neuromuscular disorders, the ability to reduce controls sensitivity, so their broader movements will result in more precise game actions and a better overall gaming experience.

Rift is a wonderful example that allows both the mouse sensitivity and camera controls to be set to levels that are extremely slow or ultrafast, which incorporates users anywhere on the spectrum.

Linage is an example of a game that does not allow sensitivity to be moved in either direction and therefore makes it difficult for anyone that does not fall in with the 'default' to move the mouse accordingly.

The problem with both sensitivity sliders and camera controls occurs when assumptions are made as to what the appropriate level of speed should be. Consider supporting a broader range of sensitivities than might seem intuitive to prevent disabled gamers from being locked out of the game entirely.

Examples:

A gamer with Muscular Dystrophy wants to play a brand-new game that just launched today. Because of his disability, he is only able to move his mouse about one inch in every direction. He fires up the game on his PC, but as soon as the splash screens are over the sensitivity levels are set to such a low level that moving the cursor around the menu screen is impossible. If the game developers have added in sensitivity settings in the game, he will be able to raise the sensitivity to a very high level, allowing that one inch of mouse movement to mean a full trip across the screen.

The other side of this, an elderly woman with tremors in her arms and hands was a gamer long before her current condition. For Mother's Day, her grandson just got her a new casual game to enjoy. Like any gamer, she hurries home and loads the game only to have her arm start acting up as soon as she sits down. She can't click on the puzzle pieces because the mouse moves too wildly and no sensitivity sliders were included in the game. If sensitivity sliders had been included, she would have been able to lower the sensitivity and make moving the mouse more manageable.

Click-to-Move/Mouse-to-Move: Extremely popular in European and Asian games, and used by disabled and non-disabled alike, Click-to-Move is when the developer allows for a gamer to click the terrain and have the character move to that position in the world. Meanwhile, Mouse-to-Move is when the designer allows for the movement of the character by pressing both the left and right mouse at the same time, using the mouse movement to steer the characters.

These two features serve different purposes. Click-to-Move allows a gamer with a strength issue to click the ground and get his character where he needs it, instead of having to press and hold the mouse buttons or WASD keys. On the other hand, Mouse-to-Move is key to gamers who do not have the arm movement to continually alternate between mouse and keyboard. Both features let players move their character where they need it, and never have to take their hands off the mouse.

An example of excellent Click-to-Move incorporation is the original Dragon Age: Origins. Depending on how you wish to control your character, you can use both mouse buttons, traditional keyboard control or Click-to-Move. This allows for gamers of any ability to control the game as they see fit.

An example of inaccessible game design is a game, which purposefully disables Click-to-Move because the community believes it allows bots. However, any program that could be used to build a bot predominantly operates by automating keyboard commands, not using click-to-move. In this situation, the game loses a useful accessibility feature for little-to-no benefit.

Example:

A disabled veteran spent a lot of time playing PvP in an MMO with his buddies when he was deployed overseas. He was injured and is now stateside. He can no longer use his arms without pain. Instead, he now uses a head mouse, which allows him to play the game without moving his arms at all. He still wants to play with his buddies. In the old days he would have to use the WASD key or hold down both mouse buttons to move his character around, but the makers of the game considered these issues during development and included Click-to-Move in the title. This means he can play the game without moving his arms, and stay connected with his old Army buddies.

Keyboard Movement: The ability to use the keyboard as the only main input device is another crossover feature that is used by both the disabled and able-bodied communities, and is part of a debate that is as old as the PC itself–some gamers swear by being 'mousers,' while others are just as convinced the only way to play is by using keyboard shortcuts.

For those that have repetitive stress injuries, carpal tunnel syndrome, and muscle or nerve injuries that cause pain when moving joints, the ability to use the keyboard to control the game may be the difference between gaming or not being able to game at all. Going back and forth from the mouse to the keyboard is a nonstarter for these players.

Example:

A woman wants to play a brand-new hack and slash game where the point of the game is little more than to run around, kill monsters, and loot new gear. However, the only way to control the character is by clicking on the ground with the mouse. Her repetitive stress injuries make it painful to move the mouse that much. If the game allowed her to rest her hand on the keyboard and use WASD, she would be able to play without pain.

Assists: There are many different types of assists spread out in the gaming sphere. Each has a varied level of interaction with different parts of the game, but each can handle functions that cause problems for certain groups of gamers, while also providing useful tools for non-disabled gamers to enjoy games as they prefer.

Comment:

Here are just a few examples of different assists developers are familiar with.

Aim-assist is one of the most highly recommended options for first-person shooters, because it can help the end-user without interfering with gameplay. Aim-assists range between automatically targeting enemies to locking onto specific parts of the target (VAT in Fallout 3).

Driving assists allow inexperienced drivers to concentrate on steering a complex course while the computer handles issues like gas and brakes. The assists can be disabled depending on the difficulty level the gamer chooses. Drive assists are important for cognitively disabled gamers who may have trouble concentrating on more than one task. It also helps mobility impaired gamers who do not have the ability to press enough buttons at one time to control multiple tasks, or those that cannot keep a button like the one controlling the gas pedal press down continuously (Forza).

Puzzle assists, often referred to as hints, give gamers progressively direct instructions on what to accomplish next in a puzzle or hidden object game. These types of assists are crucial for those with cognitive impairments and attention deficits, as well as non-disabled casual gamers.

An example of excellent an aim assist is the VATS system in Fallout 3 from Bethesda. Users can stop time completely while aiming at a select part of the target's body. This allows the user to take as much time as they need to complete intense combat situations.

A top-rated driving assist model is in the Forza series by Turn10 and Microsoft. In order to make the game accessible to players new to the racing game genre application of the brakes can be controlled by the AI automatically. This allows the user to hold down the gas during the entire race and concentrate on steering left and right while the computer controls the speed of the vehicle through the turns.

These types of assists can range from 'nice to have' to 'can't play without.' Now some developers have made comments about how these features turn off hard-core gamers from a title, an understandable concern as highlighted earlier. However, if the fear of the publisher is push back from the community over the decreased level of difficulty, allow achievements that cannot be earned while certain assists are enabled or certain game modes are played. it is imperative to remember that at the end of the day those who need these kind of assists to play the game at all do not care about achievements, they care about the ability to play. Allowing those who need the assists the option to turn them on to play the game, but keeping achievements and even certain rewards available only to those who do not use the assists is a great way to compromise in adding accessibility.

Examples:

A man logs into his favorite FPS, which he loves because he gets to play with the various types of weapons in the game. One of the features included in the game is an aim assist that allows him to lock onto a target when firing; it automatically finds a target nearest in range and slows down the action while he takes his shot. This is important to him because his disability makes targeting with any level of precision an extremely difficult activity. The assists allow him to play comfortably in a solo environment.

Who doesn't enjoy playing racing games with the family? Dreams of the fast cars and the cheering of the fans is a great addition to family game night. The young lady in this family has a form of muscular dystrophy that prevents her from being able to use the Xbox controller easily, so the game was quickly taken out of rotation on game night. She will be unable to play with her older brother, but if the game allowed the AI to control the brake for her, she could handle steering left and right, and the race would be on again.

Level Three (2 categories)

Input devices: One of the most critical elements to designing video games for disabled gamers is to include the ability for all peripherals to work adequately with each game. Most gamers in the disabled community need to use third-party software and hardware in order to mitigate some of the difficulties of playing complex games. It is important to make sure that these types of software and hardware are not disabled.

Although the list of assistive technology is long and extensive, the basic categories of assistive technology range from on-screen keyboards, voice recognition software and switch-based hardware input devices.

These types of software and hardware devices shift gaming from an impossibility to an important activity for many disabled gamers. Included in Windows by default, on-screen keyboards can be tested easily and other hardware can be white listed as programs that help the disabled community.

Examples

A gamer with Duchenne muscular dystrophy uses an on-screen keyboard and a voice control software program to play his favorite RTS. The game has a 'full-screen windowed' mode that allows the gamer to place the on-screen keyboard over top of the game and allows voice control to work properly.

Speed Settings: As a top tier option for those with mobility issues, developers should consider having the ability to slow down the game clock entirely. This allows those with dexterity, precision and strength issues to interface with the game at an easier rate of speed. It also enables those with cognitive disabilities like processing and comprehension disorders to slow the game down so they can understand the game and what is happening on screen at a pace that meets their needs.

My Football Game by VTree allows the game to be slowed down to 20 % of the default speed allowing those with difficulty moving the mouse easily or reacting quickly to complete the game at a comfortable rate.

It is important to think of the situation like this: although many of us have the ability to function in a game by a specific, universally accepted minimum of difficulty and rate of speed, some people simply can't keep up with 'normal speed.' For these individuals, if the game was slowed down to considerably, they would be able to have as much fun as everyone else.

Hearing:

Hearing-impaired gamers are the most well represented group thanks to the efforts of top-notch advocacy groups that have made subtitles mandatory for most forms of visual media. Subtitles have become a default accessibility feature, widely accepted as the standard for the entertainment industry. But there are more things than just subtitles that make a game truly accessible.

Level One (1 category)

Closed Captioning: Closed captioning for the hearing impaired has become the de facto accessibility option. It is and always will be one of the most important options to be included in games. However, many do not understand the difference between closed captioning and subtitles. Subtitles are when the speech coming from characters is written underneath as dialogue, and has become a standard feature in most of the newer games. Closed captioning on the other hand is subtitles enhanced the addition of valuable audio cues are displayed in text format.

Subtitle
- Sally: Good morning, John
Closed captioning
[Sally knocks on the door.]
Sally: Good morning, John.
[The floorboard creaks underfoot.]

If a gamer with a hearing impairment is participating in a game, it's important for the ambient noise to be captured in text on the screen, especially for horror games, spy games, and other sneaking-based games, where monsters/NPCs are making noise that is designed to be an early warning system.

Including closed captioning increases the overall experience for those with hearing impairments—example above.

Baby-Friendly Settings:

'Baby-friendly settings' reflect the idea that those with who are parents trying to play video games should be able to do so at 3 in the morning with the sound disabled and the baby sleeping right beside them. Games should be designed in a way such that the game can be completed whether the sound is enabled or not. This is another example of a feature for disabled gamers having value for non-disabled gamers.

Level Two (2 categories)

Changeable Font Sizes: Since the invention of text and graphical displays, one of the most basic options has been the ability to change the size of text. However, video games do not often provide the option. Gamers with visual difficulties including those that are legally or almost totally blind can still participate in many games, but reading text in the subtitles, directions/instructions and chats can be frustrating if the size of the text is just barely too small to read.

One example of a game that does this well is EverQuest II, which allows users to change text size.

Text size matters to more of your audience than you might think. Up to 50 % of the aging gaming population has or will have vision problems in their lifetime.

Example:

A man loves playing an online adventure game, but his eyesight is slowly fading as he ages and his macular degeneration continues. He squints at the screen with the most powerful glasses money can buy, but the text continues to be harder to read. If the game allowed him to increase the size of the text in his chat window and on quests, he would be able to continue playing the game without interference.

Changeable Text Colors: The name of this says it all. For those that have difficulty distinguishing between colors, often referred to as color deficiency, the ability to change the color of text associated with a particular function improves overall gameplay. Note that color deficiency differs from color blindness.

Color-deficient people can still see a certain color, but they cannot distinguish shades of the color, red vs maroon, or lime green vs dark green.

Often found in MMORPGs, various colors are used to indicate what type of input is being received. Green is often used for guild conversations, white is often used for local communications, light colors of pink or purple are used to indicate private message, whereas red is often used to indicate combat or enemies. This range of color usage can be challenging to players with color deficiency or color blindness.

World of Warcraft does not allow the user to change the color of any text input, but through the use of modifications, gamers can make the colors more distinguishable from one another.

In SWTOR, the individual gamer can define text colors, and this enables those with any color blindness or color deficiency to interpret the data the game is telling them more easily.

It may seem like something that is relatively simple, but for those that need the option, it's important to have. After years of development, many colors have become standardized as meaning certain things (i.e. green is almost always used for the guild chat), which is exactly why allowing colorblind or color deficient users to change colors to what is more comfortable for them is so important. As an industry, games have realized the importance of separating information based on color for easy dissemination. Therefore, it is important for those who need to alter colors to the spectrum they can see is also important. Remember this: 1 out of every 7 men are color deficient.

Example:

A woman with a strong color deficiency has trouble interpreting chats in her favorite MMORPG because the color of her guild chat is green, her group chat is a very light purple and her instant message (whispers) are pink. If the game incorporated the ability to change the colors of any text to something she could see more easily it would improve the quality of her gaming experience.

Level Three (2 categories)

Ambient Noise: Options to Include Ambient Noise as Text Output

Capturing ambient noise improves the quality of experience for those with hearing impairments. The environment is often key to the overall feel of the gaming experience, when that element is left out, the Deaf gamer may be missing the mood you are looking to set. Extending the closed captioning system above to let players turn on and off text cues for certain levels of ambient noise allows the gamer to customize the level of immersion they want

Example:

A deaf man tries out the newest zombie game, but is unable to successfully play because the developers have designed the noise the zombie makes to be an early warning system of an attack coming out of a blind spot is about to happen. If the ambient noises were captured in the subtitles—for example, [you hear a groan coming from the right]—the gamer would be able to continue playing with full situational awareness.

Alternative Reactionary Input: Often referred to as subliminal cues, these are the use of other tools available to the game developer to replicate the role of audio in indicating something important is happening. This could mean things like the screen turning red as the character is increasingly wounded, flashing when the character is low on health, or using the vibration offered by a standard console controller to have a meaning that Deaf gamers can interpret. Alternative reactionary inputs allow for the user to be more in tune with what is currently happening in the game without needing to hear the environment, or read that something is happening.

In World of Warcraft, the edges of the visual area flashes red at an increasing rate once the character reaches less than 20 % health.

In Call of Duty blood appears around the screen and the clearness of the screen blurs as the character takes additional damage. As regeneration occurs, the bloodied screen clears up. Both of these are accompanied with an audio cue as well.

These are non-audio cues that allow the gamer to interpret important game information, in this case character health, without having to hear anything. To the non-disabled gaming crowd, these types of alternate warnings are becoming more commonplace as an additional means of providing immersion and understanding of the environment. To a hearing impaired gamer, these types of warning systems are an essential feature to level the playing field.

Cognitive

Cognitive disorders are some of the most highly recognized disabilities in America today. Autism has become mainstream and as such options are needed to help those who need assistance in understanding what most of us take for granted. No matter the cognitive disorder, simple options can be included that enable those with even the most serious of disabilities.

Level One (2 categories)

Tutorial: Tutorial experiences are important for those with cognitive disorders. Many gamers with cognitive disorders experience more positive results from being shown precisely how to play a game as opposed to being left to interpret cryptic instruction pamphlets or employ trial and error. It is also not good for to expect disabled gamers to understand the "standard inputs for the genre," because many of these gamers may not remember what those were.

Most MMORPGs have so-called 'starter areas' where the basics of the game are taught through linear quests that must be completed in order and successfully in order to advance. Many times these are accompanied by special interface components that point out parts of the HUD, or lay out how combat works. The theory is that doing learns basic skills best.

It is important to include thorough tutorials for the cognitively disabled, as well as to be considered for inclusion in good game design. Some people with severe cognitive disorders simply need extra time or encouragement to continue learning.

It is important to realize that no game elements can be considered trivial if they are difficult for someone to learn. Many of these same features can also be used to welcome more casual, non-disabled players to your game, broadening your overall audience.

Example:
A woman with difficulty understanding linear steps wants to play a new game. She has no friends, caretakers or family that have the time to explain the game. There are no tutorials and as she attempts to play the game she continually does things wrong and receives ridicule from the in-game community for doing things perceived as 'simple.'

If the game offered in-game tutorials in a closed phase where she could learn the game without fear of ridicule, she would be more likely to enjoy the game and continue playing.

Difficulty Levels: Particularly pertinent to first-person shooters, strategy games, and casual timed or puzzle games, difficulty settings can mean the difference between an enjoyable experience and not being able to play the game at all.

In games such as Mass Effect 3, difficulty levels allow the user to become near invincible while experiencing the story line, traversing the game, and not worrying about dying or failing the game. Deus Ex: Human Revolution from Eidos asked gamers at the top what level they wanted to play the game at Causal, Normal, or Deus Ex (Crazy hard). It is worth noting that the more recent Madden introduced a mini game that based on the outcome took over certain actions. This allowed less able gamers to play with expert gamers and still have a good competition. This is an excellent implementation of setting difficulty level.

Starcraft II from Blizzard Entertainment and Sins of a Solar Empire from Ironclad Games allow gamers to set the level difficulty for computer opponents and thus allows the gamer to decide what difficulty of competition they can handle.

Example:
A teenager who sits down to play a brand-new strategy game for the first time gets utterly destroyed. He continues to try new games, but continually fails because there are no difficulty levels. The nature of his disability prevents them from being able to perform specific tasks at the levels demanded, and therefore he has no chance of playing the game.

Level Two (2 categories)
Training Levels: Training levels allow the user to practice playing in a multi-player environment versus computer artificial intelligence at various levels of difficulty. This allows those who may become frustrated by online gameplay to be better prepared for the coming experience of playing with other gamers.

StarSiege: Tribes by Dynamix enables people to practice its first-person shooter maneuvers with NPCs at various levels, ranges and numbers. This helps ensure gamers who have difficulty learning how to play that particular game have a chance to practice on their own without fear of ridicule, thus making the game more accessible.

Example:

A man with very little hand-eye coordination desperately wants to play his buddy's favorite first-person shooter, but every time he goes into the online environment he gets destroyed and laughed at by the competition. If the game had training levels that offered varied or increasingly difficult computer AI to fight against, the man would be better prepared at the end of training to face real competitors.

Intuitive Menus: Intuitive menus are not only good game design, but for those with cognitive disorders they provide the ability to quickly assess where certain options are without being frustrated or confused. Just as with web design, good game design includes placing menu items where they are most often found and that all options are no more than 2 levels deep.

Level Three (2 categories)

Speed Settings: This is a top-tier option for those with mobility, low vision and cognitive disorders, having the ability to slow down the game clock allows those that just need more time to process what is going on, and how they need to react to it.

A disability that is in the news a lot right now, a veteran is coping with a traumatic brain injury and has difficulty processing information. He has no issue understanding what is going on in the game, he just does not do it as fast as most others. By slowing down the game up to 80 % he can successfully play his favorite sports game that he and his friends played while over in the sands.

My Football Game by VTree LLC allows the game clock to be slowed down to 20 % for those with difficulty moving the mouse easily or reacting quickly. This enables gamers to complete the game at a comfortable rate to further particular abilities.

Enemy Marking: The ability to tell friend from foe is one of the most fundamental differentiators in all video games. Some games use naming, visual bars, symbols or color markers. For those who have trouble differentiating between similar colors, small markings can help gamers tell the difference between friend, enemy, or really difficult enemy.

In Guild Wars 2, enemies are marked by red bars with numbers representing mob level and accompanying symbols to declare the level of the faculty in either a gold or silver star.

In Battlefield 3, an option can be enabled to place small blue 'Doritos' over the heads of allies. This allows gamers to quickly recognize that those without the marking are enemies.

Examples:

A teen has trouble seeing enemies on certain backgrounds of a game with a variety of maps. Without the ability to change the colors of the enemies or mark them in some way, he will be caught unaware when an enemy walks on-screen.

Others:

Some accessibility issues don't fall anywhere else. These random thoughts may actually be some of the most important aspects of this information. It's important to understand where gamers with disabilities are coming from. Most of us don't care about achievements, goals or rewards; we simply want to be able to play the game. This section covers the issues that didn't fit into the other categories, but covers all disabilities in general at the same time.

Auto-pass: One of the more common features in video games today is the inclusion of 'quick time events.' These events are often timed obstacles that the gamer must use sequences or combinations of buttons to pass in order to advance.

The Auto-pass feature is a failsafe against frustrating gamers to the point of abandoning the game. A built-in system can recognize when a gamer fails to do a specific event certain number of times in a row, and offers a simple dialog the player to skip the event. This can be as simple as asking, "Would you like to skip this?"

There are also lighter versions of the same idea for various types of games. In a puzzle game, if someone fails to make a move, or makes the move incorrectly three times the game can offer a clue, and if they fail three more times it can offer to show the solution.

In a mandatory fight in a story-driven game, if the user fails multiple times in a row, the enemy could be weakened after each subsequent fail until the person is able to defeat the obstacle.

These types of features are failsafes that are triggered automatically by the AI when the computer notices a gamer is having difficulty. This is commonly called 'rubber band AI,' where the game will automatically increase or decrease the level of difficulty based on the gamers' performance.

Perspective: For those that have issues of vertigo, Ménière's disease or other forms of sight-based balance issues, the difference between first and third perspective can be quite daunting. Some people can simply not look down through the eyes of the character without feeling a sense of imbalance and sickness.

Whenever possible, perspective options should be offered to let players change the camera view from first person to third. In third person view, people with these types of illnesses are able to see their character and their brain can handle the environment based on a steady reference point of a character being present.

Reward System Balance: Most accessibility issues can be solved with simple accessibility options. However, there is some concern amongst the community that lowering the challenge of the game hurts the gameplay of other users by default.

Implementing reward-based systems can mitigate this fear. If an option or set of options removes much of the challenge that the game developer intended, you can present the gamer with a dialog stating that certain achievements will not be available. This allows for gamers in the disabled community to enjoy the game without enabling those who do not actually need the options to take advantage of the system. Another way to implement a balanced reward system is to look at how this was handled in the Rock Band Games from Harmonix. They have achievements at all levels, like "Hometown Throwdown," that just means you finished a set but does not care if you did it on easy or expert.

The AbleGamers Foundation, who kindly permitted the text used in this section (above and prior pages, cf [directly quoted and lightly abridged to maintain contextual meaning and clarity] http://www.includification.com) of the contribution to be extracted from their site, informs that they continuously interact with developers. They inform how they have often heard that "no one would want to play the game with X" where X is either one button, infinite life, infinite abilities, infinite power, infinite money etc." In response they state how "The truth of the matter is for most disabled gamers, they simply want to be able to play the game the best that they can. They don't particularly care about achievements, they don't care about leaderboards or being able to run around and say they're the best of the best. They simply want to play the game."

Throughout this document direct texts are used to inform undeviating from the original message meant. The direct text is used, for example in this section, from AbleGamers to illustrate the depth of the resource for supporting those with impairment. Whilst some may question the extent of quoted materials in this document the author feels it imperative to bring to the floor the important issue of accessibility and the 'players' in the field such as herein included. It is with respect to their contribution that directly quoted extracts are used. Citations and reference points determine source wherever possible without disrupting the flow of text or too many duplicate footnotes. The footnotes are used for reader immediate cross-reference to source.

The next section elaborates on CVAA exceptions and again is sourced as indicated.

14.2.9   CVAA Exemptions[50]

In line with the illuminating details supplied by AbleGamers Foundation it is notable that the CVAA's captioning requirements are for broadcast video only: where—(cf FCC publication page to maintain exactness—see section title footnote) —they do not include video in the games. The FCC does not regulate captioning of home videos, DVDs or video games; DVDs have captions most of the time because TV shows, movies, etc. must contain captions in order to be broadcast (to be in compliance). Any device that offers Advanced Communication Services (ACS) must make those communication services available to people with disabilities unless it is not "achievable" to do so. The CVAA defines ACS as:

- *interconnected VoIP service*
- *non-interconnected VoIP service (does not require connection to the public switch telephone network); for example TRS (Telecommunications Relay Service)*
- *electronic messaging service (including text messaging, instant messaging, email and two-way interactive messaging through a social networking site)*
- *interoperable video conferencing service.*

---

[50]https://www.fcc.gov/consumers/guides/21st-century-communications-and-video-accessibility-act-cvaa.

Three categories exist in the CVAA for video games: (I) game consoles, (II) game distribution and game play networks, and (III) game software. Products released before the expiration of the waiver are exempted (PS4, Xbox One, Wii U, etc. are technically exempted). Businesses and organizations with less than 30 employees are exempted. At the FCC's discretion, requirements can be waived for equipment and services that are capable of accessing Advanced Communication Services, but are designed primarily for purposes other than using ACS.

Consoles, software and online distribution services that offer broadcast video, such as television shows or movies, must provide users with a set of options so they can manipulate how subtitles are displayed. For example, YouTube, Netflix, iOS, Android, etc. currently offer full control over text size, font, and letterboxing when viewing online videos.

14.2.10   Accessibility Features on Current Consoles[51]

Xbox One:

- *screenreader*
- *magnifier (also in-game)*
- *closed caption presentation (with API)*
- *high contrast*
- *limited button remapping (at the system level)*

Playstation 4:

- *limited screenreader*
- *magnifier (also in-game)*
- *closed caption presentation*
- *high contrast*
- *bold text, large text, text speed*
- *limited button remapping (at the system level).*

Nintendo Wii U and 3DS:
Nothing as of yet—probably won't appear until the next round of consoles are released e.g. NX series.
Steam:
Nothing as of yet—this potentially puts the company out of compliance with the CVAA, but it may depend on how the exemption for preexisting devices and services is interpreted (for example, what constitutes a major update to the service?)
The access special interest group,[52] inform that things are changing and publisher-level accessibility evaluations are now offered at SCEE (Sony Computer

---

[51]cf https://igda-gasig.org.

[52]https://igda-gasig.org.

Entertainment Europe) in the form of an expert game review, player-behavior observation testing, diary studies, and general analytics. BBC (The British Broadcasting Corporation) are increasing accessibility in broadcasts. The game user research team at Sony Computer Entertainment Europe has started to offer accessibility evaluations as an internal service. The offer is for in-house expert game review, player-behavior observation testing, diary studies, and general analytics. This internal tool consists of two parts:

- A spreadsheet with a straightforward list of possible accessibility considerations, with additional columns to indicate whether each is relevant to a gameplay mechanic, how feasible it is, and recommendations for implementation
- A detailed support document with precise specifications for platform-level certification of accessibility features

This is currently an optional internal service at Sony Europe, but it will eventually be shared with the wider business and game user research community. The BBC produced a similar list/procedure this year for use across its first and third party games. The BBC is publicly funded so has a strong accessibility culture, meaning their list is a set of requirements and is not an optional service. They require games to comply with as much of the list/procedure as is reasonably possible. Other initiatives are big increases in developer accessibility considerations, in particular for accommodating epileptic and colorblind gamers. There was a strong social media reaction to the lack of colorblind friendliness in Call of Duty: Black Ops 3 and The Witness; this is compared to only two years ago where games such as Sim City and Borderlands 2 considering colorblindness was unusual enough that it received significant press coverage.

The IGDA-GASIG inform that two AAA console games are developed with intentionally patching in accessibility for completely blind gamers. But what does it cost for an established game developer, such as Rockstar, to make a game, for example Grand Theft Auto V, fully accessible to the blind. This was a thought exercise query but the developers at Rockstar took the question seriously. The short answer: $128 million as reported at the game accessibility game developers conference 2016.[53]

Also reported is how Turtle Rock Studios published an accessibility statement for its game Evolve, which details what considerations have already been made; the company also made a public commitment to further work in this area. The Witcher 3 received many accessibility patches by its developer CDProjektRed in the weeks following its launch; the developers prioritized accessibility features in a critical fix; this included full controller remapping, colorblind mode, and improving the size and contrast of text. Harmonix was actively soliciting for accessibility suggestions on their forums during development of Rock Band 4; this has resulted in more than 18 pages of suggestions from players.... Suggesting the significance of the potential market impact and possibilities for the next generation of developers to specialize.

---

[53]http://www.dailydot.com/parsec/game-accessibility-game-developers-conference-2016/.

The IGDA GA-SIG Website further informs of some small developments with Steam: now filters games that have captions, however, nothing for colorblindness, button remapping, etc. Itch.io offers these types of filters. Evaluation of how Itch.io works and filters its games would be very valuable, if anyone is interested in working on this. The organization is advocating for fixed-point mode for Switch users on iOS.

Switch accessibility means allowing access to custom controllers based on one or two simple on/off controls; for example, a sip/puff tube, headrest button, blink detector, etc. iOS and Android (to a lesser extent) offer built-in support for switch accessibility when using native interface elements. iOS has a workaround for apps that aren't developed natively (i.e. most games): the game scans the screen and the player interacts when the desired coordinates appear. IGDA GA-SIG state how there are thousands of one button mobile games that should in theory be switch-compatible, but the fixed point mode doesn't work with them, as it is incompatible with games that require any kind of timing. Barrie Ellis of One Switch/Special Effect created a video with wide backing from the game accessibility community about what works about it and what doesn't; Apple have since implemented an additional mode (switch recipes) that allows repeated presses on a single point, removing the incompatibility issue from all of those thousands of games overnight.

IGDA GA-SIG are active in presenting at conferences, both industry and academic with five accessibility talks at GDC with record attendance (three years ago, average attendance was about 30, in 2016 average attendance was about 130); six gaming sessions at CSUN (cross industry accessibility conference—a lot of web and apps).

They are also expanding accessibility at the Global Game Jam (GGJ) that have proven to be powerful awareness raisers for Game Accessibility. Included were six optional game development themes related to game accessibility (e.g. one handed controls, no visuals) with thousands of developers taking up one or more of those six optional game development themes. Aligned with this is that AbleGamers Charity is planning to offer a 24 h hotline operational during GGJ 2017 offering game accessibility support and advice.

It is reported that some gamers would not like to choose to play as a disabled gamer in a game, because they play games for escapism, whilst other gamers have strongly identified with characters that have minor or major impairments. A German indie game called 'The Unstoppables' has four different characters with different accessibility obstacles; it is a puzzle game where characters must combine their abilities to progress through a level. Other reports relate to advances during the past year, which include:

Some really significant progress in accessibility of consoles themselves is reported by the SIG, thanks to Microsoft and Sony. Also, Unity now is in the process of implementing full remapping of game controls, joining Unreal's colorblind simulator and subtitling system to form a good start of accessibility solutions in game engines to lower the threshold for all game developers. The above reported internal accessibility evaluation checklists now in use at Sony Europe (optional) and the BBC (mandatory). It is reported that charities have seen increases in donations, and been able to launch initiatives such as SpecialEffect's upgraded games room in Oxfordshire, UK, and AbleGamers' expansion packs and fellowship scheme. New accessibility websites have launched, including the re-launch of industry stalwart game-accessibility.com. Itch has introduced tagging for accessibility info, which is a good opportunity for analysis of their approach to build a case to push for Steam and others to do the same.

Accessibility in games themselves has continued to improve according to the SIG reports, in particular for colorblindness, epilepsy, and also blindness, with two current AAA console games (Killer Instinct and Mortal Kombat X) receiving blind accessibility patches. The improvements are reflected in the winners of the various game accessibility awards this year—Rocket League, MLB: The Show, Heroes of the Storm, and Ryan North's To Be Or Not To Be. How accessibility is approached has also improved, for example Evolve publishing an accessibility statement, Witcher 3 treating accessibility issues as top priority issues to be patched immediately after launch, and Rock Band 4 publicly asking their community for accessibility input during development. An engine developer wish list has been produced, opening good lines of communication with engine developers, one of which has shown interest in the wish list. In addition, an educational framework is now well underway, collaborating with the education SIG.

In closing this section it can be seen how actions are bringing about results thanks to the concentrated efforts of organizations such as IGDA GA-SIG, AbleGamers, and others. These organizations make industry and communities aware of the need. The author acknowledges their input to this document and hopes in including their findings and actions a form of dissemination outside of the usual circles may be achieved as support for the accessibility movements that warrant wide encouragement.

14.3 Discussion and Conclusions

Making all games accessible for all would seem an insurmountable challenge and one that is unreal. This is exemplified by the Rockstar example posited in the text questioning costs for making a game accessible to the blind.[54] This piece refers to Mark Barlet, president and executive director of The AbleGamers Charity, who challenged developers at the 2016 Game Developers Conference to use all the tools at their disposal to make video games accessible to as many players as possible. The extent to how which game developers can succeed in that goal is limited and constraints that developers work under is illustrated in the following. It was stated in the piece that 'making video games accessible to totally blind players involves specific sound design methodologies, like echoes that assist blind players in defining the space in which their avatar is standing. The developers at Rockstar estimated that it would cost 50 % more to make GTA V[55] accessible to totally blind players, and the game already cost $256 million to develop.' It was estimated that

---

[54]http://www.dailydot.com/parsec/game-accessibility-game-developers-conference-2016/.
[55]Grand Theft Auto V.

'Rockstar might sell around 16,000 additional copies of GTA V were it made accessible to totally blind players. That would represent a 0.03 % increase in sales. "Each copy would have to cost $8,073.61 to recoup the development costs," Barlet said. The point of the exercise was to demonstrate that making all video games perfectly accessible to all gamers across the spectrum of disabilities was impossible. Thus, the CVAA to this author's comprehension challenges the industry.

Such challenges were presented in the author's GameAbilitation + ArtAbilitation: Ludic Engagement Designs for All (LEDA) keynote at the International Conference on Accessibility, Inclusion and Rehabilitation using Information Technologies, La Havana, Cuba, 13–15 December 2011. Further elaborations of these issues were presented in the author's keynote for the VSGames2016—Virtual Worlds and Games for Serious Applications. To analogize, within the world of music customizable hardware and software interfaces for musical expression there is exhibited opportunity for all to access such that even those who are deaf can train to feel nuance of frequency and subtle changes in order to be able to express creatively. Similarly means exist that enable blind to paint by feel. Rosenbaum's publication 'See What I'm Saying: The Extraordinary Powers of Our Five Senses' highlights phenomena of overcoming sensory impairments [7]. Developers could inform themselves of such human phenomena to see if there are avenues to explore in the accessibility forum.

The apparent limited research—and knowledge—on this issue suggests major challenges for the industry in game classifications and adaptations in order to conform to the CVAA. Will it be that packaging is labeled, such as PEGI achieves, according to who can or cannot access the gameplay? Such changes were foreseen as exemplified by questions asked in the author's research asking whether an authoring game enhancement AI (impacting software plus hardware) could be implemented that adapts game variables to the player's access requirements? If so then predicted aligned with an integrated Dynamic Difficulty Adjustment (DDA) game experiences, access and impact would be optimized. Relatedly, within the field of interactive interface design early years saw the need for the human to adapt to the technology with a later reversal such that the technologies could be adapted for each user's need. Conclusions in this chapter speculate on a similar case of change whereby games software and hardware will need to evolve beyond current limitations to adapt to user's facilities/abilities, limitations, and preferences. CVAA implementations are approached herein toward introduction, promoting discussions, and for others to further investigate and build upon in future.[56] Whilst reflecting on the fact that the CVAA does not mean that popular online and other games will be totally usable by people with impairment—something that one may consider almost impossible?), it is a precedent-setting law that takes much needed steps toward greater accessibility of the communications elements of gaming, the

---

[56]In citing the CVAA the author does not posit himself as a specific expert in the act or field—reference is made to awaken discussions to the important issues surrounding accessibility and inclusion.

elements that the FCC has jurisdiction to reach [8]. If in any way this chapter has made aware the situation of accessibility and some of the key players and advocates it has contributed in a small way towards an important issue that the author believes presents many interesting challenges to overcome. It will be stimulating to see how things transpire following the January 2017 CVAA waiver deadline (now potentially extended—see not on opening page of Appendix 2) and what may follow thereafter. One thing is for sure, Virtual Worlds and Games with serious application are here, they are impactful and should be accessible, and playable by all. The author rests his case—and once again acknowledges sources used in compiling this document. The meaning should be clear in how the compilation and gathering of data has been done, i.e. towards informing rather than claiming authorship. The discussion goes on into the night... we hope for a fruitful (and accessible) outcome.

Acknowledgements Federal Communications Commission (FCC); Pan-European Game Information (PEGI) organization; The Interactive Software Federation of Europe (ISFE); NICAM (Netherlands Institute for the Classification of Audiovisual Media); The Video Standards Council (VSC), United Kingdom; AbleGamers Charity.

Disclaimer: The need to include text extracted from publicly available materials from those acknowledged above has been permitted via communication with the source. Wherever possible link and citation is given, however, the author wishes the fact known due to the form of this document and appendixes, which share in full the CVAA and waiver. Direct text is true to source (e.g. US/UK English, punctuations, and emphasis), and considered required in an attempt to minimize third-party (this author) interpreted loss of meaning.

Appendix 1

The following pages shares the Twenty-First Century Communications and Video Accessibility Act (CVAA), which is a matter of public record. The full text is included herein for direct reference for readers of the non-digital version of this chapter. Permission to use was received from The Federal Communications Commission (FCC) representative Sherita Kennedy dated Monday 22 August 2016 at 17:54 following personal communication requesting authorised approval for use. FCC is an independent U.S. government agency overseen by Congress that regulates interstate and international communications by radio, television, wire, satellite, and cable in all 50 states, the District of Columbia and U.S. territories. The commission is the United States' primary authority for communications laws, regulation and technological innovation. Web site for more information is https://www.fcc.gov

.

Requested citation by FCC is hereunder stated:

Pub. L. No. 111–260, 124 Stat. 2751 (2010)(S. 3304, 111th Cong.)(as codified in various sections of 47 U.S.C.). See also Amendment of Twenty-First Century Communications and Video Accessibility Act of 2010, Pub. L. 111–265, 124 Stat. 2795 (2010)(S. 3828, 111th Cong.), making technical corrections to the CVAA.

S. 3304

One Hundred Eleventh Congress of the United States of America

AT THE SECOND SESSION

Begun and held at the City of Washington on Tuesday, the fifth day of January, two thousand and ten

An Act

To increase the access of persons with disabilities to modern communications, and for other purposes.

Be it enacted by the Senate and House of Representatives of the United States of America in Congress assembled,

SECTION 1. SHORT TITLE; TABLE OF CONTENTS.

(a) SHORT TITLE.—This Act may be cited as the "Twenty-First Century Communications and Video Accessibility Act of 2010".

(b) TABLE OF CONTENTS.—

SEC. 2. LIMITATION ON LIABILITY.

(a) IN GENERAL.—Except as provided in subsection (b), no person shall be liable for a violation of the requirements of this Act (or of the provisions of the Communications Act of 1934 that are amended or added by this Act) with respect to video programming, online content, applications, services, advanced communications services, or equipment used to provide or access advanced communications services to the extent such person—

(1) transmits, routes, or stores in intermediate or transient storage the communications made available through the provision of advanced communications services by a third party; or

(2) provides an information location tool, such as a directory, index, reference, pointer, menu, guide, user interface, or hypertext link, through which an end user obtains access

to such video programming, online content, applications, services, advanced communications services, or equipment used to provide or access advanced communications services.

(b) EXCEPTION.—The limitation on liability under subsection (a) shall not apply to any person who relies on third party applications, services, software, hardware, or equipment to comply with the requirements of this Act (or of the provisions of the Communications Act of 1934 that are amended or added by this Act) with respect to video programming, online content, applications, services, advanced communications services, or equipment used to provide or access advanced communications services.

SEC. 3. PROPRIETARY TECHNOLOGY.

No action taken by the Federal Communications Commission to implement this Act or any amendment made by this Act shall mandate the use or incorporation of proprietary technology.

TITLE I—COMMUNICATIONS ACCESS

SEC. 101. DEFINITIONS.

Section 3 of the Communications Act of 1934 (47 U.S.C. 153) is amended—

(1) by adding at the end the following new paragraphs:

"(53) ADVANCED COMMUNICATIONS SERVICES.—The term 'advanced communications services' means—

"(A) interconnected VoIP service;

"(B) non-interconnected VoIP service;

"(C) electronic messaging service; and

"(D) interoperable video conferencing service.

"(54) CONSUMER GENERATED MEDIA.—The term 'consumer generated media' means content created and made available by consumers to online websites and services on the Internet, including video, audio, and multimedia content.

"(55) DISABILITY.—The term 'disability' has the meaning given such term under section 3 of the Americans with Disabilities Act of 1990 (42 U.S.C. 12102).

"(56) ELECTRONIC MESSAGING SERVICE.—The term 'electronic messaging service' means a service that provides real-time or near real-time non-voice messages in text form between individuals over communications networks.

"(57) INTERCONNECTED VOIP SERVICE.—The term 'interconnected VoIP service' has the meaning given such term under section 9.3 of title 47, Code of Federal Regulations, as such section may be amended from time to time.

"(58) NON-INTERCONNECTED VOIP SERVICE.—The term 'non-interconnected VoIP service'—

"(A) means a service that—

"(i) enables real-time voice communications that originate from or terminate to the user's location using Internet protocol or any successor protocol; and

"(ii) requires Internet protocol compatible customer premises equipment; and

"(B) does not include any service that is an interconnected VoIP service.

"(59) INTEROPERABLE VIDEO CONFERENCING SERVICE.—The term 'interoperable video conferencing service' means a service

S. 3304—3

that provides real-time video communications, including audio, to enable users to share information of the user's choosing."; and

(2) by reordering paragraphs (1) through (52) and the paragraphs added by paragraph (1) of this section in alphabetical order based on the headings of such paragraphs and renumbering such paragraphs as so reordered.

SEC. 102. HEARING AID COMPATIBILITY.

(a) COMPATIBILITY REQUIREMENTS.—

(1) TELEPHONE SERVICE FOR THE DISABLED.—Section 710(b)(1) of the Communications Act of 1934 (47 U.S.C. 610(b)(1)) is amended to read as follows:

"(b)(1) Except as provided in paragraphs (2) and (3) and subsection (c), the Commission shall require that customer premises equipment described in this paragraph provide internal means for effective use with hearing aids that are designed to be compatible with telephones which meet established technical standards for hearing aid compatibility. Customer premises equipment described in this paragraph are the following:

"(A) All essential telephones.

"(B) All telephones manufactured in the United States (other than for export) more than one year after the date of enactment of the Hearing Aid Compatibility Act of 1988 or imported for use in the United States more than one year after such date.

"(C) All customer premises equipment used with advanced communications services that is designed to provide 2-way voice communication via a built-in speaker intended to be held to the ear in a manner functionally equivalent to a telephone, subject to the regulations prescribed by the Commission under subsection (e).".

(2) ADDITIONAL AMENDMENTS.—Section 710(b) of the Communications Act of 1934 (47 U.S.C. 610(b)) is further amended—

(A) in paragraph (2)—

(i) in subparagraph (A)—

(I) in the matter preceding clause (i)—

(aa) by striking "initial";

(bb) by striking "of this subsection after the date of enactment of the Hearing Aid Compatibility Act of 1988"; and

(cc) by striking "paragraph (1)(B) of this subsection" and inserting "subparagraphs (B) and (C) of paragraph (1)";

(II) by inserting "and" at the end of clause (ii);

(III) by striking clause (iii); and

(IV) by redesignating clause (iv) as clause (iii);

(ii) by striking subparagraph (B) and redesignating subparagraph (C) as subparagraph (B); and

(iii) in subparagraph (B) (as so redesignated)—

(I) by striking the first sentence and inserting "The Commission shall periodically assess the appropriateness of continuing in effect the exemptions for telephones and other customer premises

S. 3304—4

equipment described in subparagraph (A) of this
paragraph."; and

(II) in each of clauses (iii) and (iv), by striking
"paragraph (1)(B)" and inserting "subparagraph
(B) or (C) of paragraph (1)";

(B) in paragraph (4)(B)—

(i) by striking "public mobile" and inserting "telephones used with public mobile";

(ii) by inserting "telephones and other customer
premises equipment used in whole or in part with"
after "means";

(iii) by striking "and" after "public land mobile
telephone service," and inserting "or";

(iv) by striking "part 22 of"; and

(v) by inserting after "Regulations" the following:
", or any functionally equivalent unlicensed wireless
services"; and

(C) in paragraph (4)(C)—

(i) by striking "term 'private radio services'" and
inserting "term 'telephones used with private radio
services'"; and

(ii) by inserting "telephones and other customer
premises equipment used in whole or in part with"
after "means".

(b) TECHNICAL STANDARDS.—Section 710(c) of the Communications Act of 1934 (47 U.S.C. 610(c)) is amended by adding at
the end the following: "A telephone or other customer premises
equipment that is compliant with relevant technical standards
developed through a public participation process and in consultation
with interested consumer stakeholders (designated by the Commission for the purposes of this section) will be considered hearing
aid compatible for purposes of this section, until such time as
the Commission may determine otherwise. The Commission shall
consult with the public, including people with hearing loss, in
establishing or approving such technical standards. The Commission may delegate this authority to an employee pursuant to section
5(c). The Commission shall remain the final arbiter as to whether
the standards meet the requirements of this section.".

(c) RULEMAKING.—Section 710(e) of the Communications Act
of 1934 (47 U.S.C. 610(e)) is amended—

(1) by striking "impairments" and inserting "loss"; and

(2) by adding at the end the following sentence: "In implementing the provisions of subsection (b)(1)(C), the Commission
shall use appropriate timetables or benchmarks to the extent
necessary (1) due to technical feasibility, or (2) to ensure the
marketability or availability of new technologies to users.".

(d) RULE OF CONSTRUCTION.—Section 710(h) of the Communications Act of 1934 (47 U.S.C. 610(h)) is amended to read as follows:

"(h) RULE OF CONSTRUCTION.—Nothing in the Twenty-First
Century Communications and Video Accessibility Act of 2010 shall
be construed to modify the Commission's regulations set forth in
section 20.19 of title 47 of the Code of Federal Regulations, as
in effect on the date of enactment of such Act.".

S. 3304—5

SEC. 103. RELAY SERVICES.

(a) DEFINITION.—Paragraph (3) of section 225(a) of the Communications Act of 1934 (47 U.S.C. 225(a)(3)) is amended to read as follows:

"(3) TELECOMMUNICATIONS RELAY SERVICES.—The term 'telecommunications relay services' means telephone transmission services that provide the ability for an individual who is deaf, hard of hearing, deaf-blind, or who has a speech disability to engage in communication by wire or radio with one or more individuals, in a manner that is functionally equivalent to the ability of a hearing individual who does not have a speech disability to communicate using voice communication services by wire or radio.".

(b) INTERNET PROTOCOL-BASED RELAY SERVICES.—Title VII of such Act (47 U.S.C. 601 et seq.) is amended by adding at the end the following new section:

"SEC. 715. INTERNET PROTOCOL-BASED RELAY SERVICES.

"Within one year after the date of enactment of the Twenty-First Century Communications and Video Accessibility Act of 2010, each interconnected VoIP service provider and each provider of non-interconnected VoIP service shall participate in and contribute to the Telecommunications Relay Services Fund established in section 64.604(c)(5)(iii) of title 47, Code of Federal Regulations, as in effect on the date of enactment of such Act, in a manner prescribed by the Commission by regulation to provide for obligations of such providers that are consistent with and comparable to the obligations of other contributors to such Fund.".

SEC. 104. ACCESS TO ADVANCED COMMUNICATIONS SERVICES AND EQUIPMENT.

(a) TITLE VII AMENDMENT.—Title VII of the Communications Act of 1934 (47 U.S.C. 601 et seq.), as amended by section 103, is further amended by adding at the end the following new sections:

"SEC. 716. ACCESS TO ADVANCED COMMUNICATIONS SERVICES AND EQUIPMENT.

"(a) MANUFACTURING.—

"(1) IN GENERAL.—With respect to equipment manufactured after the effective date of the regulations established pursuant to subsection (e), and subject to those regulations, a manufacturer of equipment used for advanced communications services, including end user equipment, network equipment, and software, shall ensure that the equipment and software that such manufacturer offers for sale or otherwise distributes in interstate commerce shall be accessible to and usable by individuals with disabilities, unless the requirements of this subsection are not achievable.

"(2) INDUSTRY FLEXIBILITY.—A manufacturer of equipment may satisfy the requirements of paragraph (1) with respect to such equipment by—

"(A) ensuring that the equipment that such manufacturer offers is accessible to and usable by individuals with disabilities without the use of third party applications, peripheral devices, software, hardware, or customer premises equipment; or

S. 3304—6

"(B) if such manufacturer chooses, using third party applications, peripheral devices, software, hardware, or customer premises equipment that is available to the consumer at nominal cost and that individuals with disabilities can access.

"(b) SERVICE PROVIDERS.—

"(1) IN GENERAL.—With respect to services provided after the effective date of the regulations established pursuant to subsection (e), and subject to those regulations, a provider of advanced communications services shall ensure that such services offered by such provider in or affecting interstate commerce are accessible to and usable by individuals with disabilities, unless the requirements of this subsection are not achievable.

"(2) INDUSTRY FLEXIBILITY.—A provider of services may satisfy the requirements of paragraph (1) with respect to such services by—

"(A) ensuring that the services that such provider offers are accessible to and usable by individuals with disabilities without the use of third party applications, peripheral devices, software, hardware, or customer premises equipment; or

"(B) if such provider chooses, using third party applications, peripheral devices, software, hardware, or customer premises equipment that is available to the consumer at nominal cost and that individuals with disabilities can access.

"(c) COMPATIBILITY.—Whenever the requirements of subsections (a) or (b) are not achievable, a manufacturer or provider shall ensure that its equipment or service is compatible with existing peripheral devices or specialized customer premises equipment commonly used by individuals with disabilities to achieve access, unless the requirement of this subsection is not achievable.

"(d) NETWORK FEATURES, FUNCTIONS, AND CAPABILITIES.—Each provider of advanced communications services has the duty not to install network features, functions, or capabilities that do not impede accessibility or usability.

"(e) REGULATIONS.—

"(1) IN GENERAL.—Within one year after the date of enactment of the Twenty-First Century Communications and Video Accessibility Act of 2010, the Commission shall promulgate such regulations as are necessary to implement this section. In prescribing the regulations, the Commission shall—

"(A) include performance objectives to ensure the accessibility, usability, and compatibility of advanced communications services and the equipment used for advanced communications services by individuals with disabilities;

"(B) provide that advanced communications services, the equipment used for advanced communications services, and networks used to provide advanced communications services may not impair or impede the accessibility of information content when accessibility has been incorporated into that content for transmission through advanced communications services, equipment used for advanced communications services, or networks used to provide advanced communications services;

S. 3304—7

"(C) determine the obligations under this section of manufacturers, service providers, and providers of applications or services accessed over service provider networks; and

"(D) not mandate technical standards, except that the Commission may adopt technical standards as a safe harbor for such compliance if necessary to facilities the manufacturers' and service providers' compliance with sections (a) through (c).

"(2) PROSPECTIVE GUIDELINES.—The Commission shall issue prospective guidelines for a manufacturer or provider regarding the requirements of this section.

"(f) SERVICES AND EQUIPMENT SUBJECT TO SECTION 255.—The requirements of this section shall not apply to any equipment or services, including interconnected VoIP service, that are subject to the requirements of section 255 on the day before the date of enactment of the Twenty-First Century Communications and Video Accessibility Act of 2010. Such services and equipment shall remain subject to the requirements of section 255.

"(g) ACHIEVABLE DEFINED.—For purposes of this section and section 718, the term 'achievable' means with reasonable effort or expense, as determined by the Commission. In determining whether the requirements of a provision are achievable, the Commission shall consider the following factors:

"(1) The nature and cost of the steps needed to meet the requirements of this section with respect to the specific equipment or service in question.

"(2) The technical and economic impact on the operation of the manufacturer or provider and on the operation of the specific equipment or service in question, including on the development and deployment of new communications technologies.

"(3) The type of operations of the manufacturer or provider.

"(4) The extent to which the service provider or manufacturer in question offers accessible services or equipment containing varying degrees of functionality and features, and offered at differing price points.

"(h) COMMISSION FLEXIBILITY.—

"(1) WAIVER.—The Commission shall have the authority, on its own motion or in response to a petition by a manufacturer or provider of advanced communications services or any interested party, to waive the requirements of this section for any feature or function of equipment used to provide or access advanced communications services, or for any class of such equipment, for any provider of advanced communications services, or for any class of such services, that—

"(A) is capable of accessing an advanced communications service; and

"(B) is designed for multiple purposes, but is designed primarily for purposes other than using advanced communications services.

"(2) SMALL ENTITY EXEMPTION.—The Commission may exempt small entities from the requirements of this section.

"(i) CUSTOMIZED EQUIPMENT OR SERVICES.—The provisions of this section shall not apply to customized equipment or services that are not offered directly to the public, or to such classes of

users as to be effectively available directly to the public, regardless
of the facilities used.

"(j) RULE OF CONSTRUCTION.—This section shall not be construed to require a manufacturer of equipment used for advanced
communications or a provider of advanced communications services
to make every feature and function of every device or service accessible for every disability.

"SEC. 717. ENFORCEMENT AND RECORDKEEPING OBLIGATIONS.

"(a) COMPLAINT AND ENFORCEMENT PROCEDURES.—Within one
year after the date of enactment of the Twenty-First Century
Communications and Video Accessibility Act of 2010, the Commission shall establish regulations that facilitate the filing of formal
and informal complaints that allege a violation of section 255,
716, or 718, establish procedures for enforcement actions by the
Commission with respect to such violations, and implement the
recordkeeping obligations of paragraph (5) for manufacturers and
providers subject to such sections. Such regulations shall include
the following provisions:

"(1) NO FEE.—The Commission shall not charge any fee
to an individual who files a complaint alleging a violation
of section 255, 716, or 718.

"(2) RECEIPT OF COMPLAINTS.—The Commission shall establish separate and identifiable electronic, telephonic, and physical receptacles for the receipt of complaints filed under section
255, 716, or 718.

"(3) COMPLAINTS TO THE COMMISSION.—

"(A) IN GENERAL.—Any person alleging a violation of
section 255, 716, or 718 by a manufacturer of equipment
or provider of service subject to such sections may file
a formal or informal complaint with the Commission.

"(B) INVESTIGATION OF INFORMAL COMPLAINT.—The
Commission shall investigate the allegations in an informal
complaint and, within 180 days after the date on which
such complaint was filed with the Commission, issue an
order concluding the investigation, unless such complaint
is resolved before such time. The order shall include a
determination whether any violation occurred.

"(i) If the Commission determines that a violation
has occurred, the Commission may, in the order issued
under this subparagraph or in a subsequent order,
direct the manufacturer or service provider to bring
the service, or in the case of a manufacturer, the
next generation of the equipment or device, into compliance with requirements of those sections within a
reasonable time established by the Commission in its
order.

"(ii) NO VIOLATION.—If a determination is made
that a violation has not occurred, the Commission shall
provide the basis for such determination.

"(C) CONSOLIDATION OF COMPLAINTS.—The Commission
may consolidate for investigation and resolution complaints
alleging substantially the same violation.

"(4) OPPORTUNITY TO RESPOND.—Before the Commission
makes a determination pursuant to paragraph (3), the party
that is the subject of the complaint shall have a reasonable
opportunity to respond to such complaint, and may include

S. 3304—9

in such response any factors that are relevant to such determination. Before issuing a final order under paragraph (3)(B)(i), the Commission shall provide such party a reasonable opportunity to comment on any proposed remedial action.

"(5) RECORDKEEPING.—(A) Beginning one year after the effective date of regulations promulgated pursuant to section 716(e), each manufacturer and provider subject to sections 255, 716, and 718 shall maintain, in the ordinary course of business and for a reasonable period, records of the efforts taken by such manufacturer or provider to implement sections 255, 716, and 718, including the following:

"(i) Information about the manufacturer's or provider's efforts to consult with individuals with disabilities.

"(ii) Descriptions of the accessibility features of its products and services.

"(iii) Information about the compatibility of such products and services with peripheral devices or specialized customer premise equipment commonly used by individuals with disabilities to achieve access.

"(B) An officer of a manufacturer or provider shall submit to the Commission an annual certification that records are being kept in accordance with subparagraph (A).

"(C) After the filing of a formal or informal complaint against a manufacturer or provider in the manner prescribed in paragraph (3), the Commission may request, and shall keep confidential, a copy of the records maintained by such manufacturer or provider pursuant to subparagraph (A) of this paragraph that are directly relevant to the equipment or service that is the subject of such complaint.

"(6) FAILURE TO ACT.—If the Commission fails to carry out any of its responsibilities to act upon a complaint in the manner prescribed in paragraph (3), the person that filed such complaint may bring an action in the nature of mandamus in the United States Court of Appeals for the District of Columbia to compel the Commission to carry out any such responsibility.

"(7) COMMISSION JURISDICTION.—The limitations of section 255(f) shall apply to any claim that alleges a violation of section 255, 716, or 718. Nothing in this paragraph affects or limits any action for mandamus under paragraph (6) or any appeal pursuant to section 402(b)(10).

"(8) PRIVATE RESOLUTIONS OF COMPLAINTS.—Nothing in the Commission's rules or this Act shall be construed to preclude a person who files a complaint and a manufacturer or provider from resolving a formal or informal complaint prior to the Commission's final determination in a complaint proceeding. In the event of such a resolution, the parties shall jointly request dismissal of the complaint and the Commission shall grant such request.

"(b) REPORTS TO CONGRESS.—

"(1) IN GENERAL.—Every two years after the date of enactment of the Twenty-First Century Communications and Video Accessibility Act of 2010, the Commission shall submit to the Committee on Commerce, Science, and Transportation of the Senate and the Committee on Energy and Commerce of the House of Representatives a report that includes the following:

S. 3304—10

"(A) An assessment of the level of compliance with sections 255, 716, and 718.

"(B) An evaluation of the extent to which any accessibility barriers still exist with respect to new communications technologies.

"(C) The number and nature of complaints received pursuant to subsection (a) during the two years that are the subject of the report.

"(D) A description of the actions taken to resolve such complaints under this section, including forfeiture penalties assessed.

"(E) The length of time that was taken by the Commission to resolve each such complaint.

"(F) The number, status, nature, and outcome of any actions for mandamus filed pursuant to subsection (a)(6) and the number, status, nature, and outcome of any appeals filed pursuant to section 402(b)(10).

"(G) An assessment of the effect of the requirements of this section on the development and deployment of new communications technologies.

"(2) PUBLIC COMMENT REQUIRED.—The Commission shall seek public comment on its tentative findings prior to submission to the Committees of the report under this subsection.

"(c) COMPTROLLER GENERAL ENFORCEMENT STUDY.—

"(1) IN GENERAL.—The Comptroller General shall conduct a study to consider and evaluate the following:

"(A) The Commission's compliance with the requirements of this section, including the Commission's level of compliance with the deadlines established under and pursuant to this section and deadlines for acting on complaints pursuant to subsection (a).

"(B) Whether the enforcement actions taken by the Commission pursuant to this section have been appropriate and effective in ensuring compliance with this section.

"(C) Whether the enforcement provisions under this section are adequate to ensure compliance with this section.

"(D) Whether, and to what extent (if any), the requirements of this section have an effect on the development and deployment of new communications technologies.

"(2) REPORT.—Not later than 5 years after the date of enactment of the Twenty-First Century Communications and Video Accessibility Act of 2010, the Comptroller General shall submit to the Committee on Commerce, Science, and Transportation of the Senate and the Committee on Energy and Commerce of the House of Representatives a report on the results of the study required by paragraph (1), with recommendations for how the enforcement process and measures under this section may be modified or improved.

"(d) CLEARINGHOUSE.—Within one year after the date of enactment of the Twenty-First Century Communications and Video Accessibility Act of 2010, the Commission shall, in consultation with the Architectural and Transportation Barriers Compliance Board, the National Telecommunications and Information Administration, trade associations, and organizations representing individuals with disabilities, establish a clearinghouse of information on the availability of accessible products and services and accessibility

S. 3304—11

solutions required under sections 255, 716, and 718. Such information shall be made publicly available on the Commission's website and by other means, and shall include an annually updated list of products and services with access features.

"(e) OUTREACH AND EDUCATION.—Upon establishment of the clearinghouse of information required under subsection (d), the Commission, in coordination with the National Telecommunications and Information Administration, shall conduct an informational and educational program designed to inform the public about the availability of the clearinghouse and the protections and remedies available under sections 255, 716, and 718.

"SEC. 718. INTERNET BROWSERS BUILT INTO TELEPHONES USED WITH PUBLIC MOBILE SERVICES.

"(a) ACCESSIBILITY.—If a manufacturer of a telephone used with public mobile services (as such term is defined in section 710(b)(4)(B)) includes an Internet browser in such telephone, or if a provider of mobile service arranges for the inclusion of a browser in telephones to sell to customers, the manufacturer or provider shall ensure that the functions of the included browser (including the ability to launch the browser) are accessible to and usable by individuals who are blind or have a visual impairment, unless doing so is not achievable, except that this subsection shall not impose any requirement on such manufacturer or provider—

"(1) to make accessible or usable any Internet browser other than a browser that such manufacturer or provider includes or arranges to include in the telephone; or

"(2) to make Internet content, applications, or services accessible or usable (other than enabling individuals with disabilities to use an included browser to access such content, applications, or services).

"(b) INDUSTRY FLEXIBILITY.—A manufacturer or provider may satisfy the requirements of subsection (a) with respect to such telephone or services by—

"(1) ensuring that the telephone or services that such manufacture or provider offers is accessible to and usable by individuals with disabilities without the use of third party applications, peripheral devices, software, hardware, or customer premises equipment; or

"(2) using third party applications, peripheral devices, software, hardware, or customer premises equipment that is available to the consumer at nominal cost and that individuals with disabilities can access.".

(b) EFFECTIVE DATE FOR SECTION 718.—Section 718 of the Communications Act of 1934, as added by subsection (a), shall take effect 3 years after the date of enactment of this Act.

(c) TITLE V AMENDMENTS.—Section 503(b)(2) of such Act (47 U.S.C. 503(b)(2)) is amended by adding after subparagraph (E) the following:

"(F) Subject to paragraph (5) of this section, if the violator is a manufacturer or service provider subject to the requirements of section 255, 716, or 718, and is determined by the Commission to have violated any such requirement, the manufacturer or provider shall be liable to the United States for a forfeiture penalty of not more than $100,000 for each violation or each day of a continuing violation, except that the amount assessed for any continuing violation shall not exceed a total of $1,000,000 for any single act or failure to act.".

(d) REVIEW OF COMMISSION DETERMINATIONS.—Section 402(b) of such Act (47 U.S.C. 402(b)) is amended by adding the following new paragraph:

"(10) By any person who is aggrieved or whose interests are adversely affected by a determination made by the Commission under section 717(a)(3).".

SEC. 105. RELAY SERVICES FOR DEAF-BLIND INDIVIDUALS.

Title VII of the Communications Act of 1934, as amended by section 104, is further amended by adding at the end the following:

"SEC. 719. RELAY SERVICES FOR DEAF-BLIND INDIVIDUALS.

"(a) IN GENERAL.—Within 6 months after the date of enactment of the Equal Access to 21st Century Communications Act, the Commission shall establish rules that define as eligible for relay service support those programs that are approved by the Commission for the distribution of specialized customer premises equipment designed to make telecommunications service, Internet access service, and advanced communications, including interexchange services and advanced telecommunications and information services, accessible by individuals who are deaf-blind.

"(b) INDIVIDUALS WHO ARE DEAF-BLIND DEFINED.—For purposes of this subsection, the term 'individuals who are deaf-blind' has the same meaning given such term in the Helen Keller National Center Act, as amended by the Rehabilitation Act Amendments of 1992 (29 U.S.C. 1905(2)).

"(c) ANNUAL AMOUNT.—The total amount of support the Commission may provide from its interstate relay fund for any fiscal year may not exceed $10,000,000.".

SEC. 106. EMERGENCY ACCESS ADVISORY COMMITTEE.

(a) ESTABLISHMENT.—For the purpose of achieving equal access to emergency services by individuals with disabilities, as a part of the migration to a national Internet protocol-enabled emergency network, not later than 60 days after the date of enactment of this Act, the Chairman of the Commission shall establish an advisory committee, to be known as the Emergency Access Advisory Committee (referred to in this section as the "Advisory Committee").

(b) MEMBERSHIP.—As soon as practicable after the date of enactment of this Act, the Chairman of the Commission shall appoint the members of the Advisory Committee, ensuring a balance between individuals with disabilities and other stakeholders, and shall designate two such members as the co-chairs of the Committee. Members of the Advisory Committee shall be selected from the following groups:

(1) STATE AND LOCAL GOVERNMENT AND EMERGENCY RESPONDER REPRESENTATIVES.—Representatives of State and local governments and representatives of emergency response providers, selected from among individuals nominated by national organizations representing such governments and representatives.

(2) SUBJECT MATTER EXPERTS.—Individuals who have the technical knowledge and expertise to serve on the Advisory Committee in the fulfillment of its duties, including representatives of—

S. 3304—13

(A) providers of interconnected and non-interconnected VoIP services;

(B) vendors, developers, and manufacturers of systems, facilities, equipment, and capabilities for the provision of interconnected and non-interconnected VoIP services;

(C) national organizations representing individuals with disabilities and senior citizens;

(D) Federal agencies or departments responsible for the implementation of the Next Generation E 9–1–1 system;

(E) the National Institute of Standards and Technology; and

(F) other individuals with such technical knowledge and expertise.

(3) REPRESENTATIVES OF OTHER STAKEHOLDERS AND INTERESTED PARTIES.—Representatives of such other stakeholders and interested and affected parties as the Chairman of the Commission determines appropriate.

(c) DEVELOPMENT OF RECOMMENDATIONS.—Within 1 year after the completion of the member appointment process by the Chairman of the Commission pursuant to subsection (b), the Advisory Committee shall conduct a national survey of individuals with disabilities, seeking input from the groups described in subsection (b)(2), to determine the most effective and efficient technologies and methods by which to enable access to emergency services by individuals with disabilities and shall develop and submit to the Commission recommendations to implement such technologies and methods, including recommendations—

(1) with respect to what actions are necessary as a part of the migration to a national Internet protocol-enabled network to achieve reliable, interoperable communication transmitted over such network that will ensure access to emergency services by individuals with disabilities;

(2) for protocols, technical capabilities, and technical requirements to ensure the reliability and interoperability necessary to ensure access to emergency services by individuals with disabilities;

(3) for the establishment of technical standards for use by public safety answering points, designated default answering points, and local emergency authorities;

(4) for relevant technical standards and requirements for communication devices and equipment and technologies to enable the use of reliable emergency access;

(5) for procedures to be followed by IP-enabled network providers to ensure that such providers do not install features, functions, or capabilities that would conflict with technical standards;

(6) for deadlines by which providers of interconnected and non-interconnected VoIP services and manufacturers of equipment used for such services shall achieve the actions required in paragraphs (1) through (5), where achievable, and for the possible phase out of the use of current-generation TTY technology to the extent that this technology is replaced with more effective and efficient technologies and methods to enable access to emergency services by individuals with disabilities;

(7) for the establishment of rules to update the Commission's rules with respect to 9–1–1 services and E–911 services

S. 3304—14

(as defined in section 158(e)(4) of the National Telecommunications and Information Administration Organization Act (47 U.S.C. 942(e)(4))), for users of telecommunications relay services as new technologies and methods for providing such relay services are adopted by providers of such relay services; and

(8) that take into account what is technically and economically feasible.

(d) MEETINGS.—

(1) INITIAL MEETING.—The initial meeting of the Advisory Committee shall take place not later than 45 days after the completion of the member appointment process by the Chairman of the Commission pursuant to subsection (b).

(2) OTHER MEETINGS.—After the initial meeting, the Advisory Committee shall meet at the call of the chairs, but no less than monthly until the recommendations required pursuant to subsection (c) are completed and submitted.

(3) NOTICE; OPEN MEETINGS.—Any meetings held by the Advisory Committee shall be duly noticed at least 14 days in advance and shall be open to the public.

(e) RULES.—

(1) QUORUM.—One-third of the members of the Advisory Committee shall constitute a quorum for conducting business of the Advisory Committee.

(2) SUBCOMMITTEES.—To assist the Advisory Committee in carrying out its functions, the chair may establish appropriate subcommittees composed of members of the Advisory Committee and other subject matter experts as determined to be necessary.

(3) ADDITIONAL RULES.—The Advisory Committee may adopt other rules as needed.

(f) FEDERAL ADVISORY COMMITTEE ACT.—The Federal Advisory Committee Act (5 U.S.C. App.) shall not apply to the Advisory Committee.

(g) IMPLEMENTING RECOMMENDATIONS.—The Commission shall have the authority to promulgate regulations to implement the recommendations proposed by the Advisory Committee, as well as any other regulations, technical standards, protocols, and procedures as are necessary to achieve reliable, interoperable communication that ensures access by individuals with disabilities to an Internet protocol-enabled emergency network, where achievable and technically feasible.

(h) DEFINITIONS.—In this section—

(1) the term "Commission" means the Federal Communications Commission;

(2) the term "Chairman" means the Chairman of the Federal Communications Commission; and

(3) except as otherwise expressly provided, other terms have the meanings given such terms in section 3 of the Communications Act of 1934 (47 U.S.C. 153).

TITLE II—VIDEO PROGRAMMING

SEC. 201. VIDEO PROGRAMMING AND EMERGENCY ACCESS ADVISORY COMMITTEE.

(a) ESTABLISHMENT.—Not later than 60 days after the date of enactment of this Act, the Chairman shall establish an advisory

S. 3304—15

committee to be known as the Video Programming and Emergency Access Advisory Committee.

(b) MEMBERSHIP.—As soon as practicable after the date of enactment of this Act, the Chairman shall appoint individuals who have the technical knowledge and engineering expertise to serve on the Advisory Committee in the fulfillment of its duties, including the following:

(1) Representatives of distributors and providers of video programming or a national organization representing such distributors.

(2) Representatives of vendors, developers, and manufacturers of systems, facilities, equipment, and capabilities for the provision of video programming delivered using Internet protocol or a national organization representing such vendors, developers, or manufacturers.

(3) Representatives of manufacturers of consumer electronics or information technology equipment or a national organization representing such manufacturers.

(4) Representatives of video programming producers or a national organization representing such producers.

(5) Representatives of national organizations representing accessibility advocates, including individuals with disabilities and the elderly.

(6) Representatives of the broadcast television industry or a national organization representing such industry.

(7) Other individuals with technical and engineering expertise, as the Chairman determines appropriate.

(c) COMMISSION OVERSIGHT.—The Chairman shall appoint a member of the Commission's staff to moderate and direct the work of the Advisory Committee.

(d) TECHNICAL STAFF.—The Commission shall appoint a member of the Commission's technical staff to provide technical assistance to the Advisory Committee.

(e) DEVELOPMENT OF RECOMMENDATIONS.—

(1) CLOSED CAPTIONING REPORT.—Within 6 months after the date of the first meeting of the Advisory Committee, the Advisory Committee shall develop and submit to the Commission a report that includes the following:

(A) A recommended schedule of deadlines for the provision of closed captioning service.

(B) An identification of the performance requirement for protocols, technical capabilities, and technical procedures needed to permit content providers, content distributors, Internet service providers, software developers, and device manufacturers to reliably encode, transport, receive, and render closed captions of video programming, except for consumer generated media, delivered using Internet protocol.

(C) An identification of additional protocols, technical capabilities, and technical procedures beyond those available as of the date of enactment of the Twenty-First Century Communications and Video Accessibility Act of 2010 for the delivery of closed captions of video programming, except for consumer generated media, delivered using Internet protocol that are necessary to meet the performance objectives identified under subparagraph (B).

S. 3304—16

(D) A recommendation for technical standards to address the performance objectives identified in subparagraph (B).

(E) A recommendation for any regulations that may be necessary to ensure compatibility between video programming, except for consumer generated media, delivered using Internet protocol and devices capable of receiving and displaying such programming in order to facilitate access to closed captions.

(2) VIDEO DESCRIPTION, EMERGENCY INFORMATION, USER INTERFACES, AND VIDEO PROGRAMMING GUIDES AND MENUS.—Within 18 months after the date of enactment of this Act, the Advisory Committee shall develop and submit to the Commission a report that includes the following:

(A) A recommended schedule of deadlines for the provision of video description and emergency information.

(B) An identification of the performance requirement for protocols, technical capabilities, and technical procedures needed to permit content providers, content distributors, Internet service providers, software developers, and device manufacturers to reliably encode, transport, receive, and render video descriptions of video programming, except for consumer generated media, and emergency information delivered using Internet protocol or digital broadcast television.

(C) An identification of additional protocols, technical capabilities, and technical procedures beyond those available as of the date of enactment of the Twenty-First Century Communications and Video Accessibility Act of 2010 for the delivery of video descriptions of video programming, except for consumer generated media, and emergency information delivered using Internet protocol that are necessary to meet the performance objectives identified under subparagraph (B).

(D) A recommendation for technical standards to address the performance objectives identified in subparagraph (B).

(E) A recommendation for any regulations that may be necessary to ensure compatibility between video programming, except for consumer generated media, delivered using Internet protocol and devices capable of receiving and displaying such programming, except for consumer generated media, in order to facilitate access to video descriptions and emergency information.

(F) With respect to user interfaces, a recommendation for the standards, protocols, and procedures used to enable the functions of apparatus designed to receive or display video programming transmitted simultaneously with sound (including apparatus designed to receive or display video programming transmitted by means of services using Internet protocol) to be accessible to and usable by individuals with disabilities.

(G) With respect to user interfaces, a recommendation for the standards, protocols, and procedures used to enable on-screen text menus and other visual indicators used to access the functions on an apparatus described in subparagraph (F) to be accompanied by audio output so that such

S. 3304—17

menus or indicators are accessible to and usable by individuals with disabilities.

(H) With respect to video programming guides and menus, a recommendation for the standards, protocols, and procedures used to enable video programming information and selection provided by means of a navigation device, guide, or menu to be accessible in real-time by individuals who are blind or visually impaired.

(3) CONSIDERATION OF WORK BY STANDARD-SETTING ORGANIZATIONS.—The recommendations of the advisory committee shall, insofar as possible, incorporate the standards, protocols, and procedures that have been adopted by recognized industry standard-setting organizations for each of the purposes described in paragraphs (1) and (2).

(f) MEETINGS.—

(1) INITIAL MEETING.—The initial meeting of the Advisory Committee shall take place not later than 180 days after the date of the enactment of this Act.

(2) OTHER MEETINGS.—After the initial meeting, the Advisory Committee shall meet at the call of the Chairman.

(3) NOTICE; OPEN MEETINGS.—Any meeting held by the Advisory Committee shall be noticed at least 14 days before such meeting and shall be open to the public.

(g) PROCEDURAL RULES.—

(1) QUORUM.—The presence of one-third of the members of the Advisory Committee shall constitute a quorum for conducting the business of the Advisory Committee.

(2) SUBCOMMITTEES.—To assist the Advisory Committee in carrying out its functions, the Chairman may establish appropriate subcommittees composed of members of the Advisory Committee and other subject matter experts.

(3) ADDITIONAL PROCEDURAL RULES.—The Advisory Committee may adopt other procedural rules as needed.

(h) FEDERAL ADVISORY COMMITTEE ACT.—The Federal Advisory Committee Act (5 U.S.C. App.) shall not apply to the Advisory Committee.

SEC. 202. VIDEO DESCRIPTION AND CLOSED CAPTIONING.

(a) VIDEO DESCRIPTION.—Section 713 of the Communications Act of 1934 (47 U.S.C. 613) is amended—

(1) by striking subsections (f) and (g);

(2) by redesignating subsection (h) as subsection (j); and

(3) by inserting after subsection (e) the following:

"(f) VIDEO DESCRIPTION.—

"(1) REINSTATEMENT OF REGULATIONS.—On the day that is 1 year after the date of enactment of the Twenty-First Century Communications and Video Accessibility Act of 2010, the Commission shall, after a rulemaking, reinstate its video description regulations contained in the Implementation of Video Description of Video Programming Report and Order (15 F.C.C.R. 15,230 (2000)), recon. granted in part and denied in part, (16 F.C.C.R. 1251 (2001)), modified as provided in paragraph (2).

"(2) MODIFICATIONS TO REINSTATED REGULATIONS.—Such regulations shall be modified only as follows:

S. 3304—18

"(A) The regulations shall apply to video programming, as defined in subsection (h), insofar as and programming is transmitted for display on television in digital format.

"(B) The Commission shall update the list of the top 25 designated market areas, the list of the top 5 national nonbroadcast networks that at least 50 hours per quarter of prime time programming that is not exempt under this paragraph, and the beginning calendar quarter for which compliance shall be calculated.

"(C) The regulations may permit a provider of video programming or a program owner to petition the Commission for an exemption from the requirements of this section upon a showing that the requirements contained in this section be economically burdensome.

"(D) The Commission may exempt from the regulations established pursuant to paragraph (1) a service, class of services, program, class of programs, equipment, or class of equipment for which the Commission has determined that the application of such regulations would be economically burdensome for the provider of such service, program, or equipment.

"(E) The regulations shall not apply to live or near-live programming.

"(F) The regulations shall provide for an appropriate phased schedule of deadlines for compliance.

"(G) The Commission shall consider extending the exemptions and limitations in the reinstated regulations for technical capability reasons to all providers and owners of video programming.

"(3) INQUIRIES ON FURTHER VIDEO DESCRIPTION REQUIRE-MENTS.—The Commission shall commence the following inquiries not later than 1 year after the completion of the phase-in of the reinstated regulations and shall report to Congress 1 year thereafter on the findings for each of the following:

"(A) VIDEO DESCRIPTION IN TELEVISION PROGRAM-MING.—The availability, use, and benefits of video description on video programming distributed on television, the technical and creative issues associated with providing such video description, and the financial costs of providing such video description for providers of video programming and program owners.

"(B) VIDEO DESCRIPTION IN VIDEO PROGRAMMING DISTRIBUTED ON THE INTERNET.—The technical and operational issues, costs, and benefits of providing video descriptions for video programming that is delivered using Internet protocol.

"(4) CONTINUING COMMISSION AUTHORITY.—

"(A) IN GENERAL.—The Commission may not issue additional regulations unless the Commission determines, at least 2 years after completing the reports required in paragraph (3), that the need for and benefits of providing video description for video programming, insofar as such programming is transmitted for display on television, are greater than the technical and economic costs of providing such additional programming.

"(B) LIMITATION.—If the Commission makes the determination under subparagraph (A) and issues additional

S. 3304—19

regulations, the Commission may not increase, in total, the hour requirement for additional described programming by more than 75 percent of the requirement in the regulations reinstated under paragraph (1).

"(C) APPLICATION TO DESIGNATED MARKET AREAS.—

"(i) IN GENERAL.—After the Commission completes the reports on video description required in paragraph (3), the Commission shall phase in the video description regulations for the top 60 designated market areas, except that the Commission may grant waivers to entities in specific designated market areas where it deems appropriate.

"(ii) PHASE-IN DEADLINE.—The phase-in described in clause (i) shall be completed not later than 6 years after the date of enactment of the Twenty-First Century Communications and Video Accessibility Act of 2010.

"(iii) REPORT.—Nine years after the date of enactment of the Twenty-First Century Communications and Video Accessibility Act of 2010, the Commission shall submit to the Committee on Energy of the House of Representatives and the Committee on Commerce, Science, and Transportation of the Senate a report assessing—

"(I) the types of described video programming that is available to consumers;

"(II) consumer use of such programming;

"(III) the costs to program owners, providers, and distributors of creating such programming;

"(IV) the potential costs to program owners, providers, and distributors in designated market areas outside of the top 60 of creating such programming;

"(V) the benefits to consumers of such programming;

"(VI) the amount of such programming currently available; and

"(VII) the need for additional described programming in designated market areas outside the top 60.

"(iv) ADDITIONAL MARKET AREAS.—Ten years after the date of enactment of the Twenty-First Century Communications and Video Accessibility Act of 2010, the Commission shall have the authority, based upon the findings, conclusions, and recommendations contained in the report under clause (iii), to phase in the video description regulations for up to an additional 10 designated market areas each year—

"(I) if the costs of implementing the video description regulations to program owners, providers, and distributors in those additional markets are reasonable, as determined by the Commission; and

"(II) except that the Commission may grant waivers to entities in specific designated market areas where it deems appropriate.

S. 3304—20

"(g) EMERGENCY INFORMATION.—Not later than 1 year after the Advisory Committee report under subsection (e)(2) is submitted to the Commission, the Commission shall complete a proceeding to—

"(1) identify methods to convey emergency information (as that term is defined in section 79.2 of title 47, Code of Federal Regulations) in a manner accessible to individuals who are blind or visually impaired; and

"(2) promulgate regulations that require video programming providers and video programming distributors (as those terms are defined in section 79.1 of title 47, Code of Federal Regulations) and program owners to convey such emergency information in a manner accessible to individuals who are blind or visually impaired.

"(h) DEFINITIONS.—For purposes of this section, section 303, and section 330:

"(1) VIDEO DESCRIPTION.—The term 'video description' means the insertion of audio narrated descriptions of a television program's key visual elements into natural pauses between the program's dialogue.

"(2) VIDEO PROGRAMMING.—The term 'video programming' means programming by, or generally considered comparable to programming provided by a television broadcast station, but not including consumer-generated media (as defined in section 3).

(b) CLOSED CAPTIONING ON VIDEO PROGRAMMING DELIVERED USING INTERNET PROTOCOL.—Section 713 of such Act is further amended by striking subsection (c) and inserting the following:

"(c) DEADLINES FOR CAPTIONING.—

"(1) IN GENERAL.—The regulations prescribed pursuant to subsection (b) shall include an appropriate schedule of deadlines for the provision of closed captioning of video programming once published or exhibited on television.

"(2) DEADLINES FOR PROGRAMMING DELIVERED USING INTERNET PROTOCOL.—

"(A) REGULATIONS ON CLOSED CAPTIONING ON VIDEO PROGRAMMING DELIVERED USING INTERNET PROTOCOL.—Not later than 6 months after the submission of the report to the Commission required by subsection (e)(1) of the Twenty-First Century Communications and Video Accessibility Act of 2010, the Commission shall revise its regulations to require the provision of closed captioning on video programming delivered using Internet protocol that was published or exhibited on television with captions after the effective date of such regulations.

"(B) SCHEDULE.—The regulations prescribed under this paragraph shall include an appropriate schedule of deadlines for the provision of closed captioning, taking into account whether such programming is prerecorded and edited for Internet distribution, or whether such programming is live or near-live and not edited for Internet distribution.

"(C) COST.—The Commission may delay or waive the regulation promulgated under subparagraph (A) to the extent the Commission finds that the application of the regulation to live video programming delivered using Internet protocol with captions after the effective date of such

S. 3304—21

regulations would be economically burdensome to providers of video programming or program owners.

"(D) REQUIREMENTS FOR REGULATIONS.—The regulations prescribed under this paragraph—

"(i) shall contain a definition of 'near-live programming' and 'edited for Internet distribution';

"(ii) may exempt any service, class of service, program, class of program, equipment, or class of equipment for which the Commission has determined that the application of such regulations would be economically burdensome for the provider of such service, program, or equipment;

"(iii) shall clarify that, for the purposes of implementation, of this subsection, the terms 'video programming distribution' and 'video programming providers' include an entity that makes available directly to the end user video programming through a distribution method that uses Internet protocol;

"(iv) and describe the responsibilities of video programming providers or distributors and video programming owners;

"(v) shall establish a mechanism to make available to video progamming providers and distributors information on video programming subject to the Act on an ongoing basis;

"(vi) shall consider that the video programming provider or distributor shall be deemed in compliance if such entity enables the rendering or pass through of closed captions and video description signals and make a good faith effort to identify video programming subject to the Act using the mechanism created in (v); and

"(vii) shall provide that de minimis failure to comply with such regulations by a video programming provider or owner shall not be treated as a violation of the regulations.

"(3) ALTERNATE MEANS OF COMPLIANCE.—An entity may meet the requirements of this section through alternate means than those prescribed by regulations pursuant to subsection (b), as revised pursuant to paragraph (2)(A) of this subsection, if the requirements of this section are met, as determined by the Commission.".

(c) CONFORMING AMENDMENT.—Section 713(d) of such Act is amended by striking paragraph (3) and inserting the following:

"(3) a provider of video programming or program owner may petition the Commission for an exemption from the requirements of this section, and the Commission may grant such petition upon a showing that the requirements contained in this section would be economically burdensome. During the pendency of such a petition, such provider or owner shall be exempt from the requirements of this section. The Commission shall act to grant or deny any such petition, in whole or in part, within 6 months after the Commission receives such petition, unless the Commission finds that an extension of the 6-month period is necessary to determine whether such requirements are economically burdensome.".

S. 3304—22

SEC. 203. CLOSED CAPTIONING DECODER AND VIDEO DESCRIPTION CAPABILITY.

(a) AUTHORITY TO REGULATE.—Section 303(u) of the Communications Act of 1934 (47 U.S.C. 303(u)) is amended to read as follows:

"(u) Require that, if technically feasible—

"(1) apparatus designed to receive or play back video programming transmitted simultaneously with sound, if such apparatus is manufactured in the United States or imported for use in the United States and uses a picture screen of any size—

"(A) be equipped with built-in closed caption decoder circuitry or capability designed to display closed-captioned video programming;

"(B) have the capability to decode and make available the transmission and delivery of video description services as required by regulations reinstated and modified pursuant to section 713(f); and

"(C) have the capability to decode and make available emergency information (as that term is defined in section 79.2 of the Commission's regulations (47 CFR 79.2)) in a manner that is accessible to individuals who are blind or visually impaired; and

"(2) notwithstanding paragraph (1) of this subsection—

"(A) apparatus described in such paragraph that use a picture screen that is less than 13 inches in size meet the requirements of subparagraph (A), (B), or (C) of such paragraph only if the requirements of such subparagraphs are achievable (as defined in section 716);

"(B) any apparatus or class of apparatus that are display-only video monitors with no playback capability are exempt from the requirements of such paragraph; and

"(C) the Commission shall have the authority, on its own motion or in response to a petition by a manufacturer, to waive the requirements of this subsection for any apparatus or class of apparatus—

"(i) primarily designed for activities other than receiving or playing back video programming transmitted simultaneously with sound; or

"(ii) for equipment designed for multiple purposes, capable of receiving or playing video programming transmitted simultaneously with sound but whose essential utility is derived from other purposes.".

(b) OTHER DEVICES.—Section 303 of the Communications Act of 1934 (47 U.S.C. 303) is further amended by adding at the end the following new subsection:

"(z) Require that—

"(1) if achievable (as defined in section 716), apparatus designed to record video programming transmitted simultaneously with sound, if such apparatus is manufactured in the United States or imported for use in the United States, enable the rendering or the pass through of closed captions, video description signals, and emergency information (as that term is defined in section 79.2 of title 47,

S. 3304—23

Code of Federal Regulations) such that viewers are able to activate and de-activate the closed captions and video description as the video programming is played back on a picture screen of any size; and

"(2) interconnection mechanisms and standards for digital video source devices are available to carry from the source device to the consumer equipment the information necessary to permit or render the display of closed captions and to make encoded video description and emergency information audible.".

(c) SHIPMENT IN COMMERCE.—Section 330(b) of the Communications Act of 1934 (47 U.S.C. 330(b)) is amended—

(1) by striking "303(u)" in the first sentence and inserting "303(u) and (z)";

(2) by striking the second sentence and inserting the following: "Such rules shall provide performance and display standards for such built-in decoder circuitry or capability designed to display closed captioned video programming, the transmission and delivery of video description services, and the conveyance of emergency information as required by section 303 of this Act."; and

(3) in the fourth sentence, by striking "closed-captioning service continues" and inserting "closed-captioning service and video description service continue".

(d) IMPLEMENTING REGULATIONS.—The Federal Communications Commission shall prescribe such regulations as are necessary to implement the requirements of sections 303(u), 303(z), and 330(b) of the Communications Act of 1934, as amended by this section, including any technical standards, protocols, and procedures needed for the transmission of—

(1) closed captioning within 6 months after the submission to the Commission of the Advisory Committee report required by section 201(e)(1); and

(2) video description and emergency information within 18 months after the submission to the Commission of the Advisory Committee report required by section 201(e)(2).

(e) ALTERNATE MEANS OF COMPLIANCE.—An entity may meet the requirements of sections 303(u), 303(z), and 330(b) of the Communications Act of 1934 through alternate means than those prescribed by regulations pursuant to subsection (d) if the requirements of those sections are met, as determined by the Commission.

SEC. 204. USER INTERFACES ON DIGITAL APPARATUS.

(a) AMENDMENT.—Section 303 of the Communications Act of 1934 (47 U.S.C. 303) is further amended by adding after subsection (z), as added by section 203 of this Act, the following new subsection:

"(aa) Require—

"(1) if achievable (as defined in section 716) that digital apparatus designed to receive or play back video programming transmitted in digital format simultaneously with sound, including apparatus designed to receive or display video programming transmitted in digital format using Internet protocol, be designed, developed, and fabricated so that control of appropriate built-in apparatus functions are accessible to and usable by individuals who are blind or visually impaired, except that the Commission may not

S. 3304—24

specify the technical standards, protocols, procedures, and other technical requirements for meeting this requirement;

"(2) that if on-screen text menus or other visual indicators built in to the digital apparatus are used to access the functions of the apparatus described in paragraph (1), such functions shall be accompanied by audio output that is either integrated or peripheral to the apparatus, so that such menus or indicators are accessible to and usable by individuals who are blind or visually impaired in real-time;

"(3) that for such apparatus equipped with the functions described in paragraphs (1) and (2) built in access to those closed captioning and video description features through a mechanism that is reasonably comparable to a button, key, or icon designated by activating the closed captioning or accessibility features; and

"(4) that in applying this subsection the term 'apparatus' does not include a navigation device, as such term is defined in section 76.1200 of the Commission's rules (47 CFR 76.1200).".

(b) IMPLEMENTING REGULATIONS.—Within 18 months after the submission to the Commission of the Advisory Committee report required by section 201(e)(2), the Commission shall prescribe such regulations as are necessary to implement the amendments made by subsection (a).

(c) ALTERNATE MEANS OF COMPLIANCE.—An entity may meet the requirements of section 303(aa) of the Communications Act of 1934 through alternate means than those prescribed by regulations pursuant to subsection (b) if the requirements of those sections are met, as determined by the Commission.

(d) DEFERRAL OF COMPLIANCE WITH ATSC MOBILE DTV STANDARD A/153.—A digital apparatus designed and manufactured to receive or play back the Advanced Television Systems Committee's Mobile DTV Standards A/153 shall not be required to meet the requirements of the regulations prescribed under subsection (b) for a period of not less than 24 months after the date on which the final regulations are published in the Federal Register.

SEC. 205. ACCESS TO VIDEO PROGRAMMING GUIDES AND MENUS PROVIDED ON NAVIGATION DEVICES.

(a) AMENDMENT.—Section 303 of the Communications Act of 1934 (47 U.S.C. 303) is further amended by adding after subsection (aa), as added by section 204 of this Act, the following new subsection:

"(bb) Require—

"(1) if achievable (as defined in section 716), that the on-screen text menus and guides provided by navigation devices (as such term is defined in section 76.1200 of title 47, Code of Federal Regulations) for the display or selection of multichannel video programming are audibly accessible in real-time upon request by individuals who are blind or visually impaired, except that the Commission may not specify the technical standards, protocols, procedures, and other technical requirements for meeting this requirement; and

"(2) for navigation devices with built-in closed captioning capability, that access to that capability through a mechanism

S. 3304—25

is reasonably comparable to a button, key, or icon designated for activating the closed captioning, or accessibility features. With respect to apparatus features and functions delivered in software, the requirements set forth in this subsection shall apply to the manufacturer of such software. With respect to apparatus features and functions delivered in hardware, the requirements set forth in this subsection shall apply to the manufacturer of such hardware.".

(b) IMPLEMENTING REGULATIONS.—

(1) IN GENERAL.—Within 18 months after the submission to the Commission of the Advisory Committee report required by section 201(e)(2), the Commission shall prescribe such regulations as are necessary to implement the amendment made by subsection (a).

(2) EXEMPTION.—Such regulations may provide an exemption from the regulations for cable systems serving 20,000 or fewer subscribers.

(3) RESPONSIBILITY.—An entity shall only be responsible for compliance with the requirements added by this section with respect to navigation devices that it provides to a requesting blind or visually impaired individual.

(4) SEPARATE EQUIPMENT OR SOFTWARE.—

(A) IN GENERAL.—Such regulations shall permit but not require the entity providing the navigation device to the requesting blind or visually impaired individual to comply with section 303(bb)(1) of the Communications Act of 1934 through that entity's use of software, a peripheral device, specialized consumer premises equipment, a network-based service or other solution, and shall provide the maximum flexibility to select the manner of compliance.

(B) REQUIREMENTS.—If an entity complies with section 303(bb)(1) of the Communications Act of 1934 under subparagraph (A), the entity providing the navigation device to the requesting blind or visually impaired individual shall provide any such software, peripheral device, equipment, service, or solution at no additional charge and within a reasonable time to such individual and shall ensure that such software, device, equipment, service, or solution provides the access required by such regulations.

(5) USER CONTROLS FOR CLOSED CAPTIONING.—Such regulations shall permit the entity providing the navigation device maximum flexibility in the selection of means for compliance with section 303(bb)(2) of the Communications Act of 1934 (as added by subsection (a) of this section).

(6) PHASE-IN.—

(A) IN GENERAL.—The Commission shall provide affected entities with—

(i) not less than 2 years after the adoption of such regulations to begin placing in service devices that comply with the requirements of section 303(bb)(2) of the Communications Act of 1934 (as added by subsection (a) of this section); and

(ii) not less than 3 years after the adoption of such regulations to begin placing in service devices that comply with the requirements of section 303(bb)(1) of the Communications Act of 1934 (as added by subsection (a) of this section).

S. 3304—26

(B) APPLICATION.—Such regulations shall apply only to devices manufactured or imported on or after the respective effective dates established in subparagraph (A).

SEC. 206. DEFINITIONS.

In this title:

(1) ADVISORY COMMITTEE.—The term "Advisory Committee" means the advisory committee established in section 201.

(2) CHAIRMAN.—The term "Chairman" means the Chairman of the Federal Communications Commission.

(3) COMMISSION.—The term "Commission" means the Federal Communications Commission.

(4) EMERGENCY INFORMATION.—The term "emergency information" has the meaning given such term in section 79.2 of title 47, Code of Federal Regulations.

(5) INTERNET PROTOCOL.—The term "Internet protocol" includes Transmission Control Protocol and a successor protocol or technology to Internet protocol.

(6) NAVIGATION DEVICE.—The term "navigation device" has the meaning given such term in section 76.1200 of title 47, Code of Federal Regulations.

(7) VIDEO DESCRIPTION.—The term "video description" has the meaning given such term in section 713 of the Communications Act of 1934 (47 U.S.C. 613).

(8) VIDEO PROGRAMMING.—The term "video programming" has the meaning given such term in section 713 of the Communications Act of 1934 (47 U.S.C. 613).

Speaker of the House of Representatives.

Vice President of the United States and
President of the Senate.

Appendix 2

A waiver extension was granted until January 2017 specific to video game software. Video Game Software is defined with sufficient specificity and shares enough common defining characteristics to be granted a class waiver that will enable platform providers and video game publishers "to continue jointly exploring novel accessibility solutions—including solutions that will facilitate access to ACS—in video game products and services.[57] P. 10022

Further to the above—On October 19, 2016, ESA filed a petition requesting an extension of the waiver for video game software for an additional 12 months—until January 1, 2018. http://transition.fcc.gov/Daily_Releases/Daily_Business/2016/db1031/DA-16-1236A1.pdf

---

[57]https://apps.fcc.gov/edocs_public/attachmatch/DA-15-1034A1_Rcd.pdf.

Before the
Federal Communications Commission
Washington, D.C. 20554

In the Matter of	)	
	)	
Implementation of Sections 716 and 717 of the	)	CG Docket No. 10-213
Communications Act of 1934, as Enacted by the	)	
Twenty-First Century Communications and Video	)	
Accessibility Act of 2010	)	
	)	
ENTERTAINMENT SOFTWARE	)	
ASSOCIATION	)	
	)	
Petition for Class Waiver of Sections 716 and 717	)	
of the Communications Act and Part 14 of the	)	
Commission's Rules Requiring Access to	)	
Advanced Communications Services and	)	
Equipment by People with Disabilities	)	

ORDER

Adopted: September 16, 2015   **Released: September 16, 2015**

By the Acting Chief, Consumer and Governmental Affairs Bureau:

I.   INTRODUCTION

1.   In this Order, the Consumer and Governmental Affairs Bureau (CGB or Bureau) of the Federal Communications Commission (FCC or Commission) addresses a petition filed by the Entertainment Software Association (ESA). ESA requests a partial extension of the class waiver of the Commission's accessibility requirements for advanced communications services (ACS) and equipment for one class of equipment – video game software. The current class waiver for video game software, granted by the Bureau in October 2012, will expire on October 8, 2015. For the reasons set forth below, the Bureau grants ESA's request for a partial extension of the class waiver for video game software until January 1, 2017.

II.   BACKGROUND

2.   On October 8, 2010, President Obama signed into law the Twenty-First Century Communications and Video Accessibility Act of 2010 (CVAA),[1] which amended the Communications Act of 1934 (Act),[2] "to help ensure that individuals with disabilities are able to fully utilize communications services and equipment and better access video programming."[3] Section 716, which was added to the Act by the CVAA, requires providers of ACS and manufacturers of equipment used for ACS

---

[1] Pub. L. No. 111-260, 124 Stat. 2751 (2010), *amended by* Pub. L. No. 111-265, 124 Stat. 2795 (2010) (containing technical corrections).

[2] *See* Title 47 of the United States Code.

[3] S. Rep. No. 111-386 at 1 (2010) (Senate Report); H.R. Rep. No. 111-563 at 19 (2010) (House Report). Congress also noted that the communications marketplace had undergone a "fundamental transformation" since it enacted section 255 of the Act in 1996. Senate Report at 1; House Report at 19. *See also* 47 U.S.C. § 255 (requiring access to telecommunications services and equipment).

to make their services and products accessible to and usable by individuals with disabilities unless doing so is not achievable.[4] The CVAA defines ACS as interconnected voice over Internet protocol (VoIP) service; non-interconnected VoIP service; electronic messaging service, such as e-mail, instant messaging, and SMS text messaging; and interoperable video conferencing service.[5] The Commission adopted rules implementing section 716 of the Act, as well as rules implementing the CVAA's recordkeeping and enforcement obligations for ACS and ACS equipment,[6] in October 2011.[7] These rules were phased in over a two-year period ending October 8, 2013.[8]

     3.   Pursuant to section 716(h)(1) of the Act,[9] the Commission may grant waivers of the ACS accessibility requirements for multipurpose equipment or services or classes of multipurpose equipment or services that are capable of accessing ACS but are nonetheless designed primarily for purposes other than the use of ACS.[10] In instances where equipment and services may have multiple primary or co-primary purposes, waivers may not be warranted, depending on the circumstances.[11]

     4.   In conducting a waiver analysis, the Commission's rules provide for a case-by-case examination of whether the equipment is designed to be used for ACS purposes by the general public and whether and how the ACS features or functions are advertised, announced, or marketed.[12] In order to make this determination, the Commission must consider "whether the ACS functionality or feature is suggested to consumers as a reason for purchasing, installing, downloading, or accessing the equipment or service."[13] The Commission may also consider the manufacturer's market research and the usage trends of similar equipment or services in order to determine whether a manufacturer or provider designed

---

[4] 47 U.S.C. § 617.

[5] 47 U.S.C. § 153(1).

[6] 47 U.S.C. § 618.

[7] *Implementation of Sections 716 and 717 of the Communications Act of 1934, as Enacted by the Twenty-First Century Communications and Video Accessibility Act of 2010; Amendments to the Commission's Rules Implementing Sections 255 and 251(a)(2) of the Communications Act of 1934, as Enacted by the Telecommunications Act of 1996; Accessible Mobile Phone Options for People who are Blind, Deaf-Blind, or Have Low Vision,* CG Docket No. 10-213, WT Docket No. 96-198, CG Docket No. 10-145, Report and Order and Further Notice of Proposed Rulemaking, 26 FCC Rcd 14557 (2011) (*ACS Report and Order*).

[8] *ACS Report and Order,* 26 FCC Rcd at 14601-14605, ¶¶ 105-113. *See also* 47 C.F.R. §§ 14.1-14.21 (applying the ACS accessibility rules to models or versions of products and services that are introduced or upgraded on or after October 8, 2013); 47 C.F.R. §§ 14.30-14.52 (implementing recordkeeping and enforcement obligations).

[9] 47 U.S.C. § 617(h).

[10] *ACS Report and Order,* 26 FCC Rcd at 14634, ¶ 181. *See also* 47 C.F.R. § 14.5. The Commission delegated to CGB the authority to act upon all such waiver requests. *ACS Report and Order,* 26 FCC Rcd at 14566, 14640-41, ¶¶ 19, 197.

[11] *ACS Report and Order,* 26 FCC Rcd at 14635, ¶ 184 (offering as an example of equipment or services that have multiple primary or co-primary purposes, smartphones that are designed for voice communications, text messaging, e-mail, web browsing, video chat, digital video recording, mobile hotspot connectivity, and several other purposes). In other words, multipurpose equipment or services that are capable of accessing ACS and are designed primarily or co-primarily for ACS, do not qualify for a waiver under this provision. 47 U.S.C. § 617(h)(1); 47 C.F.R. § 14.5(a)(1). A product or service may have co-primary purposes when it contains multiple features and functions. Conversely, as noted in the *ACS Report and Order,* the House and Senate Reports explain that "a device designed for a purpose unrelated to accessing advanced communications might also provide, on an incidental basis, access to such services. In this case, the Commission may find that to promote technological innovation the accessibility requirements need not apply." *ACS Report and Order,* 26 FCC Rcd at 14634, ¶ 181 (quoting House Report at 26; Senate Report at 8).

[12] *ACS Report and Order,* 26 FCC Rcd at 14634-35, ¶¶ 182, 183, 185. *See also* 47 C.F.R. § 14.5(a)(2).

[13] *ACS Report and Order,* 26 FCC Rcd at 14635, ¶ 185 (footnote omitted).

the equipment or service primarily for purposes other than ACS.[14] Furthermore, the following factors may be relevant to a primary purpose waiver determination: whether the ACS functionality is designed to be operable outside of other functions or aids other functions; the impact that the removal of the ACS feature has on the primary purpose for which the equipment or services is claimed to be designed, and an examination of waivers for similar products or services.[15] In addition to considering these various factors when examining a waiver request, the Commission must utilize its general waiver standard, which requires good cause to waive the rules and a showing that the particular facts of the petition make compliance with the relevant requirements inconsistent with the public interest.[16]

5.   The Commission may entertain a waiver for equipment and services individually or as a class. A waiver may apply to more than one piece of equipment or more than one service, so long as the class is carefully defined and the equipment or services in the class share common defining characteristics.[17] The Commission also may limit the time of the waiver's coverage, with or without a provision for renewal.[18] As part of this determination, the Commission will examine the extent to which the petitioner has explained in detail the expected lifecycle of the equipment or services that are part of the class.[19] To the extent a petitioner seeks a class waiver for multiple generations of similar equipment and services, the Commission will examine the justification for the waiver extending through the lifecycle of each discrete generation.[20]

6.   All products and services covered by a class waiver that are introduced into the market while the waiver is in effect will ordinarily be subject to the waiver for the duration of the life of those particular products or services – i.e., for as long as those particular products or services are sold.[21] For example, if a particular model covered by a class waiver were to be introduced to the public on the day before the expiration of the waiver period, then all products of that particular model that are sold from that point forward would be covered by the waiver.[22] For products and services already under development after a class waiver expires, the achievability analysis may take into consideration the developmental stage of the product and the effort and expense needed to achieve accessibility at that point in the developmental stage.[23]

III.   THE ESA 2012 PETITION

7.   In May 2012, ESA requested an eight-year waiver for three classes of gaming devices and services: Class I – game consoles, both home and handheld, and their peripherals and integrated

---

[14] Id. at 14635, ¶ 183.

[15] Id. at 14636, ¶ 186.

[16] Id. at 14637, ¶ 188 (citing 47 C.F.R. § 1.3; Northeast Cellular Telephone Co., L.P. v. FCC, 897 F. 2d 1164, 1166 (D.C. Cir. 1990)).

[17] ACS Report and Order, 26 FCC Rcd at 14639, ¶ 193. See also 47 C.F.R. § 14.5(b).

[18] ACS Report and Order, 26 FCC Rcd at 14638, ¶ 192. See also 47 C.F.R. § 14.5(c).

[19] ACS Report and Order, 26 FCC Rcd at 14639, ¶ 194. See also 47 C.F.R. § 14.5(c)(2).

[20] ACS Report and Order, 26 FCC Rcd at 14640, ¶ 195.

[21] Id. at 14640, ¶ 194. See also 47 C.F.R. § 14.5(c)(2).

[22] Substantial upgrades are considered new products or services for the purpose of this waiver analysis and a new waiver would be required if a substantial upgrade is made that changes the nature of the product or service. See ACS Report and Order, 26 FCC Rcd at 14639, ¶ 192. See also id. at 14609, ¶ 124 ("Natural opportunities to assess or reassess the achievability of accessibility may include, for example, the redesign of a product model or service, new versions of software, upgrades to existing features or functionalities, significant rebundling or unbundling of product and service packages, or any other significant modification that may require redesign.").

[23] ACS Report and Order, 26 FCC Rcd at 14640, ¶ 194. See also 47 C.F.R. § 14.5(c)(2).

online networks; Class II – game distribution and online game play services that distribute game software or enable online game play across a network, regardless of the device from which it is accessed; and Class III – game software used for game play.[24] On October 15, 2012, the Bureau granted the ESA 2012 Petition for a two-year period, beginning October 8, 2013 and ending October 8, 2015.[25]

 8. In addressing the ESA 2012 Petition, the Bureau determined that the classes of equipment and services for which ESA sought a waiver were defined with sufficient specificity and that the equipment and services in each class shared enough common defining characteristics to be considered a class.[26] Further, the Bureau determined that each class of equipment and services was capable of accessing ACS, but designed primarily for purposes other than using ACS, namely enabling game play, game distribution, and playing games.[27] Although the Bureau noted a "clear trend towards marketing the ACS features and functions of gaming equipment and services," it determined that, "currently, most of the marketing for these products and services emphasize[d] game playing."[28] Finally, the Bureau determined that, while good cause existed at that time to waive the Commission's rules, competing public interests were at stake that made consideration of the ESA's waiver requests a closer call than similar waivers granted to CEA and NCTA, where the inclusion of ACS in those covered products and services was not yet as prevalent.[29]

 9. The Bureau specifically recognized "the increasing role that ACS is beginning to play in online gaming systems and services – both with respect to the ability to compete effectively, and with respect to engaging in communications that are unrelated to game play."[30] Further, the Bureau believed that, "as gaming takes on an ever-present role in our society, use of online gaming systems that have ACS options may have increasing applications in the employment and educational contexts, as well as becoming a tool of social integration."[31] As a result, and considering the time it takes for a product to be developed and initially introduced in the market, the Bureau granted ESA's waiver request for a period of two years.[32] The Bureau stated that, "[g]iven the dynamism of the electronic software industry, should ESA seek to renew or extend the class waivers granted herein, it may be necessary for ESA to define with greater specificity the classes of equipment and services under consideration at that time."[33] The Bureau

---

[24] See Petition of the Entertainment Software Association, CG Docket No. 10-213, filed March 21, 2012, at 4, 21-22, 27, 34 (ESA 2012 Petition).

[25] Implementation of Sections 716 and 717 of the Communications Act of 1934, as Enacted by the Twenty-First Century Communications and Video Accessibility Act of 2010; Consumer Electronics Association; National Cable & Telecommunications Association; Entertainment Software Association; Petitions for Class Waivers of Sections 716 and 717 of the Communications Act and Part 14 of the Commission's Rules Requiring Access to Advanced Communications Services (ACS) and Equipment by People with Disabilities, CG Docket 10-213, Order, 27 FCC Rcd 12970, 12986, 12991, ¶¶ 33, 40 (CGB 2012) (CEA, NCTA, ESA Waiver Order). Two years was measured from the ACS Report and Order implementation date of October 8, 2013. See ACS Report and Order, 26 FCC Rcd at 14601-14605, ¶¶ 107-112.

[26] CEA, NCTA, ESA Waiver Order, 27 FCC Rcd at 12986-87, ¶ 34. See also 47 C.F.R. § 14.5(b); ACS Report and Order, 26 FCC Rcd at 14639, ¶ 193.

[27] CEA, NCTA, ESA Waiver Order, 27 FCC Rcd at 12987, ¶ 35. See also 47 C.F.R. § 14.5(a); ACS Report and Order, 26 FCC Rcd at 14634, ¶ 181.

[28] CEA, NCTA, ESA Waiver Order, 27 FCC Rcd at 12987, ¶ 35.

[29] Id., 27 FCC Rcd at 12987-88, ¶ 36. See also 47 C.F.R. § 1.3; ACS Report and Order, 26 FCC Rcd at 14637, ¶ 188.

[30] CEA, NCTA, ESA Waiver Order, 27 FCC Rcd at 12988, ¶ 36 (footnotes omitted).

[31] Id., 27 FCC Rcd at 12990, ¶ 39 (footnotes omitted).

[32] Id., 27 FCC Rcd at 12989-90, 12991-92, ¶¶ 38, 40.

[33] Id., 27 FCC Rcd at 12992, ¶ 41.

also noted that "the Commission's analysis of any future requests to extend the waivers [can] be informed not only by the evidence specific to game products and services, but more generally by the development of accessibility features throughout ACS industries."[34] In this respect, the Bureau believed that, during the waiver period, gaming equipment manufacturers and service providers would benefit from and be able to utilize the experience gained in making ACS accessible in other contexts.[35]

IV.   THE ESA 2015 PETITION

10.   On May 22, 2015, ESA filed a petition for a partial extension of the class waiver of the Commission's ACS accessibility requirements for one class of equipment – video game software – for a period of just under 15 months, from October 8, 2015 to January 1, 2017, without prejudice to requests for additional extensions.[36] ESA did not request an extension of the waivers previously granted by the Bureau for the other two classes of equipment: video game consoles; and video game distribution platforms.[37] On June 9, 2015, the Bureau placed the ESA 2015 Petition on public notice.[38] No comments or reply comments were filed.

11.   ESA asserts that it has defined the class of "video game software" with sufficient specificity and that this type of equipment shares enough common defining characteristics to be considered a class.[39] According to ESA, video game software means "playable games on any hardware or online platform . . . includ[ing] game applications that are built into operating system software."[40] ESA further explains that video game software is "a clearly defined category separate from other forms of software, Internet services, entertainment media, and consumer electronics."[41]

12.   ESA also asserts that video game software that is capable of accessing ACS is designed primarily for purposes other than using ACS, namely game play.[42] In support of this assertion, ESA

---

[34] *Id.*, 27 FCC Rcd at 12992, ¶ 40, n.185 (suggesting that the 2014 CVAA biennial report to Congress would include a review of the development of accessibility features and functions in new ACS products and services).

[35] *Id.*, 27 FCC Rcd at 12988, ¶ 36.

[36] Petition of the Entertainment Software Association for Partial Extension of Waiver, CG Docket No. 10-213, filed May 22, 2015 (ESA 2015 Petition) at 1-2, 17. ESA describes itself as "the U.S. association exclusively dedicated to serving the business and public affairs needs of companies that publish computer and video games for video game consoles, handheld devices, personal computers, and the Internet." *Id.* at 1, n.1.

[37] *Id.* at 1. As a result, the waivers of the Commission's ACS rules for covered video game consoles and video game distribution platforms will expire October 8, 2015.

[38] *Request for Comment; Petition for Partial Extension of Class Waiver of Commission's Rules for Access to Advanced Communications Services and Equipment by People with Disabilities*, CG Docket No. 10-213, Public Notice, 30 FCC Rcd 6141 (CGB 2012) (*ESA 2015 Petition Public Notice*).

[39] ESA 2015 Petition at 6.

[40] *Id.* at 6.

[41] *Id.* at 6.

[42] *Id.* at 6. ESA also notes that, even when video game software is capable of accessing ACS, "many gamers opt not to engage in communications for reasons unrelated to accessibility," such as to enjoy the game in "story mode," to complete the game in the fastest time possible, to avoid conversations with strangers, or because they do not have or prefer not to use the necessary peripheral devices (*e.g.*, headset and microphone). *Id.* at 10-11. However, ESA offers no data to support this assertion. While the Commission may consider the usage trends of equipment in determining whether a manufacturer designed equipment primarily for purposes other than ACS, assertions about such trends must be supported by more than anecdotal evidence. *See, e.g., Implementation of Sections 716 and 717 of the Communications Act of 1934, as Enacted by the Twenty-First Century Communications and Video Accessibility Act of 2010, Coalition of E-Reader Manufacturers' Petition for Class Waiver of Sections 716 and 717 of the Communications Act and Part 14 of the Commission's Rules Requiring Access to Advanced Communications Services and Equipment (ACS) by People with Disabilities*, CG Docket. No. 10-213, 30 FCC Rcd 396, 407, ¶ 24

(continued....)

Federal Communications Commission   DA 15-1034

describes and provides examples of marketing materials for three video game genres (action games, sports games, and role-playing games), all of which, ESA maintains, emphasize game play and not ACS features or functions.[43] According to ESA, "ACS still plays only a peripheral role in video game software,"[44] and "video game software continues to be designed primarily for the purpose of game play, and not for providing ACS."[45] ESA asserts that "the role of ACS as a complement to game play is no more significant today than it was in 2012."[46]

13.   According to ESA, while "the video game industry has rapidly evolved to incorporate technological advances . . . [that] have enabled new features to facilitate game play for individuals with disabilities," this evolution has not impacted the ACS features and functions in video games.[47] Contrary to the Bureau's expectations in 2012,[48] ESA maintains that the gaming industry has not yet been able to benefit from and utilize the experience gained in making ACS accessible in other contexts.[49] The reason for this, ESA states, is that "[v]ideo games present dramatic technical challenges and a unique – and difficult – environment in which to deliver accessibility solutions."[50] Among other things, ESA describes these challenges to include the fast paced nature of the gaming environment and the frequent involvement of multiple gamers communicating simultaneously during gaming chat sessions,[51] which can make the use of speech-to-text technologies impractical or impossible.[52] ESA also asserts that, because video games make use of specialized hardware and specialized software application program interfaces (APIs), accessibility solutions developed for other environments cannot be re-used in the video game environment.[53] For example, on-screen text cannot be accessed readily by screen readers due to differences in the display technologies used by video games.[54]

14.   In addition, ESA identifies other factors that have hindered the development of accessibility solutions for the gaming environment. For example, ESA claims assistive technology developers tend to focus on technology in employment, education, and core communications contexts, rather than the video game market, which has a short product life cycle.[55] Further, because of the variability of text and voice chat features within games and their implementation, "there is no one

---

(Continued from previous page)
(CGB 2015) (citing studies that had been conducted by the petitioners to support their argument that the low usage of browsers for any purpose, including ACS, demonstrated that ACS was not a primary or co-primary function of basic e-readers).

[43] See ESA 2015 Petition at 6-11, Appendix A (presenting copies and descriptions of screenshots of websites advertising video game software, packaging or "box art," and other advertising or "sell sheets").

[44] Id. at 1.

[45] Id. at 11.

[46] Id. at 4.

[47] Id. at 4. See also id. at 16 (describing an increase from 50 to 350 games with accessibility features and improved accessibility in game consoles since 2012).

[48] See ¶ 9, supra.

[49] ESA 2015 Petition at 11.

[50] Id.

[51] Id.

[52] Id. at 12-13.

[53] Id. at 12, Appendix B (Declaration of Mike Paciello) at ¶ 5.

[54] Id. at 12, Declaration of Mike Paciello at ¶ 5(d).

[55] Id. at 13-14, Declaration of Mike Paciello at ¶¶ 5(e), 6-7.

solution that can be rapidly deployed to address chat accessibility across the broad spectrum of games."[56] Nonetheless, "ESA anticipates that as platforms develop accessibility solutions for the ACS systems that they may offer, these features may be available for reuse by game publishers, and could help inform other solutions to publishers for in-game ACS."[57]

15.   Finally, ESA states that extending the waiver for video games would continue to serve the public interest because it would permit the industry to continue releasing innovative games and allow video games that have ACS to compete with other video game products.[58] In addition, ESA claims that granting a waiver extension will enable platform providers and video game publishers "to continue jointly exploring novel accessibility solutions – including solutions that will facilitate access to ACS – in video game products and services."[59]

V.   DISCUSSION

16.   We extend the class waiver from the Commission's ACS accessibility rules for video game software that is capable of ACS, but designed primary for the purpose of game play, until January 1, 2017. As an initial matter, we agree that, as required by the *ACS Report and Order*, the class of equipment for which ESA seeks a waiver extension – video game software – is defined with sufficient specificity and shares enough common defining characteristics to be granted a class waiver.[60] Specifically, for purpose of this class waiver, we define video game software to include playable games on any hardware or online platform, including game applications that are built into operating system software.[61]

17.   Next, we consider whether video game software is capable of accessing ACS but is nonetheless designed primarily for purposes other than the use of ACS.[62] As noted above, among the factors used to determine whether ACS is a primary or co-primary use in video game software is the extent to which the ACS functionality is advertised, announced, or marketed to consumers as a reason for purchasing, installing, downloading, or accessing the software.[63] The Bureau previously concluded that video game software was capable of accessing ACS, but designed primarily for playing games.[64] We are persuaded by ESA's claims that "ACS still plays only a peripheral role in video game software,"[65] and "the role of ACS as a complement to game play is no more significant today than it was in 2012."[66] ESA provides examples of current marketing materials for action, sports, and role-playing games that emphasize game play and not ACS features or functions.[67] For example, the marketing website for the video game "Call of Duty: Advanced Warfare" focuses on game play features such as characters,

---

[56] *Id.* at 13.

[57] *Id.* at 5. *See also id.* at 14.

[58] *Id.* at 15 (citing *CEA, NCTA, ESA Waiver Order*, 27 FCC Rcd at 12988-89, ¶ 36).

[59] *Id.* at 4. *See also id.* at 15.

[60] *See* 47 C.F.R. § 14.5(b); *ACS Report and Order*, 26 FCC Rcd at 14639, ¶ 193.

[61] ESA 2015 Petition at 6. *See also* ¶ 11, *supra*.

[62] *ACS Report and Order*, 26 FCC Rcd at 14634, ¶ 181. *See also* 47 C.F.R. § 14.5.

[63] *ACS Report and Order*, 26 FCC Rcd at 14635, ¶ 185.

[64] *CEA, NCTA, ESA Waiver Order*, 27 FCC Rcd at 12987, ¶ 35.

[65] ESA 2015 Petition at 1.

[66] *Id.* at 4.

[67] *See id.* at 6-11, Appendix A.

movement, tools, and weapons, rather than the voice chat function.[68] Similarly, the "features" section of the website for "FIFA 15," a soccer simulation game, touts the emotional intensity of the characters and does not mention the voice chat function.[69] Likewise, the website for the "Diablo III" video game advertises game play features, such as the types of characters that can be played in the game, and does not mention the game's text or voice chat features.[70] These and other examples provided by ESA demonstrate that video game software marketing currently emphasizes game playing, not ACS. Accordingly, we find that the equipment defined by this class is capable of accessing ACS, but, at present, is designed primarily for the purpose of game play, which meets the waiver criteria of section 716(h)(1)(A) and (B) of the Act and section 14.5(a) of the Commission's rules.[71]

18.   We must next determine whether good cause exists to extend the waiver and that granting the requested waiver is in the public interest.[72] We remain mindful of the competing public interests at stake – *i.e.*, the ability of consumers with disabilities to use ACS to both communicate with others in the video game environment to compete effectively and to engage in communication that is unrelated to game play,[73] versus the gaming industry's interest in releasing innovative games and allowing video games that have ACS to compete with other video game products.[74] After reviewing the record, we conclude that good cause exists and granting the limited extension requested by ESA is in the public interest at this time.

19.   When the Bureau granted ESA's first waiver request, it concluded that gaming equipment manufacturers and service providers would "benefit from and utilize the experience gained in making ACS accessible in other contexts, to develop and implement ACS accessibility in the equipment and services that are subject to the class waiver in a more efficient and cost-effective manner."[75] This approach appears to have been successful for two of the three classes of gaming equipment and services for which ESA previously received waivers: video game consoles and video game distribution platforms.[76] For example, ESA reports that console manufacturers have released firmware upgrades that have successfully improved ACS accessibility for people with disabilities.[77]

20.   The record demonstrates, however, that eliminating accessibility barriers to the ACS features and functions in video game software remains challenging.[78] For example, ESA asserts that the use of speech-to-text technologies may be impractical or impossible in the context of a fast-paced video

---

[68] *Id.* at 7 (citing Call of Duty: Advanced Warfare Game Home Page, Activision Publishing, Inc., https://www.callofduty.com/advancedwarfare/game and https://www.callofduty.com/advancedwarfare/mp).

[69] *Id.* at 9, Appendix A at 2 (citing FIFA 15 Game features Page, Electronic arts, Inc., https://www.easports.com/fifa/features).

[70] *Id.* at 9-10, Appendix A at 8 (citing Diablo III Game Home Page, Blizzard Entertainment, Inc., http://us.battle.net/d3/en/console/).

[71] 47 U.S.C. § 617(h)(1)(A) and (B); 47 C.F.R. § 14.5(a).

[72] *See* ¶ 4, *supra* (citing 47 C.F.R. § 1.3; ACS *Report and Order*, 26 FCC Rcd at 14637, ¶ 188).

[73] *See* ¶ 9, *supra*.

[74] *See CEA, NCTA, ESA Waiver Order*, 27 FCC Rcd at 12988-89, ¶ 36.

[75] *Id.*, 27 FCC Rcd at 12989, ¶ 36.

[76] *See* ¶ 10, *supra* (noting that ESA did not request an extension of the waivers previously granted by the Bureau for video game consoles and video game distribution platforms).

[77] ESA 2015 Petition at 16. By way of illustration, ESA notes that a recent PlayStation 4 update includes a screen-reader for messages received through the console's text-chat feature and for console menus, as well as the ability to adjust the size and contrast of on-screen text. *Id.* at 16-17.

[78] *See* ¶¶ 13-14, *supra*.

game played by multiple gamers who are communicating simultaneously.[79] Similarly, ESA claims that accessibility solutions developed for other environments, such as screen readers, cannot always be re-purposed for the video game environment due to differences in the display technologies used by video games.[80] In addition, according to ESA, a single solution across the spectrum of video games may not be possible because of the variability of text and voice chat features within the games and their implementation.[81]

21.   ESA states that the modest waiver request it now makes – solely for the class of gaming software, for a period of less than 15 months, in contrast to its original waiver request of eight years for three classes of gaming equipment and services[82] – is necessary to enable the industry to leverage the progress already made with respect to incorporating ACS accessibility into gaming consoles and at the platform level.[83] We are persuaded by ESA's assertions that the accessibility challenges that continue to exist for the ACS features on gaming software warrant a short waiver for an additional period of time. Specifically, we agree with ESA that the increased availability of accessible console and platform-level ACS features is likely to help game developers and publishers address their ACS compliance obligations if such additional time is granted to achieve such compliance.[84] We further agree that "[t]he greater economies of scale associated with game consoles . . . may spur further innovation in assistive technologies" that can help achieve accessible ACS features in the gaming software that use those consoles.[85] As ESA notes, taking this cooperative approach can provide an attractive entry point for assistive technology developers.[86]

22.   Moreover, the limited nature of ESA's waiver request – in both scope and time – minimizes the potential harm that the lack of access to ACS in video games will present to individuals with disabilities. That there were no comments opposing ESA's extension request further suggests that any potential harm to consumers in granting the waiver would be minimal. On balance then, we believe that the circumstances presented weigh in favor of extending the waiver for the additional period of time requested by ESA. Accordingly, we conclude that good cause exists and that it is in the public interest to extend the waiver of the Commission's ACS accessibility rules for video game software, to enable video game platform providers and video game publishers to build upon the accessibility solutions that have been developed by other segments of the gaming industry. Like ESA, we are hopeful that the accessibility solutions developed for video game consoles and distribution platforms, and the ACS they may offer, will be available for use by video game publishers to make the ACS features and functions in their games accessible to individuals with disabilities.[87]

23.   Finally, with respect to the duration of a waiver, we consider the expected lifecycle of the class of video game software.[88] Although ESA previously described the development cycle for video games as five to seven years,[89] in its 2015 petition, ESA refers to the "especially short product life cycle"

---

[79] ESA 2015 Petition at 11-13.

[80] *Id.* at 12, Declaration of Mike Paciello at ¶ 5.

[81] *Id.* at 13.

[82] *See* ¶¶ 7, 10, *supra.*

[83] ESA 2015 Petition at 15.

[84] *Id.*

[85] *Id.*

[86] *Id.* at 13-15, Declaration of Mike Paciello at ¶¶ 5(e), 6-7. *See also* ¶ 14, *supra.*

[87] *See* ¶ 14, *supra.*

[88] *See* ¶ 5, *supra.*

[89] ESA 2012 Petition at 19.

of video games.[90] Notwithstanding this unexplained discrepancy, as noted above, ESA seeks a waiver only until January 1, 2017, which is less than 15 months more than its current waiver.[91]

24.   We find that it is both reasonable and in the public interest to grant ESA's request to extend the waiver of the Commission's ACS accessibility rules until January 1, 2017, to enable video game platform providers and video game publishers to continue exploring ACS accessibility solutions. We expect that extending the waiver for this period of time will permit the industry to continue releasing innovative games, allow video games that have ACS to compete with other video game products, and encourage the gaming industry to incorporate accessibility designed to benefit people with disabilities.[92] During the period of the waiver, we will not require video game software covered by the waiver to comply with the obligations of section 14.20, the performance objectives of section 14.21, and the recordkeeping obligations of section 14.31 of the Commission's rules.[93]

25.   The action we take herein is without prejudice to ESA exercising its right to come back to the Commission at a later time to request another extension of the waiver.[94] However, manufacturers or providers of video game software will be expected to plan for accessibility during the waiver period and to consider accessible design early during the development stages of the next generation of video games to better enable them to eliminate accessibility barriers when the class waiver expires on January 1, 2017.[95] The Commission will take a careful look at industry developments to determine whether any further extension, if requested, is justified.[96]

VI.   ORDERING CLAUSES

26.   Accordingly, IT IS ORDERED that, pursuant to the authority contained in sections 4(i), 4(j) and 716 of the Communications Act of 1934, as amended, 47 U.S.C. §§ 154(i), (j) and 617, and sections 0.361, 1.3 and 14.5 of the Commission's rules, 47 C.F.R. §§ 0.361, 1.3 and 14.5, this Order IS ADOPTED.

27.   IT IS FURTHER ORDERED that the Petition of the Entertainment Software Association for Partial Extension of Waiver IS GRANTED.

---

[90] Specifically, ESA states that the product life cycle of video game software is "especially short . . . compared with software whose core purpose is facilitating communications." *Id.* at 13.

[91] *See* ¶¶ 10, 16, *supra.*

[92] *See* ¶ 15, *supra. See also* ESA 2015 Petition at 4, 15 (citing *CEA, NCTA, ESA Waiver Order*, 27 FCC Rcd at 12989, ¶ 36).

[93] 47 C.F.R. §§ 14.20, 14.21 and 14.31. The waiver of these rules also includes a waiver of the obligation to conduct an achievability analysis during the period of the waiver. *See ACS Report and Order*, 27 FCC Rcd at 14607-14619, ¶¶ 119-148.

[94] *See* ¶ 16, *supra.*

[95] *See ACS Report and Order*, 26 FCC Rcd at 14609, ¶ 124 (wherein the Commission noted that "in many instances, accessibility is more likely to be achievable if covered entities consider accessibility issues early in the development cycle"). We recognize that the achievability analysis conducted for products and services already under development at the time when the class waiver expires may take into consideration the developmental stage of those products or services and the effort and expense needed to achieve accessibility at that point in the developmental stage. *See id.*, 26 FCC Rcd at 14640, ¶ 194. *See also* 47 C.F.R. § 14.5(c)(2). However, if a manufacturer or provider of video game software attempts to demonstrate, in response to an enforcement action, that accessibility is not achievable for software introduced after January 1, 2017, the manufacturer or provider would also need to demonstrate that it has conducted accessibility planning throughout the time period of the class waiver, as early as possible during the design process for the video game software. *ACS Report and Order*, 26 FCC Rcd at 14602, ¶ 108.

[96] *CEA, NCTA, ESA Waiver Order*, 27 FCC Rcd at 12973, ¶ 4.

28.   IT IS FURTHER ORDERED that this Order SHALL BE EFFECTIVE upon release.

29.   To request materials in accessible formats for people with disabilities (Braille, large print, electronic files, audio format), send an e-mail to fcc504@fcc.gov or call the Consumer and Governmental Affairs Bureau at 202-418-0530 (voice), 202-418-0432 (TTY).

FEDERAL COMMUNICATIONS COMMISSION

Alison Kutler
Acting Chief
Consumer and Governmental Affairs Bureau

References

(NB—extensive use of footnotes supplements this reference list—this is purposeful to offer readers immediate linkage to source)

1. Miller M (2015) New requirements for video games this fall. Interact Access. http://www.interactiveaccessibility.com/news/new-requirements-video-games-fall#.V8blk2WIcgU
2. Powers GM, Nguyen V, Frieden LM (2015) Video game accessibility: a legal approach. Disab Stud Q. The Ohio State University Libraries in partnership with the Society for Disability Studies 35:1. http://dsq-sds.org/article/view/4513/3833
3. Lewis DB (1998) The professional artist: grants and fellowships. National Forum on Careers in the Arts for People with Disabilities: Concept Papers. The John F. Kennedy Center for the Performing Arts. http://artsedge.kennedy-center.org/forum/papers/lewis.html
4. Purcell J (2004) WaveRider operating manual. MindPeak, USA
5. Brooks AL, Camurri A, Canagarajah N, Hasselblad S (2002) Interaction with shapes and sounds as a therapy for special needs and rehabilitation. ICDVRAT. The University of Reading, pp 205–212
6. Brooks AL, Petersson E (2005) Play therapy utilizing the Sony EyeToy®. In: Proceedings of the presence 2005, pp 303–314. http://astro.temple.edu/~lombard/ISPR/Proceedings/2005/Brooks%20and%20Petersson.pdf
7. Rosenbaum LD (2010) See what i'm saying: the extraordinary powers of our five senses. Norton, New York
8. Richert M (2012) FCC pushes back on gaming industry accessibility waiver request, consumer voices tip the scales. AFB Access World Magazine. http://www.afb.org/afbpress/pub.asp?DocID=aw131205. Accessed 20 Oct 2014

Printed in the United States
By Bookmasters